Ecological Studies, Vol. 184

Analysis and Synthesis

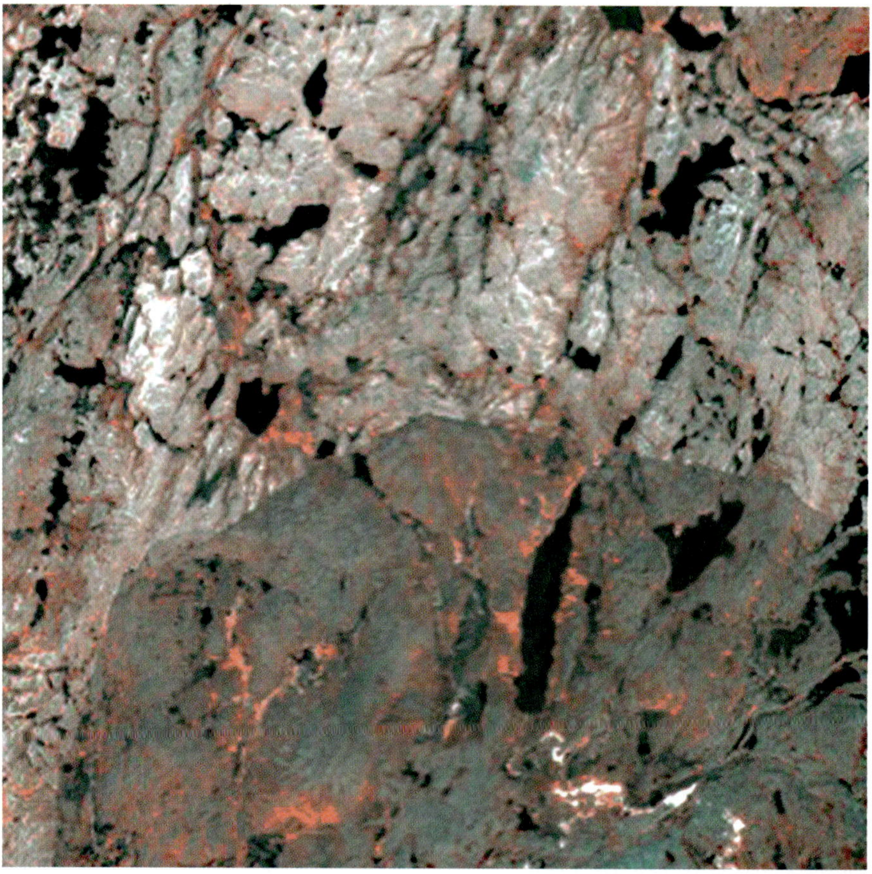

IKONOS-2 false color satellite image from Jauristunturit, acquired 28 June 2001. The reindeer fence on Finnish–Norwegian border is visible because of higher reindeer lichen coverage in Norway (*upper part* of the image). For details see Chap. 8 and 9. lmage coverage is 10 km^2

B.C. Forbes M. Bölter L. Müller-Wille J. Hukkinen
F. Müller N. Gunslay Y. Konstantinov (Eds.)

Reindeer Management in Northernmost Europe

Linking Practical and Scientific Knowledge in Social-Ecological Systems

With 71 Figures, 2 in Color, and 32 Tables

Springer

Prof. Dr. Bruce C. Forbes
Arctic Centre
University of Lapland
P.O. Box 122
96101 Rovaniemi, Finland

Dr. Felix Müller
Ecology Center
Christian-Albrechts-University Kiel
Olshausenstr. 75
24118 Kiel, Germany

Prof. Dr. Manfred Bölter
Institute for Polar Ecology
Christian-Albrechts-University Kiel
Wischhofstr. 1-3, Bldg. 12
24148 Kiel, Germany

Dr. Nicolas Gunslay
Arctic Centre
University of Lapland
P.O. Box 122
96101 Rovaniemi, Finland

Prof. Dr. Ludger Müller-Wille
Department of Geography
McGill University
805 Sherbrooke Street West
Montreal H3A 2K6, Canada

Prof. Dr. Yulian Konstantinov
New Bulgarian University
Montevideo Street 21
1233 Sofia, Bulgaria

Prof. Dr. Janne Hukkinen
Laboratory of Environmental Protection
Helsinki University of Technology
P.O. Box 2300
02015 Espoo, Finland

Cover illustration: Reindeer herd at calf marking near Näkkälä, Finland (Photo: B. Burkhard)

ISSN 0070-8356
ISBN-10 3-540-26087-0 Springer Berlin Heidelberg New York
ISBN-13 978-3-540-26087-5 Springer Berlin Heidelberg New York

Springer is a part of Springer Science+Business Media
springeronline.com

Printed in Germany

Editor: Dr. Dieter Czeschlik, Heidelberg, Germany
Desk editor: Dr. Andrea Schlitzberger, Heidelberg, Germany
Cover design: *design & production* GmbH, Heidelberg, Germany
Typesetting and production: Friedmut Kröner, Heidelberg, Germany

31/3152 YK – 5 4 3 2 1 0 – Printed on acid free paper

Preface

The findings presented in this volume represent a concerted effort to develop a more inclusive form of reindeer management for northernmost Europe. Our guiding principle has been to foster a new paradigm of participatory research. We wish to move beyond the historical reliance on western approaches to basic and applied science. These have been concerned primarily with interactions between herded animals and the various components of their biophysical environment, e.g., plants, insects, predators, climate, and others. In our view, sociocultural and economic drivers, along with herders' experience-based knowledge, gain equal currency in the effort to understand how management may mitigate against the negative aspects of the challenges modern herding faces, while also exploring concepts of sustainability from different perspectives (see also Jernsletten and Klokov 2002; Kankaanpää et al. 2002; Ulvevadet and Klokov 2004).

This broadening of the pool of disciplines and local, national, and international stakeholders in policy-relevant research invariably complicates virtually all aspects of the research process. Multidisciplinary or, in our sense, transdisciplinary approaches also require extraordinary effort from all participants if they are to succeed. As such, those approaches should not be undertaken lightly, nor without personnel who possess appropriate experience in cooperating with those of different disciplines and, preferably, also with relevant practitioners and public social and administrative institutions. In such settings the potential for misunderstandings is quite high.

For these reasons, the education of new generations of young scientists, herders and administrators who can work together, learn from and trust each other, is an important long-term objective. The alternative – business as usual – necessarily excludes potentially constructive partnerships necessary for handling the range of issues confronting contemporary reindeer herding. Another common consequence is the generation of results that are not accepted by all stakeholders.

The collected contributors to this volume cannot claim to have suddenly solved any of the urgent problems that have been facing reindeer management for decades. However, as a group we do feel that the effort has been more than worthwhile. It is a particularly significant achievement that such a diverse coalition of partners from a cross section of northern European nations, cultures, and academic backgrounds were able to forge a common mission. The group undertook not only high-level scientific scholarship across international borders, but also tackled the difficult task of bringing collectively the biological, physical, and social sciences, as well as practitioners' knowledge, to bear in a domain traditionally dominated by process-oriented studies. The trust of the respective reindeer-herding individuals, families, and communities involved has led to their explicit participation in planning, executing, and reporting the research to form a variety of social capital new to the region. Its importance should not be underestimated. In this sense, the project has succeeded – and so hopefully presents an attractive alternative template for future research.

Such a large and complex research undertaking would not have been possible without the strong and sustained efforts of a great many individuals and institutions. To acknowledge them all is not possible, but a few deserve special mention. The input of Johan Mathis Turi and other members of the Association of World Reindeer Herders proved particularly invaluable in motivating, framing, and executing participatory research in the study region. Several individual herders and their families, who do not appear as authors in this book, were vital contributors. Among them are: Tarja Arttijeff; Ola Furmark; Harry Grape; Karl-Erik Gunnari; Vladimir, Grigoriy, and Vassiliy Hatanzei; Jaakko Kantola; Reijo Kyrö; Jari Heikki Länsman; Antti J. Magga; Petri Mattus; Pekka Nevalainen; Antti Näkkälä; Antero Paldan; Taisto Ristimella; Oskar Rosendahl; Martti Sainmaa; Karl-Gunnar Söderholm; Piotr Terent'iev; Mikko Vehkaoja, to name but a portion of the many helpful herders throughout the reindeer districts of northern Finland, Norway, Russia, and Sweden.

And the project never would have gotten off the ground or managed to continue without the sound advice of our successive EU contact officers Xabier Goenaga and John Claxton of the Directorate General Research.[1] The personnel of the Coordination Office at the Arctic Centre, University of Lapland (Rovaniemi, Finland), were professional and a pleasure to work with on a daily basis. Special thanks to its director, Paula Kankaanpää, for consistently sound administrative guidance and Tuija Holm for skillfully negotiating the labyrinthine subtleties and seemingly weekly challenges, otherwise

[1] The main project from which the reported findings are derived was funded by the EU 5th Framework Programme under the key action Quality of Life and Management of Living Resources (Contract No. QLK5-CT-2000-00745).

known as the EU financial reporting requirements. Finally, we are thankful to Otto Lange, this series' editor, and to the personnel at Springer-Verlag for thoughtfully embracing the challenges presented by including a volume clearly distinct from the usual content encompassed by the Ecological Studies series.

Bruce C. Forbes, Manfred Bölter, Ludger Müller-Wille, Janne Hukkinen, Felix Müller, Nicolas Gunslay, and Yulian Konstantinov

September 2005 Rovaniemi, Finland

References

Jernsletten J-L, Klokov K (2002) Sustainable reindeer husbandry. Arctic Council/Centre for Saami Studies, Tromsø, 157 pp

Kankaanpää S (ed) (2002) The 2nd World Reindeer Herders' Congress Anár 2001. Arctic Centre Reports 36, University of Lapland, Rovaniemi, 182 pp

Ulvevadet B, Klokov K (eds) (2004) Family-based reindeer herding and hunting economies, and the status and management of wild reindeer/caribou populations. Arctic Council/Centre for Saami Studies, Tromsø, 170 pp

Contents

Contributors

P. Aikio

Unarintie 17, 99600 Sodankylä, Finland (pekka.aikio@samediggi.inet.fi)

A. Aschfalk

Dept. of Arctic Veterinary Medicine, Norwegian College of Veterinary Medicine, Stakkevollveien 23B, 9212 Tromsø, Norway

H. Beach

Institute for Cultural Anthropology and Ethnology, Uppsala University, Box 631, 751216 Uppsala, Sweden (hugh.beach@antro.uu.se)

M. Bölter

Institute for Polar Ecology, Christian-Albrechts-University Kiel, Wischhofstr. 1-3, Bldg. 12, 24148 Kiel, Germany (mboelter@ipoe.uni-kiel.de)

B. Burkhard

Ecology Center, Christian-Albrechts-University Kiel, Olshausenstr. 75, 24098 Kiel, Germany (bburkhard@ecology.uni-kiel.de)

B.C. Forbes

Arctic Centre, University of Lapland, P.O. Box 122, 96101 Rovaniemi, Finland (bforbes@ulapland.fi)

N. Gunslay

Arctic Centre, University of Lapland, P.O. Box 122, 96101 Rovaniemi, Finland (ngunslay@ulapland.fi)

L. Heikkilä

Ounastie 2060, 99410 Rovaniemi, Finland (lydia.heikkila@pp.inet.fi)

D. Heinrich

Institute for Zoology, Christian-Albrechts-University Kiel, Olshausenstr. 40, 24098 Kiel, Germany (dheinrich@ifh.uni-kiel.de)

T. Helle

Finnish Forest Research Institute METLA, P.O. Box 16, 96301 Rovaniemi, Finland (timo.helle@metla.fi)

C. Höller

Bavarian Health and Food Safety Authority, Veterinärstr. 2, 85764 Oberschleißheim, Germany (christiane.hoeller@lgl.bayern.de)

R. Horn

Institute for Plant Nutrition and Soil Science, Christian-Albrechts-University Kiel, Olshausenstr. 40, 24098 Kiel, Germany (rhorn@soils.uni-kiel.de)

J. Hukkinen

Helsinki University of Technology, Laboratory of Environmental Protection, P.O. Box 2300, 02015 Espoo, Finland, (janne.hukkinen@hut.fi)

O. Jääskö

Angeli, 99870 Inari, Finland (outi.jaasko@pp.inet.fi)

J.-L. Jernsletten

Institute for Cultural Anthropology and Ethnology, Uppsala University, Engelska parken, Humanistiskt centrum Thunbergsv. 3H, 751 26 Uppsala, Sweden
(johnny-leo.jernsletten@antro.uu.se)

N. Kemper

Institute of Animal Breeding and Husbandry, Christian-Albrechts-University Kiel, Olshausenstr. 40, 24098 Kiel, Germany (nkemper@tierzucht.uni-kiel.de)

H. Kitti

Arctic Centre, University of Lapland, P.O. Box 122, 96101 Rovaniemi, Finland (heidi.kitti@metla.fi)

I. Kojola

Finnish Game and Fisheries Research Institute, Oulu Game and Fisheries Research, Tutkijantie 2 A, 90570 Oulu, Finland

Y. Konstantinov

Bulgarian Society for Regional Cultural Studies, P.O. Box 59, Montevideo Street 21, 1233 Sofia, Bulgaria (yulian1@bitex.com)

T. Kumpula

Department of Geography, University of Joensuu, Yliopistokatu 7, 80101 Joensuu, Finland (timo.kumpula@joensuu.fi)

A. Laakso

Arctic Centre, University of Lapland, P.O. Box 122, 96101 Rovaniemi, Finland (alaakso@ulapland.fi). Current address: Institute of Biology, University of Tromsø, 9037 Tromsø (ari.laakso@ib.uit.no)

K. Lakkala

Finnish Meteorological Institute, Arctic Research Centre, Tähteläntie 62, 99600 Sodankylä (kaisa.lakkala@fmi.fi)

V.-P. Lehtola

Giellagas, Sámi Studies, University of Oulu, P.O. Box 1000, 90014 Oulu, Finland

H. Magga

Lappi Reindeer Herding Cooperative, Taivalammentie, 99690 Vuotso, Finland

J. Magga

Näkkälä Reindeer Herding Cooperative, 99401 Hetta, Finland (juha.magga@pp2.inet.fi)

R. Möller

Institute for Polar Ecology, Christian-Albrechts-University Kiel, Wischhofstr. 1-3, Bldg. 12, 24148 Kiel, Germany (rmoeller@ipoe.uni-kiel.de)

F. Müller

Ecology Center, Christian-Albrechts University Kiel, Olshausenstr. 75, 24118 Kiel, Germany (felix@ecology.uni-kiel.de)

L. Müller-Wille

Department of Geography, McGill University, 805 Sherbrooke St West, Montreal H3A 2K6, Canada (ludger.muller-wille@mcgill.ca)

S. Nevalainen

Lomajärventie 1035, 99280 Tepsa, Finland

S. Peth
Institute for Plant Nutrition and Soil Science, Christian-Albrechts-University Kiel, Olshausenstr. 40, 24098 Kiel, Germany (s.peth@soils.uni-kiel.de)

O. Pokuri
Lappi Reindeer Herding Cooperative, Taivalammentie, 99690 Vuotso, Finland

K. Raitio
Kalliojärventie 119, 80330 Reijola, Finland (raitio@cc.joensuu.fi)

P. Soppela
Arctic Centre, University of Lapland, P.O. Box 122, 96101 Rovaniemi, Finland (paivi.soppela@ulapland.fi)

M.-L. Sutinen
Finnish Forest Research Institute, P.O. Box 16, 91500 Muhos, Finland (marja-liisa.sutinen@metla.fi)

J.-M. Turi
Association of World Reindeer Herders, P.O. Box 109, 9521 Kautokeino, Norway (office@reindeerworld.org)

M. Turunen
Arctic Centre, University of Lapland, P.O. Box 122, 96101 Rovaniemi, Finland (minna.turunen@ulapland.fi)

C. Uhlig
The Norwegian Crop Research Institute, 9292 Tromsø, Norway (christian.uhlig@holt.planteforsk.no)

V. Vladimirova
Bulgarian Society for Regional Studies, P.O. Box 59, Montevideo Street 21, 1233 Sofia, Bulgaria (vladi.vladimirova@antro.uu.se)

N. West
Kittilänratsutie 13B5, 99870 Inari (nwest@ulapland.fi)

A. Zink
Institute for Plant Nutrition and Soil Science, Christian-Albrechts-University Kiel, Olshausenstr. 40, 24098 Kiel, Germany

Prologue

Historically, due to environmental and human circumstances, reindeer herding has experienced and coped with dramatic changes and also losses in both reindeer and pasture. Although nature includes regeneration, modern reindeer management can only be sustained in the future if people will make decisions that secure conditions for the continuation of human–reindeer relations in both environmental and human terms. In Northern Fennoscandia and Northwest Russia, the issue is not just the survival of a northern livelihood but that of a northern indigenous people, the Sámi.

After the war, the welfare society solidified itself alongside the traditional social system maintained by reindeer management. In the welfare society, everything is measured in money. A reindeer herder with a family must today have extra jobs that are difficult to correlate with the annual cycle of reindeer management. Permanent jobs are out of the question. Temporary jobs are available only during the busy seasons of reindeer work. The spouse must also work elsewhere. Should the reindeer owner travel back in time and switch over to a subsistence economy? What changes are required to prevent the struggle between the traditional livelihood and the pressures of modern society?

Protecting and preserving the sociocultural system of a minority is much more difficult than protecting the natural environment, which can be easily described and measured. New technologies make reindeer work easier but at the same time they replace old practices rich with knowledge about the reindeer, the weather, the land, and the values that go with them. Cultures and communities have coexisted and influenced each other before, but today the largest cultures penetrate all aspects of life and overwhelm the smaller ones. In the euphoria of globalization, we have forgotten to take care of the rights of local cultures.

The RENMAN project represents research that can respond to these challenges. Traditional know-how should be accepted more broadly as a legitimate part of official research. Research must be targeted also on topics that have been chosen by reindeer herders and will most rapidly improve their

Ecological Studies, Vol. 184
B.C. Forbes, M. Bölter, L. Müller-Wille, J. Hukkinen,
F. Müller, N. Gunslay, Y. Konstatinov (Eds.)
Reindeer Management in Northernmost Europe

prospects. Research must focus on issues relevant to reindeer management without excluding parallel livelihoods. Lost jobs and other damage caused by other land uses to reindeer management have to be compensated justly. More research such as RENMAN is urgently needed if we wish to preserve the diversity of cultures, livelihoods, and nature in Europe.

Outi Jääskö, Reindeer Herder, Muotkatunturi Reindeer Herding Cooperative
Hannu Magga, Reindeer Herder, Lappi Reindeer Herding Cooperative
Satu Nevalainen, Reindeer Herder, Kuivasalmi Reindeer Herding Cooperative
Johan Mathis Turi, Reindeer Herder, Association of World Reindeer Herders

Part I

Herders and Reindeer: The Cultural and Socio-Economic Dynamics of Human–Animal Relations

Photo B. Burkhard

Preface

Part I includes chapters that deal with the cultural and socioeconomic dynamics of keeping and herding reindeer by peoples in northernmost Europe (Finland, Norway, Sweden, and Russia). In these contributions particular emphasis is put on the intricate relation between herders and reindeer, its evolution over time, and the ways in which they have been shaped under rapid changes at the beginning of the 21st century.

Chapter 1 provides an overview of the book's structure. It introduces the research problems, rationale, setting, and semantics and future outlook. Reflections on prehistoric, historic and contemporary practices of reindeer herding and the dynamics of human–reindeer relations in Northernmost Europe are given in Chapter 2 to aid in understanding the anthopological and sociological background. Evolutionary lines are drawn from the prehistoric and historic developments to the present, highlighting crucial changes in reindeer herding. However, the focus is on contemporary issues that are at the core of the discussion over the continuation of reindeer herding and management among the Sámi in northern Finland and Northwest Russia. Their livelihoods are under considerable economic and political pressures to adapt to the principles of the free market. Chapter 2 concludes with the following challenges for today's reindeer management and policy: first, local population should have access to and control over the territory and specific resources; second, local decision-making should take place within a framework of fair negotiations; and third, the nation states of Northernmost Europe should recognize reindeer management as a livelihood that signifies cultural and linguistic diversity. Subsequent chapters of Part I elaborate these challenges and how they might be tackled in practice.

Enhancing cultural diversity, local control, and decision making requires significant changes in the governance of northern natural resources. The research reported in Chapter 3 began by asking how to set in motion the required changes in the institutional and administrative structures that govern reindeer management today. The chapter is an introduction to the application of workshops, or roundtables, which was used as a contemplative and deliberative methodology to bring together herders and other stakeholders with the purpose of developing participatory institutions for reindeer management in Finland. This contribution presents a diagnosis of the deliberation and the emerging integration of knowledge by practitioners (herders), managers, and scientists that would contribute to sustainable reindeer management.

To appreciate the cultural and linguistic diversity that reindeer management embraces, we need a clear understanding of the elements of that diversity. In Chapter 4, comparisons of the indigenous, i.e., Sámi, and scientific, i.e. Western European, perceptions of reindeer management are made and analyzed. The differences in the basic premises and assumptions are exposed and discussed in relation to the development and application of herding strategies formulated by either side. The key finding is that the scientific tradition has not yet fully accepted the validity of indigenous knowledge. Therefore misunderstandings and misinterpretations continue to occur in the management of reindeer herding.

Cultural diversity is also reflected in specific reindeer herding practices. In Chapter 5, the specific case of participation and management in the Torneå Valley in northern Sweden is presented. It discusses the challenges and dilemmas of a particular form of herding practise called "concession" reindeer management. By looking at the environmental and socioeconomic conditions of this practise, the pros and cons of the case are highlighted with respect to the broader context of reindeer management in Sweden.

Finally, in Chapter 6, the situation of the Sámi herders in Lovozero (Murmansk District, Northwest Russia) is presented based on intensive fieldwork. It is an analysis of the changes in property regimes and management practices of herding collectives during the post-Soviet period (since the early 1990s). These changes were the introduction of new agrarian legislation, the continuation of practices inherited from the former state farms (*sovkhoz*), and the establishment of companies with limited liability or agricultural production cooperatives.

These studies present a diverse picture of modern reindeer herding practices and highlight the continuing changes that reindeer management experiences. The research results indicate an urgent need for the development of participatory institutions capable of coping with the expectations and demands of herders within the context of regional development, nation states, and trans-national entities such as the European Union. The presentation of scenarios and recommendations is an earnest contribution for fostering constructive and positive plans for the continuation of reindeer herding.

The Reindeer: A Brief Natural History[1]

Taxonomy and Characteristic Features. The reindeer is the sole representative of the genus *Rangifer* within the deer family Cervidae and is classified by the subfamily Rangiferinae. Reindeer differ from other deer in a number of characters. They are the only cervid species in which both sexes carry antlers; stags shed them in late autumn after the rutting season, females and calves in spring. Noteworthy is the sex dimorphism in the antlers. Antlers of males are by far the larger ones; their prongs are partly formed to shovel, especially the basal ones (brow tine, bez tine). Several features of the species are adaptations to the cold climate of its habitat. The two hooves, which are broad and rounded at the lateral margin, are unusually deeply divided. The accessory metacarpal bones carry small hooves that touch the ground; around and among the hooves reindeer are densely hairy. The reindeer's steadiness on snow and ice is augmented by these characteristics. The coat of reindeer is very thick and dense and the guard hairs are hollow to increase insulation; a further protection from the cold is the covering of the nose with hairs. The coat color of wild reindeer is greyish-brown with white belly and throat mane; the legs are dark. In summer the coat is usually darker than in winter. However, the variability in color is remarkable; this is true especially for herded reindeer. It is not only the color of the coat of wild reindeer that is variable, but they also differ to a certain degree in shape and size of the body and shape of the antlers.

Range. The distribution area of reindeer, wild or herded, was in former times primarily the tundra and taiga of the Holarctic, where their range extended from Ellesmere Island (84° N) in the North to Sakhalin (46° N) in the South. However, today the range is reduced and partly disjoint, especially in Europe. The great variation in reindeer especially concerning the antlers within this large area has led to many attempts to classify them into different subspecies and even species – attempts that have been at least partly unsatisfactory. However, reindeer can be divided into four main types. These are: (1) the relatively small, short-legged polar type; (2) tundra and taiga reindeer, which occupy the central parts of the range, (3) those of the woodlands, which are relatively large of a darker color, and finally (4) the largest reindeer, which live in the southern forest and mountain regions. It needs to be mentioned that the word "caribou", used in North America for wild reindeer, does not signify a special type or a subspecies, but it is derived from the Mik'maq Indian word "xalibu", which means "shovel" thus probably referring to the shape of the antlers of moose, elk, or wild reindeer.

[1] This section was written by D. Heinrich.

Ecology. Reindeer live in different landscape types of the circumpolar North, but especially in the tundra. The reindeer is gregarious; this is especially true for the animals that live in the tundra, mainly in arctic North America. During the summer they prefer windswept terrains in order to avoid the plague of insects. Their winter range is mainly dependent on food requirements. However, there are probably different reasons for the migratory instinct. In spring and summer the females and their calves congregate in small herds like the males only a few of them remain single. In the autumnal rutting season the animals live in mixed herds; finally, a male drives some females together, thus founding harems. After the rut, reindeer again form mixed herds. The wandering herds comprise large numbers, even thousands of animals. For instance, in late winter, when the food supply is still sufficient only in few areas, the animals gather in these before migration starts. Reindeer are able to dig – mainly with their hooves – for their winter food, mostly lichens (*Cladonia* and *Cetraria* species) if the snow cover is not too thick. In the other seasons reindeer feed on a great variety of plants such as sedges, grasses, leaves, and forbs; they prefer soft new sprouts and parts of the plants with the highest nutritional value. Sometimes they pursue lemmings and voles and gnaw on shed antlers to obtain scarce protein and mineral substances (compilation after Heptner et al. 1966; Clutton-Brock 1981; Skjenneberg 1984; Herre 1986; Benecke 1994; Starck 1995).

Fig. 1. Reindeer in summer pasture in the forest (Photo: B. Burkhard)

References

Benecke N (1994) Der Mensch und seine Haustiere. Die Geschichte einer jahrtausendealten Beziehung. Theiss, Stuttgart

Clutton-Brock J (1981) Domesticated animals from early times. Heinemann, London

Heptner VG, Nasimovič AA, Bannikov AG (1966) Paarhufer und Unpaarhufer. In: Heptner VG, Naumov NP (eds) Die Säugetiere der Sowjetunion 1. Fischer, Jena

Herre W (1986) *Rangifer tarandus* – Ren, Rentier. In: Niethammer J, Krapp F (eds) Handbuch der Säugetiere Europas 2/II Paarhufer – Artiodactyla. Aula, Wiesbaden, pp 198–216

Skjenneberg S (1984) Reindeer. In: Mason I L (ed) Evolution of domesticated animals. Longman, London, pp 128–138

Starck D (1995) Lehrbuch der Speziellen Zoologie, Bd II: Wirbeltiere. Teil 5/2: Säugetiere. Fischer, Jena

1 The Challenges of Modernity for Reindeer Management in Northernmost Europe

B.C. Forbes

1.1 Introduction: The Research Setting

This book presents the combined results of a research and development project funded by the European Union's 5th Framework Programme of 2001–2004. The full title of the project was "The challenges of modernity for REiNdeer MANagement: integration and sustainable development in Europe's subarctic and boreal regions (RENMAN)"[1]. Within this Framework Programme, RENMAN was funded under the Key Action "Quality of Life and Management of Living Resources" (Contract QLK5-CT-2000-0745). This portion of the program mandated that funded research aim toward (1) "the design and negotiation of policies that will have a direct impact on the local conditions and prospects for securing an acceptable level of quality of living", and (2) "the continuation of appropriate management regimes for the utilization of living resources locally and regionally within the European context". As such, it called for a direct role for local stakeholders – in this case reindeer herders – in the process of policy-relevant research and for the use of historical trends to facilitate adaptation to future changes. In these aims, RENMAN is quite similar to other EU-funded research projects that focus on particular species in northernmost or Arctic Europe to provide stakeholders and policy makers with a predictive framework of outcome for land use in fragile ecosystems, such as the projects "HIBECO" (Wielgolaski 2005) and "FRAGILE".

The purpose of RENMAN was therefore to develop new tools and models of participatory research and planning in reindeer management that would foster integrated and sustainable use of reindeer (*Rangifer t. tarandus* L.) and related living resources in northernmost Europe (for the natural history of the reindeer see Preface to Part I). Reindeer management is among the most

[1] The acronym "REN" ("ren" = "reindeer" in German and Scandinavian languages) stands for "reindeer" and "MAN" for "management".

Ecological Studies, Vol. 184
B.C. Forbes, M. Bölter, L. Müller-Wille, J. Hukkinen, F. Müller, N. Gunslay, Y. Konstatinov (Eds.)
Reindeer Management in Northernmost Europe

important mutually competing activities using natural resources and the environment in the Barents Euro-Arctic Region (BEAR). It is also one of the oldest and most resilient forms of livelihood in the region. Reindeer are a living resource central to the culture of many northern circumpolar indigenous peoples, including the Sámi, as well as non-indigenous peoples in some areas, e.g., in Finland and Russia.

RENMAN was conceived as a 36-month project comprised of 12 tightly integrated subprojects or "workpackages" (referred to as WPs in this book) divided among nine partner institutes in five countries.[2] The project investigated both the human dimensions and natural conditions of reindeer management in order to formulate sustainable future scenarios. To achieve efficient integration of the workpackages, an integrative systems analysis was conducted to assist the whole project in problem distinction, system identification, interdisciplinary understanding, and sustainability evaluation (Chap. 16). The results presented here therefore provide a complex picture of the position and role of reindeer management in both the physical environment and cultural landscape of the research region.

Local study areas were selected in subarctic and boreal Fennoscandia and the adjoining Kola Peninsula (Fig 1.1), with special emphasis on the traditional homeland of the Sámi, Sápmi.[3] Multiple research localities allowed for comparisons through the application of standardized research methods. The precise study sites were selected in conjunction with herders during interviews and field excursions at the start of the project in winter/spring 2001 and aimed to represent the variability in management within a representative portion of the reindeer herding area. In fact, during each phase of the project – conceptualization, implementation, and reporting – herders were integral participants as practitioners and researchers alongside external scientists (Fig. 1.2). The project involved herders from northern Finland, Sweden, Nor-

[2] Arctic Centre, University of Lapland, Finland (coordinator, principal contractor); Department of Geography, University of Oulu, Finland; Finnish Forest Research Institute, Rovaniemi, Finland; Helsinki University of Technology, Espoo, Finland; Christian-Albrecht University, Kiel, Germany; The Norwegian Crop Research Institute, Tromsø, Norway; Dept. of Social Anthropology, University of Tromsø, Norway; Institute for Cultural Anthropology and Ethnology, Uppsala University, Sweden; Institute for Anthropological Field Research, New Bulgarian University, Sofia, Bulgaria.

[3] We gratefully acknowledge the active participation and tremendous input of the many reindeer herders, their families, communities and organizations that are not listed in the formal authorship of this book,, but who contributed to our collective understanding of reindeer management in many important ways. Without this generous assistance and mutual trust the project would not have been possible. We are pleased that this experiment in participatory research has been deemed a success in the local communities involved and sincerely hope that the new approaches to research tested here, and the resulting management proposals, can lead to further discussion and implementation in the future.

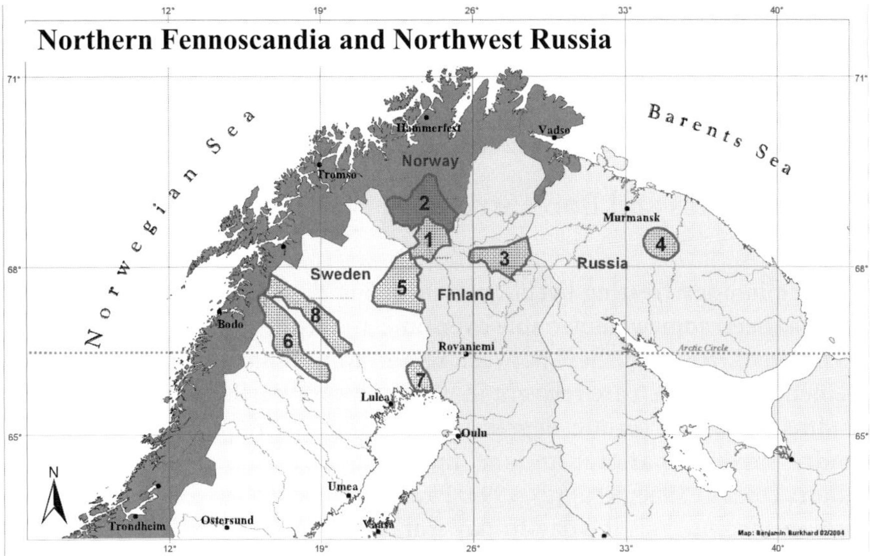

Fig. 1.1. Map of the intensive RENMAN study areas in Sweden, Norway, Finland and Russia are as follows: *1.* Näkkälän paliskunta; *2.* Kautokeino common winter pastures district 30; *3.* Lapin paliskunta; *4.* Lovozero region; *5.* Muonio Sameby; *6.* Tuorpon Sameby; *7.* Liehittäjä concession Sameby; and *8.* Sirkas Sameby. Many other districts within Finland also took part in the project via a series of annual participatory workshops (Chap. 3)

Fig. 1.2. RENMAN project participants in Oteren, Norway, discussing the selection of study sites along the Norway/Finland border in May 2001. In the photo from left are as follows: Johan Mathis Turi, [former] President of the Association of World Reindeer Herders, Kautokeino; Dr. Christian Uhlig, researcher from the Norwegian Crop Research Institute; Heidi Kitti, Ph.D. student from Arctic Centre, University of Lapland; and Timo Kumpula, Ph.D. student from Department of Geography, University of Joensuu. *Photo: B.C. Forbes*

way, and on Kola Peninsula, Russia. We are not aware of previous projects in Europe of this magnitude that have involved local stakeholders to such an extent in all aspects of the research.

1.2 Semantics and Book Structure

The terminology treating livelihoods involving the keeping and use of reindeer has been somewhat variable over time and from country to country. We attempt to differentiate here between reindeer husbandry, herding, and management, all of which are relevant in contemporary northernmost Europe. To be sure, there is a certain degree of conceptual overlap in the terms husbandry, herding, pastoralism, and management as discussed by different authors (Ingold 1980; Krupnik 1993; Paine 1994; Anderson 2005). In general, herding encompasses the day-to-day work with a herd. It concerns the relationship between a herd and its pasture as directed by herders for the welfare of the animals and, if necessary, to the exclusion of the comfort of the herders themselves. Husbandry, on the other hand, has more to do with a herd as the "harvestable" resource of its owners. According to Paine (1994), "herding" is about the control and nurture of reindeer within their terrain and "husbandry" is more related to the growth of capital ("on the hoof") and profit. In addition, he argues that husbandry includes social mechanisms such as dowry and inheritance. To this we would add that husbandry clearly also deals with reproduction and breeding.

In this book we generally adhere to the terms herding, as defined above, and management. Management is used in the broadest possible sense, and is intended to encompass all practices pertaining to the keeping of reindeer. This term is not perfect. For example, individual herders may not feel they are actually "managers" of their own herds, which are allowed to range freely much of the time. However, in a comprehensive discussion treating ecological, sociocultural, political, and economic issues within four different countries, we are certainly dealing with management at several scales, from local *siida* and landscapes to the wide region of northernmost Europe. Furthermore, the RENMAN project was funded in part to tackle issues of policy relevance within Europe. As such, we adopted a participatory approach in an attempt to address management holistically.

The book is structured in three main parts, reflecting the various research topics and approaches. Part I covers first the geographic and social evolution of reindeer herding as a form of livelihood (Chap. 2), and the development of participatory approaches to management (Chap. 3). The next few chapters address, respectively, the different conceptions of "environmental management" (Chap. 4), the dilemmas of so-called concession reindeer management in modern-day Sweden (Chap. 5), and post-Soviet changes in property regimes on the Kola Peninsula (Chap. 6). Together, the chapters in this book section outline the main sociocultural and economic conflicts that lie at the heart of contemporary reindeer management within the region.

Turning to the natural and physical sciences in Part II, Chapter 7 provides a suitable transition as it treats a question right at the nexus of the natural and social sciences – what is the best way to characterize pasture "quality"? Assessments of forage quantity and quality derived primarily from satellite surveys have provided the basis for pasture inventories and policies concerning population regulation in recent years (e.g., Kumpula et al. 1997, 1999, 2004). Yet, herders have their own sets of indicators that vary in space and time, combining ecology, geography, and social relations, all of which bear heavily on their patterns of pasture use. Chapters 8 and 9 contrast vegetation cover and structure patterns on either side of the Finnish–Norwegian border, where strongly divergent management systems have been in place for nearly 50 years. Chapter 10 focuses on common forage species in wetlands and reveals their individualistic short-term responses to ultraviolet radiation, an important component of global change. The remaining contributions address questions surrounding: the long term effects of grazing and trampling in mineral (Chap. 11) and organic (Chap. 12) soils, as well as microbial ecology (Chap. 13), and the hygienic status of both soils and surface waters (Chap. 14).

The final section, Part III, provides an integrative overview, beginning with an analysis of the various drivers underlying recent population trends of semi-domestic reindeer in Fennoscandia (Chap. 15). A systems analysis of reindeer management balances some of the different social, economic, and ecological interactions within reindeer management. Such analysis also simulates future scenarios based on trade-offs that develop when certain inputs are altered (Chap. 16). Finally, Chapter 17 provides a multidisciplinary synthesis, drawing conclusions from the combined findings of all the authors.

1.3 Research Problems, Questions, and Regions and Sites

Changes in reindeer systems that adversely impact small and dispersed communities involved with herding include, among others, reductions in grazing land and forage resources. Such losses may occur as a result of large-scale exploitation by the expanding forestry and hydroelectric power industries, mining, and competition with tourism or crop-based agriculture, as well as of future habitat alteration due to global warming. Reindeer themselves have long been critical constituents of the northern portions of the BEAR, with marked capacities for affecting ecosystem structure and function (Oksanen 1978; Helle and Aspi 1983; Fox 1998; Helle 2001; Moen and Danell 2003). Large areas in northern Norway, Sweden, and Finland and portions of the adjoining Kola Peninsula have been grazed for millennia (Staaland and Nieminen 1993;

Forbes 2005; Chap. 2). As the aforementioned competition over space and resources has increased and its effects have become visible, in particular over the past 20 years or so, there have been recurrent and widespread reports of depletion of plant resources that some scientists have termed "overgrazing" and calling for significant reductions in the number of reindeer in the Nordic countries (Torp 1999; Riseth 2000; Tahkolahti 2000; Tuomi–Nikula2001). The combined effect of these various trends is that political discussion about reindeer management policy and its relationship with other uses of the environment continues to intensify (Jernsletten and Klokov 2002). Relevant recommendations regarding land management and human rights have already been promulgated in the context of World Heritage Sites (Agenda 21), of the EU's Northern Dimension (in particular Protocol No. 3 on the Sámi people attached to the EU accession treaties of Finland and Sweden), and of the EU General Secretariat's "Resolution on Indigenous Peoples".

Intended in part as a response to this trend, the International Arctic Science Committee (IASC) workshop on the Human Role in Reindeer Systems was held at the Arctic Centre in Rovaniemi, Finland, in 1999 (Forbes and Kofinas 2000). This was a firm step toward the integration of international, multidisciplinary research pertaining to both semi-domestic and wild reindeer. The workshop brought together biologists, social scientists, and indigenous practitioners to discuss priorities in reindeer-related research for the first decade of the new millennium. As such, it was a watershed event in terms of bringing reindeer herders into the research process in a meaningful way. During the workshop it was noted that Sámi herders perceive that there has been inordinate emphasis on meat production and the ostensibly poor condition of lichen-dominated winter pastures. According to some Sámi, much of the research addressing these issues has been conducted without acknowledgment of the importance of practitioners' knowledge and the sociocultural context of reindeer management.

The IASC workshop built upon a theme earlier put into action via a more circumscribed participatory planning workshop held in Inari, Finnish Lapland, in 1997. That workshop was organized as a concerted action of the "Human Environmental Interactions" theme of the European Commission's Arctic–Alpine Terrestrial Ecosystems Research Initiative (ARTERI). The mandate of ARTERI was to develop research themes and encourage their implementation in a way that would allow for discussions among local residents, natural and social scientists, and policy makers concerned with environmental protection and management in arctic and alpine regions of Europe. Although the participants at the ARTERI workshop identified concerns in diverse fields of human environmental interactions in upper Finnish Lapland, their future scenarios shared common topics. Reindeer management, forestry, and tourism were the dominant livelihoods that discussed (Burgess 1999; Müller-Wille and Hukkinen 1999). Subsequently, a successful application for management improvement was submitted to the Academy of

Finland by J. Hukkinen "Reindeer Management as Environmental Management in Finland", which brought together a large number of scientists with the express purpose of integrating herders. The project was not implemented at that time for administrative reasons. However, ensuing discussions among scientists and herders contributed greatly to the RENMAN project proposal to the EU, which then developed over a two or three year period and finally came into focus during the 1999 Rovaniemi workshop.

The critical research questions for RENMAN included:

- Is "overgrazing" the only problem and reducing the number of reindeer the only solution in reindeer management?
- What are the key issues in reindeer management for the local reindeer herding communities?
- What is the future of reindeer management in northern Fennoscandia and adjacent northwest Russia?
- Much is known about winter pastures, but what about the quality of summer pastures?
- Can herders participate meaningfully in the research process?
- How can reindeer management be best integrated with other uses of the northern environment, such as tourism, forestry, and hydropower?

From the beginning, herders urged us to pay special attention to summer pastures because they felt that the emphasis on winter pastures in recent years had been at the expense of an appreciation and understanding of summer pastures. Summer pastures are important for calf growth and overall herd production (Kumpula et al. 1999). Other key reasons for selecting the chosen study areas were: (1) the demonstrated willingness of the herders and their representative organizations[4] to participate; (2) the existence of previous studies that provided necessary background biological data; and (3) the various investigators' familiarity with these regions, several through long term studies. For reasons of broader interpretations in anthropological and socioeconomic terms, the general reference region has been all of northernmost Fennoscandia. The overall objectives of the social science workpackages within RENMAN were to: (1) establish a novel and thoroughly participatory mode of analyses for human environmental interactions in Europe; (2) develop future scenarios of reindeer management in northern Fennoscandia; and (3) create a framework for interactive reindeer management planning that is grounded in the knowledge base of reindeer herders and supported with scientific studies (Chaps. 3 to 6). The results presented here provide a

[4] Svenska Samernas Riksförbund; Suoma Boazosámit (Sámi Reindeer Herders of Finland); Paliskuntain Yhdistys (Association of Finnish Reindeer Herding Districts); and the Association of World Reindeer Herders, based in Norway.

complex picture of the position and role of reindeer management in both the physical environment and cultural landscapes of the research region.

Our starting point in the project was that ecology itself is subject to cultural perceptions (of both indigenous and dominant society) and that reindeer management policies therefore have cultural implications. An explanatory model that is based on range biology alone without consideration for ethnic and cultural sustainability as well is destined to fail. Further investigations are therefore needed of the relationship between indigenous and non-indigenous reindeer herding practices and external factors influencing these applications. Central to a better understanding of this relationship is a systematic comparison between practitioners' experience-based knowledge and science-based knowledge concerning, for example, the quality of pastures. Adding to the sense of urgency has been the critical state of reindeer herding in the Kola Peninsula of northwest Russia. In the context of recent food crises in that region, there has been strong interest in enhancing the role of reindeer herding in the local economy, thereby increasing regional food production and improving living standards for the herders and their families (Chap. 6).

1.4 Issues and Concerns

In general, northernmost Fennoscandia is subject to extensive and often intensive land use and ecosystem fragmentation in contrast to much of the boreal zone in Russia and North America (Forbes et al. 2004). For Norway, Sweden, and Finland, boreal forests and forestry are an important source of economic wealth and, relative to neighboring portions of Russia, highly managed for pulp and timber. There is concern that the sustainability of ecosystem resources is already threatened by this intensive use and that future policy changes may turn out to be ineffective to protect resilience (Kankaanpää et al. 2002).

Part of the problem is the lack of an agreed structure to monitor the effectiveness of management strategies, beyond matters pertaining to forest and pulp production, and questions about the efficacy of current monitoring protocols. One example comes from the availability of arboreal lichens, an important source of winter food for reindeer. Such lichens are plentiful within the canopies of older coniferous forests but are typically scarce or even absent in clear-cuts and the young and medium-aged regenerating stands, which characterize large portions of Lapland. In a recent inventory of arboreal lichens, a detailed ground survey conducted by one of the region's reindeer herding cooperatives (Sipilä et al. 2000) found significantly fewer lichen resources than those reported for the same areas by satellite survey (Korkalo et al. 1997). This is not surprising, given the scales of resolution of the different methods, but the fact that the lichen resources tend to be overestimated by satellite surveys, which form the foundation for the government's reindeer management policy, results in distrust of official statistics at the local level.

In northern Finland ground lichens have also been greatly reduced in reindeer herding districts in recent decades (Helle and Aspi 1983; Kumpula et al. 1997; Kumpula 2001). This has occurred through a combination of factors, including generally more intensive reindeer management and "double" grazing use (summer and winter) in some areas, resulting in both winter consumption and summer trampling of lichens. This has been coupled with incremental spatial, economic, and administrative encroachment on reindeer "pasture" resources by diverse land uses such as forestry, hydropower, mining, tourism, and increased off-road vehicle traffic by the herders themselves (Olli 1995; Jernsletten and Klokov 2002; Kankaanpää et al. 2002). Here again, disagreement over the root causes of the problem, invariably characterized as "overgrazing" with most of the blame attributed to the practices of the reindeer herders, means that long term solutions are delayed or avoided while ecosystems continue to degrade under the synergistic effects of multiple and overlapping land uses (Torp 1999; Jernsletten and Klokov 2002; Kankaanpää et al. 2002).

To characterize this trend as overgrazing is to oversimplify a complex suite of issues driven by political imperatives, socioeconomic and demographic changes, and agricultural norms. Already in the 1970s it was clear that summer trampling was having an impact as significant as that of winter consumption of lichens alone, if not more so (Oksanen 1978). The problem is especially acute in Finland, where fences have truncated former migration routes. Summer and winter pastures in some areas are thus tightly interspersed in a patchwork mosaic, whereas formerly they would have been in distinct, widely separated areas (Chap. 7). The result is that unattended animals are able to move back and forth across winter lichen grounds during summer to reach preferred habitats such as riparian zones and wetlands. When dry, the brittle lichens are crushed into small pieces that are then readily blown or washed away. Through this process, large areas can become denuded of lichens even when animals are subsisting mainly on deciduous shrubs, graminoids and other summer forage (Chap. 9). This simple but important concept is rarely encompassed in the discussion and recommendations concerning "overgrazing".

In Russia, renewable and non-renewable resource management has been somewhat chaotic at least since the break-up of the Soviet Union, and arguably long before that. Fennoscandia presents a relatively benign situation by comparison, due to its economic and political stability since World War II, as well as a strong regulatory framework for environmental management (Forbes et al. 2004). However, this can create a false perception that the major problems are now behind us, thus threatening adaptive management. The aforementioned example of "overgrazing" as an overriding concern in contemporary reindeer management points to the lack of cooperative management of natural resources in northernmost Europe. Co-management is defined as a shared decision-making process, formal or informal, between a government authority and a user group for managing a species of fish, wildlife, or other resources. Ideally, co-management serves to incorporate elements of scientific and local or practitioners' knowledge to sustain viable pools of resources that may, in turn, secure the livelihoods that depend on them. In practice, there are great difficulties in actually integrating the two types of knowledge (Kendrick 2003), although there are instances in North America of different user groups and scientists each benefiting by using information derived from the other source (Klein et al. 1999). According to Kendrick (2003: ii), "...[W]ithout venues where biologists and caribou hunters and elders can share their knowledge in direct and regular interactions, the social learning involved in linking knowledge systems is not likely". A similar recommendation is warranted in the context of European reindeer management with semi-domestic animals and, in fact, was included among the management proposals made by the RENMAN project (Chap. 3).

1.5 Challenges in Modern Reindeer Management

This is a time of great change for reindeer management in northernmost Europe. Serious challenges threaten the future of reindeer management, and the transition from relatively low-tech management prior to the "snowmobile revolution" (Pelto 1973) has often been difficult. To begin with, since World War II, and long before that in the Soviet Union, nation states have appropriated important aspects of decision making. The international borders with their fences separating the Nordic countries, and restricting large cross-border migrations of herds, are just one example of this. During the 1980s heavy grazing by relatively unsupervised herds during summer left many winter ranges in poor condition. The discussion has focused on the absence of favored forage species of lichens and the establishment of less palatable ruderal or "weedy" species. This is in part because of the dramatic disappearance of once-thick lichen mats coupled with the exposure and, in some places, erosion of organic and mineral substrates (Väre et al. 1996; Olofsson 2001). However, even less apparent changes in soils and surface waters may also be critical for long-term ecosystem dynamics (Stark 2002; Chaps. 13–15). At the same time, global change has raised concerns about pasture accessibility (icing events, heavier winter snow accumulations with corresponding longer spring snowmelt), quality, and animal population dynamics (Turi 2000; Helle and Timonen 2001; Nuttall et al. 2005; Chaps. 11, 12). As well, there are economic factors that are largely beyond herders' control. For example, in autumn 2003 the price of reindeer meat plummeted as, among other factors, local supermarkets throughout Fennoscandia became flooded with comparatively inexpensive deer and ostrich meat imported from central Europe and New Zealand (Chap. 3; Valtavaara 2003). As one herder noted during the course of the project, "it is not enough to simply be a good herder anymore. Now we have to be experts in a number of other things as well, like marketing, in which we don't really have much experience".

Yet it is necessary to point out that there are also excellent opportunities (Turi 2002). One pertinent for a wide spectrum of disciplines and applications is the advent of participatory models of policy-relevant research in Europe. The RENMAN project has been a test case in participatory research and in valuing the local knowledge of herders (Chaps. 3, 7). Participatory studies, and the cooperative management of renewable resources, have been underway for several decades in North America (Treseder 1999). These are relatively new and innovative tools in terms of ecosystem management in Europe (Sandström et al. 2003; Chaps. 3–5). Another promising, if costly, tool is high-resolution satellite imagery (Allard 2003; Chap. 8), which we have found useful for developing timelines of visible patterns and changes in the land at scales meaningful for herders.

Traditionally, Nordic and Russian reindeer-related research has been of high quality yet functionally isolated within the various disciplines, e.g., zoology, agronomy, cultural anthropology, and human geography. Also, funding has tended to derive from national sources so that international comparisons have been difficult to achieve. There is a need to broaden the scope of the discussion and the range of research topics to better address herders' own needs concerning their future, rather than focusing narrowly on agronomic indicators like meat production and "carrying capacity" within individual nations. One example is the definition of "pasture quality", which herders in different countries themselves define much more comprehensively than is possible employing standard western scientific indicators such as vegetation composition, cover, biomass, nutrition, digestibility, etc. (Chaps. 7, 9). As it is, the bounds of debate are inherently limited by rather strict adherence to the prevailing agricultural norms (Laakso 2002). As a result, herders are often marginalized as players within the regional and national dialogue on reindeer management policy, even more so when other forms of land use are under discussion. At a minimum, this decoupling of interconnected social–ecological systems (sensu Berkes and Folke 1998) effectively excludes herders from the dialogue and inevitably leads to conflict (Laakso 2002; Sandström et al. 2003). In the worst case scenarios, herders are vilified in the media as irresponsible stewards of the land and, therefore, considered incapable as managers in the contemporary milieu of modern commercialized and industrialized land uses within which reindeer management must function (Torp 1999; Aikio 2002). In such a strained atmosphere, concepts of "sustainability" are limited in their applicability and the adoption of new measures to embrace a more holistic form of ecosystem management is difficult at best.

1.6 Conclusions: Outlook for Resolutions to Conflicts in Reindeer Management

Even if the direct transfer of North American-style co-management as a model to alleviate conflicts between aboriginal peoples and western-based governments does not seem practical for northernmost Europe, in the short term it is necessary to acknowledge at least that we are dealing with complex systems that transcend simple agricultural paradigms and models of "carrying capacity". Reindeer management functions now, as it always has, at the nexus of diverse social and ecological systems (Folke et al. 2002; Berkes et al. 2003). In Fennoscandia, the overlaying of national and international borders and the extensions of post-World War II political and economic agendas on the region have invariably complicated the study of holistic solutions to what are simultaneously local and regional issues and concerns. Across the border in Russia, the situation has rightly been characterized as a "crisis" since the collapse of the Soviet Union (Krupnik 2000). While the problems are often quite different from those faced by herders in the Nordic countries, for example, widespread poaching of privately owned herds (Konstantinov 2002), there are still lessons to be learned by comparisons between the two regions (Chap. 6).

To weigh all of these different factors properly in a manner that allows us to develop possible scenarios for the future is a daunting task. Models based solely on "ecological" criteria are clearly insufficient. Other methods can include integrative systems analysis (Chap. 16), in which the ramifications of different choices are balanced against the entire suite of socioeconomic and ecological interactions that characterize modern reindeer management. In general, the RENMAN project scientists felt it was important to return a measure of power in decision making to herders. Already there have been further developments in this direction in Finland, where the Ministry of Agriculture and Forestry has held participatory workshops, using the RENMAN template, expanding the discussion beyond matters of ecology in envisioning different future scenarios for reindeer management (Saijets 2004; Talvitie 2004). At the same time, there needs to be more balance in power relations with other players in terms of regional land use: forestry, hydropower, mining, and tourism (Chap. 2). Internationally, as practitioners of reindeer management, herders themselves are increasingly well represented by the Association of World Reindeer Herders (WRH; Fig. 1.1). WRH not only seeks to advance the sustainability of reindeer management as a viable livelihood, but also makes great efforts to coordinate and integrate their own activities with those of researchers in both natural and social sciences (Kankaanpää 2002). It is in this spirit of shared interest in the future of reindeer management that the RENMAN study has functioned (Chap. 17). We hope that the findings presented here are of broad interest to scientists, administrators, and herders, and that the participatory model of research can gain newfound traction in the European North to better link practical and scientific knowledge.

References

Aikio P (2002) The prospects of Sámi reindeer herding in Finland. In: Kankaanpää S (ed) The 2nd World Reindeer Herders' Congress Anár 2001. Arctic Centre Rep 36:36-40
Allard A (2003) Detection of vegetation degradation on Swedish mountainous heaths at an early stage by image interpretation. Ambio 32:510-519
Anderson DG (2005) Reindeer pastoralism. In: Nuttall M (ed) Encyclopedia of the Arctic, vol 3. Routledge, New York, pp 1752-1754
Baskin L (2000) Reindeer husbandry/hunting in Russia in the past, present and future. Polar Research 19:23-29
Berkes F, Folke C (eds) (1998) Linking social and ecological systems. Cambridge Univ Press, Cambridge, 459 pp
Berkes F, Mathias J, Kislalioglu M, Fast H (2001) The Canadian Arctic and the Oceans Act: the development of participatory environmental research and management. Ocean Coastal Manage 44:451-469
Berkes F, Colding J, Folke C (eds) (2003) Navigating social-ecological systems. Cambridge Univ Press, Cambridge, 393 pp
Burgess P (1999) Human environmental interactions in Upper Lapland, Finland. Arctic Centre Rep 27:1-80
Colpaert A, Kumpula J, Nieminen M (2003) Reindeer pasture biomass assessment using satellite remote sensing. Arctic 56:147-158
Finstad GL, Bader H, Prichard AK (2002) Conflicts between reindeer herding and an expanding caribou herd in Alaska. Rangifer 13:33-37
Folke C, Carpenter S, Elmqvist T et al (2002) Resilience and sustainable development: building adaptive capacity in a world of transformations. Ministry of the Environment, Stockholm, 74 pp
Forbes BC (2005) Reindeer. In: Nuttall M (ed) Encyclopedia of the arctic, vol 3. Routledge, New York, pp 1750-1752
Forbes BC, Kofinas G (eds) (2000) The human role in reindeer and caribou grazing systems. Polar Research 19(1):1-142
Forbes BC, Fresco N, Shvidenko A, Danell K, Chapin FS III (2004) Geographic variations in anthropogenic drivers that influence the vulnerability and resilience of high latitude social-ecological systems. Ambio 33:377-382
Fox J (1998) Finnmarksvidda: reindeer carrying capacity and exploitation in a changing pastoral ecosystem - a range ecology perspective on the reindeer ecosystem in Finnmark. In: Jentoft S (ed) Commons in cold climate: coastal fisheries and reindeer pastoralism in north Norway. Parthenon Publ, Paris, pp 17-39
FRAGILE (2003-2006) Fragility of Arctic Goose habitat: Impacts of Land use, conservation, and Elevated temperature. http://www.fragile-eu.net/index.php (accessed 2 March 2005)
Helle T (2001) Mountain birch forests and reindeer husbandry. In: Wiegolaski FE (ed) Nordic mountain birch ecosystems. UNESCO/The Parthenon Publishing Group, Paris, France, pp 279-291
Helle T, Aspi J (1983) Effects of winter grazing by reindeer on vegetation. Oikos 40:337-343

Helle T, Timonen M (2001) The North Atlantic Oscillation and reindeer husbandry. In: Kankaanpää P et al (eds) Arctic Flora and Fauna: Status and Conservation, Edita, Helsinki, p. 25

Ingold T (1980) Hunters, pastoralists and ranchers. Cambridge Univ Press, Cambridge, Cambridge, 336 pp

Jernsletten J-L, Klokov K (2002) Sustainable reindeer husbandry. Arctic Council/Centre for Saami Studies, Tromsø, 157 pp

Kankaanpää S (ed) (2002) The 2nd World Reindeer Herders' Congress Anár 2001. Arctic Centre Rep 36:1-182

Kankaanpää S, Müller-Wille L, Susiluoto P, Sutinen M-L (eds) (2002) Northern timberline forests: environmental and socio-economic issues and concerns. Finnish Forest Research Institute, Jyväskylä, 289 pp

Kendrick A (2003) Caribou co-management and cross-cultural knowledge sharing. PhD thesis, University of Manitoba, Winnipeg, 266 pp

Klein DR, Moorehead L, Kruse J, Braund SR (1999) Contrasts in use and perceptions of biological data for caribou management. Wildlife Society Bulletin 27:488-498

Konstantinov Y (2002) Soviet and post-Soviet reindeer-herding collectives: transitional slogans in Murmansk Region. In: Kasten E (ed) People and the land: pathways to reform in post-Soviet Siberia. Reimer, Berlin, pp 171-188

Korkalo M, Pouta T, Valle P, Kokkonen P, Korpivuoma J, Itkonen P, Ollila T (1997) Peurakaira's ecological-area plan (in Finnish). National Board of Forestry Forest Economy Publication 9, 79 pp

Krupnik I (1993) Arctic adaptations: native whalers and reindeer herders of northern Eurasia. University Press of New England, Hanover, 355 pp

Krupnik I (2000) Reindeer pastoralism in modern Siberia: research and survival in the time of crash. Polar Research 19:49–56

Kumpula J (2001) Productivity of the semi-domesticated reindeer (*Rangifer t. tarandus* L.) stock and carrying capacity of pastures in Finland during 1960-1990´s. PhD Thesis, University of Oulu, Oulu, 44 pp

Kumpula J, Colpaert A, Kumpula T, Nieminen M (1997) Winter pasture resources of the Finnish reindeer management area (in Finnish w/English summary). Finnish Game and Fisheries Research Institute, Kaamanen, 42 pp

Kumpula J, Colpaert A, Nieminen M (1999) Summer pasture resources of the Finnish reindeer management area (in Finnish w/English summary). Finnish Game and Fisheries Research Institute, Kaamanen, 54 pp

Kumpula J, Colpaert A, Anttonen M, Nieminen M (2004) The repeated reindeer pasture inventory in the northernmost part (13 districts) of the reindeer management area during 1999-2003 (in Finnish w/English summary). Finnish Game and Fisheries Research Institute, Helsinki, 39 pp

Laakso AM (2002) The anomalies of contemporary reindeer herding management and supporting reindeer research: the need for a new paradigm. Pro gradu (Masters) thesis, University of Lapland, Rovaniemi, 127 pp

Moen J, Danell Ö (2003) Reindeer in the Swedish mountains: an assessment of grazing impacts. Ambio 32:397-402

Müller-Wille L, Hukkinen J (1999) Human environmental interactions in Upper Lapland, Finland: development of participatory research strategies. Acta Borealia 16(2):43-61

Nuttall M, Berkes F, Forbes BC, Kofinas G, Vlassova T, Wenzel GW (2005) Hunting, herding, fishing and gathering. In: Arctic Climate Impact Assessment. Cambridge Univ Press, Cambridge, pp. 649–690

Oksanen L (1978) Lichen grounds of Finnmarksvidda, northern Norway, in relation to summer and winter grazing by reindeer. Rep Kevo Subarctic Res Station 14:64-71

Olli A (1995) Off-road vehicle traffic and possibilities to control it in the Pöyrisjärvi wilderness. In: Sippola A-L, Alaraudanjoki P, Forbes BC, Hallikainen V (eds) Northern wilderness areas: ecology, sustainability, values. Arctic Centre Publ 7:354-364

Olofsson J (2001) Long term effects of herbivory on tundra ecosystems. PhD Thesis, Department of Ecology and Environmental Science, Umeå University, Umeå, 49 pp

Paine R (1994) Herds of the tundra: a portrait of Saami reindeer pastoralism. Smithsonian Institution Press, Washington, DC, 242 pp

Pelto PJ (1973) The snowmobile revolution: technology and social change in the Arctic. Cummings Publishing Company, Menlo Park, CA, 225 pp (revised Waveland Press, Prospects Heights, IL 1987, 234 pp)

Riseth JÅ (2000) Sámi reindeer management under technological change 1960-1990: implications for common-pool resource use under various natural and institutional conditions. PhD Thesis, Department of Economics and Social Sciences, Agricultural University of Norway, Ås, 237 pp

Saijets J (2004) Reindeer economy should not be like a museum (in Finnish). Lapin Kansa, 7 Oct 2004, p 5

Sandström P Granqvist Pahlen T, Edenius L et al (2003) Conflict resolution by participatory management: remote sensing and GIS as tools for communicating land-use needs for reindeer herding in northern Sverige. Ambio 32:557-567

Sipilä P, Magga H, Aikio P (2000) Looking for arboreal lichen: Lapland Cooperative area's arboreal lichen inventory 1999-2000 (in Finnish w/English summary). Lapland Reindeer Cooperative, Vuotso, 24 pp

Staaland H, Nieminen M (1993) World reindeer herding: origin, history, distribution, economy. Proceedings World Conference on Animal Production, Edmonton, pp 161-203

Stark S (2002) Reindeer grazing and soil nutrient cycling in boreal and tundra ecosystems. PhD Thesis, University of Oulu, Oulu, 31 pp

Tahkolahti J (2000) Researcher would reduce reindeer by one third (in Finnish). Helsingin Sanomat, 14 Nov 2000, p A11

Talvitie M (2004) Reindeer economy could move into two camps (in Finnish). Lapin Kansa, 16 Sept 2004, p 21

Torp E (1999) Reindeer herding and the call for sustainability in the Swedish mountain region. Acta Borealia 16(1):83-95

Treseder L (1999) The evolution and status of wildlife co-management in Canada. In: Treseder L, Honda-McNeil J (eds) Northern Eden: community based wildlife management in Canada. University of Alberta, Edmonton, pp 7-18

Tuomi-Nikula H (2001) Overgrazing threatens reindeer management (in Finnish). Lapin Kansa 30 Dec 2001, p 4

Turi JM (2000) Native reindeer herders' priorities for research. Polar Res 19:131-133

Turi JM (2002) The world reindeer livelihood - current situation: threats and possibilities. In: Kankaanpää S, Müller-Wille L, Susiluoto P, Sutinen M-L (eds) Northern timberline forests: Environmental and socio-economic issues and concerns. Finnish Forest Research Institute, Jyväskylä, pp 70-75

Valtavaara M (2003) Reindeer breeders look to southern Finland for help as producer-prices fall (in Finnish). Helsingin Sanomat 6 Oct 2003, p A7

Väre H, Ohtonen R, Mikkola K (1996) The effect and extent of heavy grazing by reindeer in oligotrophic pine heaths in northeastern Fennoscandia. Ecography 19:245-253

Wielgolaski FE (ed) (2005) Plant ecology, herbivory and human impact in Nordic mountain birch forest. Ecological studies 180. Springer, Berlin Heidelberg New York, 365 pp

2 Dynamics in Human–Reindeer Relations: Reflections on Prehistoric, Historic and Contemporary Practices in Northernmost Europe

L. Müller-Wille, D. Heinrich, V.-P. Lehtola, P. Aikio, Y. Konstantinov, and V. Vladimirova

2.1 Introduction: People and Reindeer in Northern Environments

Reindeer (*Rangifer t. tarandus* L.) have been hailed as an icon in the past and present, representing northern peoples and environments in the circumpolar regions (Beach 1990). As a species reindeer are well adapted to their natural northern habitats (Jacobi 1931; Banfield 1961). In northernmost Europe, the Sámi, the aboriginal people of these tiers, have lived with reindeer as the thread of life that has connected environment, culture, and livelihood for times immemorial; the neighboring Finns to the south have integrated this animal as an additional element into their emerging agricultural cycles in a more recent era (Chap. 1). Here at the beginning of the 21st century, keeping and herding reindeer is under considerable pressure to adapt and change internally and externally to being the processes of modernization, globalization, and mechanization (Pelto 1973; Heikkinen 2002; Jernsletten and Klokov 2002; Riseth 2003). These rapid changes involve reindeer to a degree that these animals have become iconoclastic and emblematic symbol in national and global image building. Within this context the Nordic countries (particularly Finland) as members of the European Union appear to have emerged globally as representatives – along with the ever-so-popular Santa Claus and his acolytes – of a mythical "Northern Wilderness" attracting worldwide tourism. Recently, Finnish corporate businesses and their products, be it Finlandia Vodka inviting enjoyment of a hardy drink or Nokia connecting people, have invoked images of reindeer and their herding peoples.

This chapter presents reflections on prehistoric and historic processes, as well as on contemporary developments, in the intricate human–reindeer relationships of northern environments. Emphasis is placed on the practices of

Ecological Studies, Vol. 184
B.C. Forbes, M. Bölter, L. Müller-Wille, J. Hukkinen, F. Müller, N. Gunslay, Y. Konstatinov (Eds.)
Reindeer Management in Northernmost Europe

reindeer hunting, keeping, and herding among the Sámi of northernmost Europe. It is not the intention of this chapter to provide a complete picture of human–reindeer relations as they also have evolved among Finns, Komi, Nenets and other peoples in the northwestern portion of the Eurasian continent (Jernsletten and Klokov 2002). Elsewhere in this book (Chaps. 3–7) specific aspects of different types of modern reindeer management are discussed in various sociocultural, economic, political and ecological settings in northernmost Europe. This chapter presents the reindeer–human relations first through prehistoric and early historic times (Heinrich, Sect. 2.2), second by outlining the historic evolution of herding practices among the Sámi (Lehtola, Sect. 2.3), third by discussing the current conflicts between Sámi herders and governmental authorities and their management policies in Sápmi on the Finnish side (Aikio, Sect. 2.4), and fourth by assessing the emerging post-Soviet developments in the Kola Peninsula on the Russian side (Konstantinov and Vladimirova, Sect. 2.5). These reflections highlight the complexity of human–reindeer interactions over time and in space and the importance of these animals to the peoples in the circumpolar north in the context of rapid changes that are taking place and seeming to threaten the very foundation of this relationship.

2.2 Remarks on Prehistoric and Early Historic Dimensions of Human–Reindeer Relations

In Europe the relation between the human, then as a beast of prey, and the reindeer dates back to the Middle Paleolithic. Especially in the Magdalenian (ca. 13,500–12,000 BC) and the Hamburgian culture (ca. 12,500–12,000 BC) of the Upper Paleolithic, it was the most important game. This was also true for the Ahrensburgian culture at the end of the last glaciation (younger Dryas or Dryas III, ca. 10,700–9,600 BC). Reindeer are known even from Magdalenian sites in northern Spain (Asturia, Cantabria, Basque Country) and, judging from the number of their remains, the reindeer were abundant in southwestern France at that time (Herre 1986; Altuna 1999), thus representing the southernmost distribution in Europe during the last glaciation (see Fig. 2.1). Obviously, the late Paleolithic Magdalenian people of West and Central Europe as well as those of the Hamburgian culture and, later, the Ahrensburgian one in northern Central Europe relied on reindeer hunting. It was thought that the hunters would have migrated with the reindeer between their winter ranges and summer pastures according to the postulation of herd control (Sturdy 1975), and even perhaps as a pre-stage to domestication. However, although reindeer were the main game animals based on the large number of remains of this species – in some materials they amount to 99 % of all fragments –, the faunal assemblages from several sites indicate seasonally

Fig. 2.1. Engraving of a reindeer on an antler bone from Kesslerloch/Thayngen; Magdalénien, approx. 12,000 BC; Rosengartenmuseum, Konstanz

changing priorities in hunting patterns. Reindeer were hunted during migrations at special ambush sites during autumn and spring.

Other prehistoric sites, at least Magdalenian ones, were dominated in their remains by horses (*Equus ferus*). This points to summer or winter hunting (Bratlund 1994, 1996; Baales 1996) and indicates that special resources were exploited at specific dwelling places. For instance, the seasonally limited Hamburgian samples from Meiendorf and Stellmoor, situated at a tunnel valley in the utmost north of today's Hamburg (Ahrensburg), bear direct evidence of specialized reindeer hunting in the autumn. The other species identified – such as molting and therefore flightless swans (*Cygnus cygnus*) and geese (*Anser* spp., *Branta leucopsis*) – were perhaps partly hunted at an unknown site in late summer before reindeer hunting started. Some species such as grouse (*Lagopus lagopus*) and hare (*Lepus* spec.) might have been hunted in autumn as well. It is not known what happened during the other months of the year. Perhaps the tunnel valley was settled in spring, too, when reindeer had left the wintering grounds in contemporary Schleswig-Holstein and unknown lands of today's North Sea area. In summer and winter, hunters went to a different eco-zone such as, for instance, the relatively nearby Elbe valley to exploit alternative resources (Bokelmann 1979; Bratlund 1994).

Reindeer must have been of overwhelming importance for the late glacial hunters in the northern German Lowlands, as can also be shown by a study of the development of the ungulate fauna for the time in question. Except for the Allerød stage (ca 12,000–10,700 BC) there is unambiguous evidence for the presence of only two species, mainly reindeer and to a lesser degree wild horse. During the Allerød, a period of increased warmth, elk (*Alces alces*) was the dominant ungulate species; remains of giant deer (*Megaloceros giganteus*) and horse were also found, and reindeer were identified for the Allerød of Denmark and southern Sweden (Benecke and Heinrich 2003). Recently, it has also been proved that reindeer could survive in Schleswig-Holstein during this time (an antler axe was found in Klappholz in this region; Clausen 2004) and even until the Preboreal as is shown by a single antler from Brodersdorf near Kiel (Find: 9746 calBC, KIA 12858; Clausen 2004).

That reindeer were exclusively hunting prey in the late glacial period is also shown by hunting lesions with embedded pieces of flint in some specimens of reindeer bone from the Hamburgian sites of Meiendorf and Stellmoor, especially from the Ahrensburgian layers of Stellmoor. With these discoveries it was possible to reconstruct the hunting methods to a certain extent, particularly by analysis of impact angles. While these samples in the Hamburgian specimens point to stalking of individual animals or small flocks, mainly bagged from a lateral location, the shot position in the Ahrensburgian material was variable, probably resulting from organized battues with hunters on both sides of the narrow passage (Bratlund 1990). On the other hand, there were no marks to be seen of slaughtering as Herre (1955) had already pointed out, countering suggestions of earlier investigators.

Another feature of the relationship between humans and reindeer is the keeping of reindeer, which is connected with the domestication of the species. The oldest archaeological evidence may be the frozen horse discovered in a tomb from Pazyryk in the Altai Mountains near the headwaters of the river Ob dated to the 5th century BC. The animal wore a mask, which probably was to "convert" it into a reindeer, certainly a hint that the reindeer was likewise ridden. Rock pictures in the upper Yenisei region showing reindeer drawing sledges or mounted by humans are believed to date back to the beginning of the Christian era. Furthermore, in the Sajan Mountains a wooden figure was found showing a reindeer wearing a harness; it is dated to the 2nd century AD. These earliest archaeological hints point to a Southern Siberian origin of domestication. It was thought that keeping and herding reindeer in this region was an imitation of such a practice pursued with cattle and horses by other peoples nearby.

According to ethnographic sources, several differences in the original mode of reindeer keeping and herding existed among different peoples in Eurasia. The Sámi in northernmost Europe, for instance, used their animals for driving and transport of people and goods, but they also practiced milking and used dogs for herding, whereas the Chukchi and Koryak in the Siberian Far East had very large herds, and they neither milked their animals nor used dogs in herding. Some Finno-Ugrian peoples used reindeer for pulling sledges, milking was not practiced, and dogs were important in herding. These and other differences may point to an evolution of reindeer herding, inviting postulates that it emerged independently in different locations. Elements also exist, however, in reindeer keeping that are similar among all peoples of the whole area. For instance, the custom of castration by biting the testicles, and that of using a lasso and skis, point to a diffusion of reindeer herding practices from one area to another.

The question is still open whether reindeer keeping developed independently in northern Fenno-Scandia or was introduced, for instance, by contacts with eastern peoples (Nenets/Samoyeds). The taming of decoy animals, which may be the earliest step and type of reindeer keeping, was known among the Sámi from about 890 AD, as indicated by the report of the Norwegian chieftain Othar to King Alfred of England. Othar mentioned keeping tame animals, of which some were decoys used for catching wild reindeer. However, it is stipulated that large-scale, intensive reindeer herding was developed by the Sámi in the Middle Ages, when the population of wild reindeer was diminished severely by non-sustainable hunting practices (Skjenneberg 1984; Benecke 1994).

2.3 The Emergence of Sámi Practices in Reindeer Nomadism and Large-Scale Herding

In the forested and tundra areas of northernmost Europe, the traditional *siida* or village system of the early Forest Sámi, with its relationship to wild reindeer, was the prototype for Sámi society (Lehtola 2002). It was a village unit that provided for community activities, and it was the territory wherein the members of the society had usage rights of land and water. The siida, owned a certain area that usually had well defined borders. The siida system was a permanent socioeconomic and political institution and had been functional for centuries (Aikio 1992). These Sámi were usually thought to have had no permanent settlements or organizational structure before the Scandinavian and Finnish agricultural colonists expanded into Sápmi after the Middle Ages. That is why they were often referred to as "vagabond Lapps". According to current research this picture is wrong. Sámi migrated for their livelihood, but only within their own carefully defined areas. In fact, the Swedish Crown recognized the territorial boundaries and ownership rights of the siida until the 1700s.

The siida's communal lands and waters were divided into usage areas that were exclusive for kin groups or families. No one else had any business in these areas except those dealing with marriage negotiations or trade. Special assemblies, *goahti* ("dwelling" in Northern Sámi), managed the area. These institutions constituted the central management that decided and watched over the use areas, decided on usage rights, and carried out administrative duties and other community business, such as presiding over marriages and accepting new members. All members of the siida had voting rights.

The organization of the siida was based on annual migration. Kin groups and families stayed in their own areas from spring to fall, moving from one settlement site to another according to available resources and fishing and hunting needs. At the height of winter the whole community gathered in the winter village, where foreign merchants, tax collectors, and church ministers were known to frequent. The multi-based Sámi livelihood depended on various sources: freshwater fishing, hunting wild reindeer, snaring birds, sea fishing, picking berries, and gathering – each a sensitive use of renewable resources. The siida system, which took on different forms depending on natural conditions, underwent many kinds of change from the 1500s onward (Aikio 1992). The external reason was the expanding southern central states' influence and the spread of colonization. The internal reason was the intensity change in livelihood related to the beginning of a new economic sector, intensive nomadic reindeer herding, at the turn of the 1500s to the 1600s. Large-scale reindeer nomadism meant that reindeer herds dictated lifestyle, movements, and land use and occupancy.

Reindeer keeping is divided into several developmental stages. Long before economically viable reindeer herding began, reindeer were tamed for use as draft and pack animals, and as decoys to attract other reindeer. *Intensive reindeer keeping* had been practiced, at least in the mountains of Sweden, probably since the 1400s. In conjunction with herding small groups of reindeer, reindeer milk was used for subsistence and reindeer calving was controlled in specific locations. *Extensive reindeer herding* or full herding economy developed later. According to records, the number of tamed reindeer grew significantly in the early 1600s, and on the Swedish side their numbers exploded within just a few decades. The increase in tamed reindeer may be only statistical; perhaps the tax collectors only then began to become interested in reindeer numbers. But it is obvious that larger changes are also linked to this issue (Lehtola 2002). It is clear that full nomadism spread into different directions from the combined central areas of Sápmi on the Swedish side. It is thought to have spread southwestward and northeastward from the Caledonian Ridge. It reached the present area of Finnish Lapland at the end of the 1600s, and arrived in the Kola Peninsula only in the 1800s (Aikio 1992).

It is argued that the birth of *nomadic reindeer herding* is connected with the decrease of wild reindeer. In the mountains between the shores of the Arctic Ocean and Swedish Lapland, reindeer hunting had changed toward the end of the Middle Ages, mostly due to overhunting because the European market consumed endless quantities of furs. Pits and corrals for hunting constantly became bigger and on some mountains the corrals became gigantic. An entire herd would be driven into the corral and all the animals would be killed. With the decrease of the wild reindeer stock and the opening of pastures, reindeer keeping became a sensible way to continue the old livelihood. Hunting wild reindeer provided the most important basis for the development of reindeer herding. The knowledge about wild reindeer, the terminology and techniques, tools and hunting systems were transferred to reindeer herding as special skills.

The change to reindeer herding meant a complete change in the way of life. It became full time work to enlarge the herds and to watch over one's animals. The increase in numbers of reindeer led to a decrease in lichen vegetation in the traditional siida areas. It became necessary to travel far with the herd and to move the home along with it, taking the family in order to keep it together. The herd had to be taken along on trips to the market places, as well, because it could not be left unguarded.

In summer, the migration route ended at the Arctic Ocean coast where there were fewer mosquitoes and more pleasant conditions for the reindeer. The forest pastures were left in peace for the summer. When autumn came the Sámi and their herds migrated back to the forest areas at the edge of the altitudinal tundra where the snow was softer than at the coast. The migration was suited to the rhythm of the seasons – on the cool coast in summer; shielded in the forest from the wind in winter – and lifewas made easier in other ways. Slaughtering was done at the coast in the fall, so the carcasses did not have to be carried long distances to market. It was also easier to migrate with smaller herds. The Sámi herders saved on the externally imposed tax rolls while remaining in the winter villages and pastures. In winter, reindeer (and thus taxes) were fewer, but they increased in the fall for the markets.

These changes also caused a breakdown in the siida system. The Sámi economic model, based on distribution of surplus, changed in the mountain areas to a system favoring private ownership and private enterprise. Nomadic reindeer herding expanded and rapidly spread, and lands were freely claimed for pastures, disturbing the situation of other Sámi groups. A "reindeer village" was not now being founded on the traditional siida practice of exclusive kin group communities; it could be formed of independent individuals or units and families. Under the influence of nomadic reindeer herding, the physical structure of the siida also changed. Small rectangular areas used by the Sámi and located next to the community became greatly elongated Summer residences now could be hundreds of kilometers from the winter settlements.

Reindeer nomadism had many consequences. The expansionist nature caused competition for pastures among siidas as well as among other Sámi groups, for instance, the Sea Sámi. The emphasis on private ownership led to the emergence of differences in wealth and even to capitalism, which had the trait of following "the principle of constant growth and the expectation of unlimited growth" (as expressed in modern terminology). On the other hand, it is said that the emerging Reindeer Sámi had the most secure social system of all the Sámi and of the colonizing settlers. Their food supply exceeded their own needs, and poorer members were provided with reindeer meat for survival. Nomadism encouraged specialization because herders did not make all of their equipment such as the *gieres* (reindeer sled).

Regardless of the fact that full reindeer herding is a historically "young" phenomenon in Sámi culture and that the reindeer herding Sámi way of life has continued as only one of the basic modes of Sámi society, over the years reindeer nomadism has emerged as their defining global image. This is true because external observers took a look and promoted it as the central characteristic and dominant livelihood of all Sámi people. As a widespread form it nevertheless did not persist for more than 200 years. The demise of nomadic reindeer herding was particularly felt on the Finnish side of Sápmi because of externally imposed national border arrangements during the 1700s and 1800s that led to spatial interruption of migratory patterns and curtailed access to previously used pastures. The border closures to Norway in 1852 and Sweden in 1889 by Russia, rulers of Finland at the time, effectively divided the Sámi reindeer communities (Lehtola 2002). The closures caused relocation of people and changes in the daily life of the siida, as well as reduction in the size of reindeer herds by the early 1900s. On the Norwegian and Swedish sides the migratory patterns have continued under legal conventions between the two countries allowing for the crossing of the international well into the 21st century.

Going back to the early 19th century, by that time economic development policies and their applications by the Finnish, Norwegian, and Swedish states in their respective northern regions, i.e., Sápmi, had led to the establishment of separate national laws and regulations dealing with reindeer herding, fishing, hunting, and gathering under the umbrella of agricultural practices. From that time on, the Sámi system of reindeer herding developed into different directions in these three countries. Reindeer herding attained a "national flair". This process has gradually eroded the cultural, socioeconomic, and political conventions, norms, and identity of the Sámi siida system.

2.4 Sámi Reindeer Herding in Crisis: Reflections on Practices in Sápmi, Finland

Here at the beginning of the 21st century, conditions in open-pasture reindeer herding among the Sámi have reached crisis proportions in Sápmi on the Finnish side, the Sámi Home Region in northernmost Finland. Unfolding developments threaten to destroy the very foundations of Sámi reindeer herding. In a short time, the situation has culminated in an economic emergency. In late 2003, the meat price in Finland's reindeer herding area dropped by

25% compared with 2002. Furthermore, in Sápmi on the Finnish side the annual average income of reindeer herders had dropped to 5,045 per annum per capita, well below the European Union poverty level set at 5,800. In comparison Finnish farmers in the southern parts of the reindeer herding area had an annual gross income of 11,600 (all figures are from Statistics Finland 2004). This crisis has not been created within the reindeer herding culture, it is rather the consequence of rationally developed and clearly directed policies by Finnish governmental authorities.

Among reindeer herders the notion exists that the Finnish State intends to keep Sámi herders at the lowest income level and, moreover, encourages the discontinuation of open-pasture reindeer herding and blocks the Sámi struggle for aboriginal land rights in their home region. If Sámi herders "voluntarily" relinquish or are forced by "economic laws" to give up open-pasture grazing, they are no longer protected by international conventions such as the Indigenous and Tribal Peoples Convention (ILO 1989). They will then lose their rights to land and resources unless they continue the practice of open-pasture reindeer herding (Aikio and Hyvärinen 2004). By 2004, Finland had not yet ratified the convention. Similar tendencies toward centrally directed control of reindeer herding are noticeable in other areas of Sápmi on the Norwegian or Swedish side (Chaps. 4–7; see also discussions in Beach 1990; Paine 1994; Hukkinen et al. 2003; Riseth 2003).

This imminent crisis relates to the discussions over the application of sustainable development as they have propagated since the Report by the Commission on Environment and Development (WCED 1987). The principles of sustainable development touch upon the needs of contemporary populations with a view of how future generations will be able to maintain appropriate levels of resource utilization and quality of life (Hukkinen 2003). For reindeer herders, sustainable development means the maintenance of sociocultural identity supported by economic activities and land use rights in relation to externally introduced expanding and competing interests such as, for example, globalized industrial forestry and large-scale tourism. In this context, sustainable development clearly contains and combines ecological, economic, and sociocultural dimensions.

Current circumstances in reindeer herding indicate that chaos reigns at various levels due to a multitude of interests, which try to exert control over herding activities and open-pasture. Semi-domestic reindeer, like wild reindeer, are migratory animals that move freely from pasture to pasture depending on ecological conditions and annual cycles. Open-pasture reindeer herding in northern reaches has always reacted and adjusted substantially to ecological requirements (Chap. 7), in contrast to reindeer herding in southern areas, which has been attached to agricultural practices. This northern practice has always been based on presupposed pasture rotation and has required only light technology. Nonetheless, laws, regulations, and administrative activities dealing with reindeer herding have been established in Finland based on introduced sedentary settlement patterns modeled after the requirements of crop and cattle farming.

Reindeer herding, if defined as a meat-producing economy, could be seen as one possibility or element of general agricultural production, which, in essence, aims at profitability. However, the total volume of reindeer meat production in Finland is only about two million kilograms per year. In comparison, Finns consume six million kilograms of pork during the Christmas holidays alone (Statistics Finland 2004). Reindeer herding is not a free-market enterprise, rather it is a livelihood based on limited renewable resources. However, despite this fact, herding is a tightly regulated field of activity. Subsidies are not paid on the same principles as in agriculture and, if they are, then they are proportionally lower in comparison with conventional agriculture. Sámi reindeer herding is very much culturally based and thus does not compare well with agricultural activities aimed at maximizing meat production, and thus profitability.

In the 1990s, the discussions concerning the fate of reindeer herding within the process of European integration have emphasized (1) the restructuring of "northern or arctic agriculture", (2) the strengthening of reindeer herding productivity and profitability as an element of agriculture, as well as (3) the European Union's demand for the protection of large predators in Finland. The rhetoric used is taken straight from concepts propagated by the business community and the industrial complex. Links between powerful economic demands and the human values of Sámi culture are not visible – are not even discussed. Still, after considerable lobbying, the European Union has thus far looked favorably at Sámi reindeer herding and its development based on Sámi cultural reasoning. Protocol No. 3 on the Sámi People attached to the 1994 accession treaty between Finland, Sweden, and the European Union provides a strong commitment to aboriginal rights including "reindeer husbandry" (European Union 1994).

In this connection, it is necessary to examine sustainable development for reindeer herding through its central elements. In herding, people, reindeer, pasture and social norms, conventions, and rights related to land use practices are a totality in themselves, within which all elements are linked and on whose prevailing balance Sámi sustainable reindeer herding is based. In fact, nature has provided the basis for sustainable use. The well-developed relationship between humans and reindeer is the foundation of Sámi philosophy, culture, and economy (Lehtola 2002). This holistic condition has shaped basic, original knowledge through cultural and linguistic development over hundreds and even thousands of years.

Under contemporary conditions Sámi herding is strictly monitored and limited in expressing its practitioners' points of view. The keystone of reindeer herding as a livelihood and practice is completely misunderstood from the outside when tight norms of the market economy are applied to assess and measure the productive value of reindeer herding and its economic potential. In this situation the acceptance and recognition of the sociocultural dimension of sustainable development by people outside the herding community is still very much lacking.

The Sámi Home Region is about 10 % of Finland's national territory and is located in its northernmost extension (Chap. 1, Fig. 1.1). This area consists of open fell and timberline forests. The average annual mean temperature is around minus one centigrade and thus is not considered favorable for agricultural practices. The fell region is also less productive biologically than the more southerly agricultural zone dominated by pine forests. It is estimated that the annual biomass production of the fell or tundra areas can only support 5 % of the people are sustained by the pine forest belt. This means that it would require an area 20 times larger if the same number people should live in the fell region as in the pine forests (Kallio 1971).

Under current circumstances, the ecological dimension of reindeer herding is defined and tightly restricted by other competing land use practices such as commercial forestry, industrial extraction of mineral and hydroelectric resources and, to a lesser degree, agriculture. In the public debate, the viewpoint is expressed that reindeer are too numerous in relation to the existing pasture resources and thus their numbers need to be decreased. However, the impact of other land users is not taken into commensurable account. For decades now the number of reindeer has been tightly controlled yearly by quota, and the systematic decrease of reindeer has become a prevailing theme in the development regime. The principle of setting quotas emerged in administrative regulations during the 19th century. Since that time reindeer have been seen as a nuisance or even pest, causing damage to and competing with agriculture and the emerging industrial forestry. Reindeer herding has continuously been forced to relinquish pasture to other land users against whom its competitive position was weak due to unfavorable laws and regulations (Aikio 1978).

The impact of climate change on reindeer grazing has only been more recently assessed. Changing levels in temperature and precipitation and their influence on snow cover and, in extension, on reindeer grazing habits are more and more recognized as decisive factors in herding (Turi 2000). In addition, the killing of reindeer by predators needs to be considered, particularly with respect to the principles of conservation, but also in view of the difficulties of securing exact numbers of lost animals, the base capital of reindeer herders.

In economic terms, sustainable development in reindeer herding is a conflict between livelihood and free-market enterprise and business. In fact, reindeer herding is a small factor in the national and regional economy. Reindeer meat production in Finland has been around half a kilogram per capita or approximately 2.2 million kilograms per year. The total turnover per year was calculated at 33.6 million in 2002; this is less than a year's profit at the recreational complex of Saariselkä in northern Finland (Statistics Finland 2004).

Sámi reindeer herding is founded on direct physical benefits derived from hides, antlers, meat, heart, liver, blood, etc., with very small monetary income. Still herders are expected to participate in the market economy. However, reindeer herding has not been transformed from livelihood to player in the market economy. Although it is seen as part of agriculture on the national level, programs for subsidies and social security have barely been expanded to benefit Sámi herders.

Sámi reindeer herding is a collective livelihood defined by Sámi culture and society. The income levels of reindeer herders are very low and cannot be influenced by the practitioners themselves since they are dependent on natural conditions, high expenditures, lack of protective regulations covering land use, controlled slaughter quotas and, finally, lagging policies on subsidies and social welfare. Reindeer herders live today under tight economic conditions with a low level of engagement and a growing risk of marginalization. In 1998, the national subsidy per farm homestead was almost 16,600 in Finland. At the same time, herders had to be content with a subsidy of 20 per year per animal, which amounts to barely 12 % of subsidies paid per sheep (166.50 per year). These figures illustrate very well the negative effects of the State's subsidization policies on reindeer herding (Saamelaistoimikunnan mietintö 2001). The reindeer is a small animal providing a low amount of edible meat. The setting of quotas by the State constantly reduces the number of reindeer. Thus, herding cannot expand its economic activity in the meat-producing sector.

In general, international law tends to favor aboriginal peoples and their rights. International conventions protect Sámi herders against assimilation, and Protocol No. 3, mentioned above, includes the rights of Sámi to aboriginal status and reindeer herding in their home region. Under this protocol Sámi have been able to access European funds for specific development projects. However, the Finnish Ministry of Agriculture and Forestry has steadfastly argued that reindeer herding is part of Finnish agriculture and that Protocol No. 3 has no implications for national policies.

So what is the future for Sámi reindeer herding in Finland? First, as part of general agricultural development policies it is not tenable under current circumstances. Second, under these policies open-pasture herding would come to an end and the collective Sámi right to land use would be meaningless. Third, newly proposed regulations would result in "reindeer farming" comparable to beef cattle farming. Herding would become commercialized with the side effect of catering to tourism. Fourth, such developments would lead to further rapid linguistic and cultural assimilation. And finally, in this atmosphere, the northern forest and tundra areas would come under stronger regulatory control favoring industrial and recreational exploitation. In the worst case scenario, such a vision could be realized within a couple of years.

On the other hand, a proposal can be envisioned aiming to revitalize herding for the Sámi. Such a proposal would re-establish their sovereignty over reindeer as a resource, an age-old mainstay of northern peoples. Therefore, a new profile of open-pasture reindeer herding needs to be projected to include sustainable development with a holistic approach to ecological, socioeconomic, and cultural conditions appropriate and acceptable to Sámi society. The legal position of reindeer herding would need to be secured in Finland's constitution and by international conventions based on historical aboriginal rights to lands and its resources including the grazing by reindeer. Under such legal protection, administration and management of Sámi reindeer herding would be transferred from the Ministry of Agriculture and Forestry to the Sámi Parliament in Finland, right into the heartland of the Sámi and the reindeer (Aikio and Hyvärinen 2004). The result of this new profile would be the separation of reindeer herding from agriculture and industrial meat production. In the Sámi Home Region herding would legally be declared *Sámi reindeer herding* and would include all herders working in this sector. Moreover, Sámi reindeer herding would receive full recognition as a separate economic and cultural activity by the European Union, thus strengthening its foundations. Furthermore, the herding sector could develop cooperative programs with nature conservation and tourism and therefore contribute to the particular cultural air and flair of the northernmost edge of the European Union.

2.5 Human–Reindeer Relationships in the Post-Soviet Context: Kola Peninsula (Northwest Russia)

The period since 1991 has seen the dissolution of the Soviet Union and the introduction of market-oriented economic policies by its successor. For reindeer herding in the Kola Peninsula (Murmansk Region) this has meant the transformation of the two existing state reindeer herding farms – "*Tundra*" of Lovozero, and "*Pamiat Lenina*" (in Memory of Lenin) of Krasnoshchelye – into herding cooperatives (Chap. 6).

This transformation from *sovkhoz* (state farm) to cooperative can be considered relatively superficial in terms of the relations between two crucial groups of actors, namely, the administors of the farms on one hand, and the herders on the other. The herders' status as salaried employees of the former state farms did not change. Also the strict hierarchical division between administration and workers has remained. On the popular discursive level this is reflected in the fact that, despite all superficial changes, the herding community still refers to the farm as "the sovkhoz", while the various new appellations have a life only in official documents and are rarely mentioned in the regional press. *TOO* is the Russian acronym equivalent to a company with limited liability and *SKhPK* refers to an agricultural cooperative. The former state farms became TOOs in 1992 and were changed to SKhPKs in 1997 (see detailed explanations in Chap. 6).

While the herders' status vis-à-vis the superimposed bureaucracies has been essentially preserved, the economic state of the two reindeer farms has been seriously jolted by new forces. Most significantly, the state ceased to take care of buying the production of the farms and thus relieving them from the worries of finding customers. In accordance with the new pro-market course, the state also withdrew from subsidizing the whole productive process. An indirect part of such subsidizing was connected with the low prices of fuel before 1992. In the Soviet era this allowed unproblematic mechanized transportation, both surface and airborne. As the herders remember those days: "Helicopters would come whenever we asked for them, sometimes for trifles".

The huge gap opened by the withdrawal of the state has been partly filled by the operation – again since 1992 – of a Swedish-owned slaughtering facility. This company, "*Norfrys*", has since become the main buyer of meat from the two cooperatives, renamed "SKhPK *Tundra*" and "SKhPK *Olenevod*".

What might be called a *quasi-sovkhoz* mode of existence could define the current status quo. At the same time, the herding community and its leading actors, i.e., the herders currently employed by the cooperatives, have introduced internal management changes of their own. It can be said that these changes have far reaching effects on the further development of the farms as well as on the relationship of humans and reindeer in this particular part of the herding universe (Chap. 6). Some of these changes are highlighted briefly.

The people, predominantly men, who are presently actively engaged in herding in the Kola Peninsula do not exceed 200. They include herders and veterinarians, both of various ranks, camp cooks, drivers, and corral workers. In terms of ethnic composition this group is numerically dominated by the Komi, the Sámi come second, and third are other groups, mainly Komified Nenets, Ukrainians and Russians. These ethnic lines of division are highly arbitrary, however, since a great part of the local population is of mixed descent through a number of generations.

The type of reindeer herding practiced also represents a fact requiring adjustment to current realities. To speak of "Sámi reindeer herding" in Kola does not bear close examination. Before the arrival of the Komi in the Kola Peninsula in 1883/1884 (Konakov 1993), Sámi reindeer herding was characterized by keeping relatively small herds, by close handling of and caring for reindeer ("intensive herding"), and, most significantly, by free grazing during the summer, when the Sámi engaged in fresh and salt water fishing (Jernsletten and Klokov 2002). The most radical changes introduced by the Komi upon arrival was a considerable increase of herd size and year-round controlled herding, both very much against Sámi traditions and interests. It should be noted that Soviet herding management, especially after the Second World War, did everything possible to promote year-round control of reindeer herds, favoring Komi management practices and aiming at higher numbers (Chap. 6).

At present, what can be observed is a system that does not answer to either of these ideal types – neither to intensive Sámi herding (Beach 1990) combined with free summer grazing, nor to Komi year-round controlled herding (nor to the Soviet modification of the latter, for that matter). Present management is characterized by very loose and sporadic control over herds, and, significantly, by very late slaughtering extending into last months of pregnancy of females (i.e., March and April). In comparison to Soviet-time herding, the greatest change has been the mixing of herds and the parallel mixing of territories. Soviet state farm herding had imposed a strictly defined herding team (brigade), with the division of herds and their grazing range corresponding roughly to the Sámi *pogost* (siida) arrangements of earlier times. Post-Soviet herding is characterized, in contrast, by allowing former brigade herds to mix into oversized composite herds, which in some observed cases reach 10,000 or even 12,000 head. The animals roam more or less freely over parts of the Kola Peninsula and are loosely referred to as the "left side" (Brigade No. 3, 4, 6, and 7 of "*Tundra*"), the "right side" (Brigade No. 1, 2, 8, and 9) and the "Krasnoshchelye side", which extends all the way from Lake Kalmozero to the southeastern coast of the Barents and the White Sea.

The driving force behind these changes can be seen in the desire of the herding community to avoid the risks of private herding practices by clinging to the current quasi-sovkhoz mode of existence (Beach 1992). As a worldview and practice, this disposition is seen as being dominant in the community, both at its administrative and its rank-and-file levels, a state of affairs that can be called "sovkhoism" (Chap. 6).

Meanwhile, the current relationship between humans and reindeer on the Kola Peninsula can be defined as one of growing alienation (Chap. 6). This is characterized by minimal or even totally absent control over herds, and the "going wild" (*odichanie*) of herds, particularly of the females. Contact with humans is established mainly for slaughtering purposes and is thus comparable to predation (Ingold 1986). On the other hand, and in an opposite direction, a close bond is noted between herders and draft animals (sled geldings), a characteristic dependence that exists throughout the Russian Far North and Siberia. In the absence of control and the existence of informal privatization as a type of property transfer, alienation seems to be the process defining the human-*Rangifer* bond in this region.

2.6 Conclusion: The Future of Human-Reindeer Relations and its Practices

Since the last deglaciation in northernmost Europe, human–reindeer relations have been a mainstay for the continuation of human existence and occupancy under these arctic and subarctic conditions. Human reliance on this circumpolar species has allowed the flourishing of intricate human–environmental interactions, an interplay has resulted in continuous adaptive processes as a response to changes in natural, socioeconomic, cultural and political conditions. The above reflections on these processes give an indication that the human–reindeer relationship has remained fundamental in the lives of the Sámi in northernmost Europe, as well as other northern peoples, since prehistoric times.

During the last 50 to 60 years, conditions have changed rapidly and dramatically. These changes touch upon the very foundation of the precarious relationship between environment, animals, and people, the latter having based their sustenance and well-being on the continuation of this link (Chap. 3). Until recently, reindeer keeping and herding was almost self-contained and did not face major competitors in the use of land and water resources. These circumstances have been altered irreversibly by the expanding external economic and political interests extending in to these northern tiers by focusing specifically on the utilization of its evident mineral, wood, hydroelectric resources, and on its potential for recreation. This emerging constellation has led to competition and conflicts between the local, aboriginal and immigrant populations as well as the central institutions of nation-states whose sovereignty has evolved spatially and administratively by carving boundaries of their respective national territories into the North.

In this volatile situation, the question arises how reindeer herding can continue to function under evolving modern management practices in coexistence with other land uses, competing for space and attention, to be represented fairly in public policies that purport to deal with collective or common resources – territory (i.e., "wilderness") and animals (Chap. 4). In the authors' opinion, realizing the current state of prevailing political conditions in the North, the following issues are crucial to the negotiation and creation of a new balance among the interests and claims of multiple users of land and its resources, in particular concerning the future viability of reindeer herding.

The issues are also spelled out in international conventions to which states have already signed up or still intend to sign fulfilling its requirements.

1. Continued access to and control of territory by the local population such as the Sámi is a prerequisite for the application of the principles of sustainable development to safeguard the foundation for reindeer herding. Such a solution does not mean that other users are excluded, rather a negotiated management regime will allow for the coexistence of different interests (Chap. 3, 16).
2. The rights to and the protection of the utilization of specific resources by aboriginal peoples need to be accepted, particularly in the circumpolar North with low human populations and severe environmental conditions. The efforts of the European Union to enhance the continuation of specific

cultural landscapes and cultural traits and elements are steps to recognize the value and maintenance of human diversity.

3. Local decision-making powers within a framework of fair negotiations over existing interests and related management policies are crucial to secure a continued basis for, and shape the future of, reindeer management for Sámi herders. It must be understood that such powers contribute fundamentally to the physical, mental, and spiritual well-being and the viability of small communities in northernmost Europe (Chap. 5).
4. A prerequisite for the above points is the acceptance of reindeer herding as a livelihood versus an activity within the free-market economy, which latter is based on strong demands for success and profitability. Reindeer herding cannot be economically expanded significantly due to its limitations in resources (pasture) and reproductivity (reindeer).
5. The final issue concerns the recognition of cultural and linguistic diversity, which is closely related to livelihood. Culture and language are of dynamic application and wide-ranging in contribution to human diversity and humanity. Its value can only be maintained by affording mutual respect despite the inevitable conditions of minority–majority relations. After all, the lasting strength of a democracy comes not from enforcing "the will of the majority", but rather from honoring the equal rights of the minority.

In the modern context of northernmost Europe, in the polar fringe of the European Union, reindeer herding is a minute economic activity whose viability and feasibility are questioned in light of the radical processes of modernization and globalization. The reflections expressed in this and other chapters of this book show that there exists vitality in reindeer herding communities that give an indication of the determination to continue herding practices. However, this determination needs to be recognized and accepted by the external forces to allow it continue.

Acknowledgments. The English translation and adaptation of the original Finnish version was provided for Sect. 2.3 by Linna Weber Müller-Wille and for 2.4 by Ludger Müller-Wille. Kustannus-Puntsi Publisher (Aanaar/Inari, Finland) granted permission to republish passages from "The Sámi People. Traditions in Transition" (Lehtola 2002); letter by Veli-Pekka Lehtola dated May 27, 2004.

References

Aikio P (1978) The breakdown of a Lappish ecosystem in northern Finland. In: Müller-Wille L, Pelto PJ, Müller-Wille Li, Darnell R (eds) Consequences of Economic Change in Circumpolar Regions. Univ Alberta Boreal Institute for Northern Studies Occas Publ 14:91–104

Aikio P, Hyvärinen H (2004) Valtio heikentää poronhoidon kannattavuutta (in Finnish; The State weakens the profitability of reindeer herding). Suomen Kuvalehti 8:75

Aikio S (1992) Olbmot ovdal min. Sámiid historjá 1700-jogu rádjái (in Northern Sámi; The people before us. Sámi History to the 1700s). Girjegiisá Oy, Ohcejohka

Altuna J (1999) Mammal changes between the Dryas and the Holocene in Northern Spain. In: Benecke N (ed) The Holocene history of the European Vertebrate Fauna. Modern Aspects of Research. Leidorf, Rahden/Westf, pp 1–8

Baales M (1996) Umwelt und Jagdökonomie der Ahrensburger Rentierjäger im Mittelgebirge. Römisch-Germanisches Zentralmuseum Monographien, Bd 38. Habelt, Bonn

Banfield AWF (1961) A Revision of the Reindeer and Caribou, Genus *Rangifer*. National Museum of Canada Bulletin 177. Biol Ser 66:1–137

Beach H (1990) Comparative Systems of Reindeer Herding: In: Galaty JG, Johnson DJ (eds) The world of pastoralism. Herding systems in comparative perspective. Belhaven Press, London, pp 255–298

Beach H (1992) Reindeer Herding on the Kola Peninsula – report of a visit with Saami Herders of Sovkhoz Tundra. In: Kvist R (ed) Readings in Saami history, culture and language III. University of Umeå, Umeå, pp 113–142

Benecke N (1994) Der Mensch und seine Haustiere. Die Geschichte einer jahrtausendealten Beziehung. Theiss, Stuttgart

Benecke N, Heinrich D (2003) Neue Daten zur Entwicklung der Huftierfauna im Tieflandgebiet zwischen Elbe und Oder im Spätglazial und Altholozän. Archeozoologia 21:19–36

Bokelmann K (1979) Rentierjäger am Gletscherrand in Schlesig-Holstein? Offa 36:12–22

Bratlund B (1990) Rentierjagd im Spätglazial. Eine Untersuchung der Jagdfrakturen an Rentierknochen von Meiendorf und Stellmoor, Kreis Stormarn. Offa 47:7–34

Bratlund B (1994) A survey of the subsistence and settlement pattern of the Hamburgian Culture in Schleswig-Holstein. Jahrb Röm Germ Zentralmus Mainz 41:59–93

Bratlund B (1996) Archaeozoological Comments on Final Palaeolithic Frontiers in South Scandinavia. In: Larsson L (ed) The Earliest Settlement of Scandinavia and its relationship with neighbouring areas. Acta Archaeol Lundensia Ser 8(24):23–33

Clausen I (2004) The reindeer antler axe of the Allerød period from Klappholz LA 63, Kreis Schleswig-Flensburg/Germany. Is it a relict of the Federmesser, Bromme or Ahrensburg culture? In: Terberger T, Eriksen BV (eds) Hunters of a changing world – environment and archaeology of the Pleistocene–Holocene transition (ca. 11000–9000 B.C.) in Northern Europe. Workshop of the U.I.S.S.P. Commission XXXII at Greifswald in Sept 2002. Leidorf, Rahden/Westfalen, pp 141–164

European Union (1994) Protocol 3 – on the Sami people, Act concerning the conditions of Accession to the Treaties on which the European Union is founded, 11994N/PRO/03. Official Journal C 241, 29/08/1994, p 0352

Heikkinen H (2002) Sopeutumisen mallit. Poronhoidon adaptaatio jälkiteolliseen toimintaympäristöön Suomen läntisellä poronhoitoalueella 1980 – 2000 (in Finnish; Models of adaptation – adaptation of reindeer herding to the post-industrial environment in the western Finnish reindeer herding area 1980 – 2000). Finnish Literature Society, Helsinki

Herre W (1955) Das Ren als Haustier. Eine zoologische Monographie. Geest and Portig, Leipzig

Herre W (1986) *Rangifer tarandus* – Ren, Rentier. In: Niethammer J, Krapp F (eds) Handbuch der Säugetiere Europas 2/II Paarhufer – Artiodactyla. Aula, Wiesbaden, pp 198–216

Hukkinen J (2003) Sustainability indicators for anticipating the fickleness of human-environmental interaction. Clean Technol Environ Policy 5:200–208

Hukkinen J, Müller-Wille L, Heikkinen H (2003) Development of participatory institutions for reindeer management in northern Finland: Preliminary synthesis and report. Technol Soc Environ 6:1–29
ILO (International Labour Organization) (1989) Indigenous and Tribal Peoples Convention (C.169). International Labour Organization, Geneva
Ingold T (1986) Reindeer economies and the origins of pastoralism. Anthropol Today 2(4):5–10
Jacobi A (1931) Das Rentier. Eine zoologische Monographie der Gattung *Rangifer*. Zool Anz 96 [Suppl]:1–258
Jernsletten J-L, Klokov K (2002) Sustainable reindeer husbandry. Centre for Saami Studies, Tromsø
Kallio P (1971) Ihminen Lapin ekosysteemin osana (in Finnish; Humans as part the ecosystem in Lapland). Kalevalan vuosikirja 51:4–113
Konakov ND (1993) Ecological adaptations of Komi resettled groups. Arctic Anthropol 30:92–102
Lehtola V-P (2002) The Sámi People. Traditions in Transition. Translated by Linna Weber Müller-Wille. Kustannus-Puntsi, Aanaar-Inari
Paine R (1994) Herds of the Tundra. A portrait of Saami Reindeer Pastoralism. Smithsonian Institution Press, Washington, DC
Pelto PJ (1973) The Snowmobile revolution: technology and social change in the arctic. Cummings Publishing Company, Menlo Park, CA (revised Waveland Press, Prospects Heights, IL 1987)
Riseth JÅ (2003) Sámi Reindeer Management in Norway: Modernization Challenges and Conflicting Strategies. In: Jentoft S, Minde H, Nielsen R (eds) Indigenous peoples: resource management and global rights. Eburon, Delft, pp 229–247
Saamelaistoimikunnan mietintö (2001) Ehdotus maahan, veteen, luonnonvaroihin ja perinteisiin luontaiselinkeinoihin liityvien oikeuksien järjestämisesta saamelaistenkotiseutualueella valtion mailla (in Finnish; Proposal for the establishment of rights on state land in the Sámi Home Region related to land, water, natural resources and traditional livelihoods). Komiteanmietintö 14:1–140
Skjenneberg S (1984) Reindeer. In: Mason IL (ed) Evolution of domesticated animals. Longman, London, pp 128–138
Statistics Finland (2004) www.stat.fi. Accessed 30 May 2004
Sturdy DA (1975) Some reindeer economies in prehistoric Europe. In: Higgs ES (ed) Palaeoeconomy. Cambridge Univ Press, London, pp 55–95
Turi JM (2000) Native reindeer herders' priorities for research. Polar Res 19:131–133
WCED (World Commission on Environment and Development; 1987) Our common future. Oxford Univ Press, Oxford New York

3 Development of Participatory Institutions for Reindeer Management in Finland: A Diagnosis of Deliberation, Knowledge Integration and Sustainability

J. HUKKINEN, L. MÜLLER-WILLE, P. AIKIO, H. HEIKKINEN, O. JÄÄSKÖ, A. LAAKSO, H. MAGGA, S. NEVALAINEN, O. POKURI, K. RAITIO, and N. WEST

3.1 Introduction

The objective of this chapter is to diagnose the mechanisms by which a focused effort to integrate knowledge based on various professional and disciplinary backgrounds can result in emergent participatory institutions for resource management – in this case reindeer herding management in northern Finland. Considering knowledge integration as a process of institution-building makes sense, since in environmental governance literature institutions are widely understood as working rules that are common knowledge to resource users (Ostrom 1990). In this context, the crucial point is the process of knowledge transfer through networking and communication. Three stages of knowledge integration characterize the entire research effort, which the authors conducted as Workpackage 1 (WP1) under the RENMAN project: (1) pioneer networking, (2) translational networking, and (3) modular networking (Bruun et al. 2002; Hukkinen et al. 2003b; Forbes et al. 2004).

Our typology of knowledge integration draws from recent literature on organizational learning in innovation (Bruun et al. 2002; Langlais et al. 2004). The project became a prime example of a focused effort of knowledge integration, because every stage of it was based on participatory processes. We go on to show that the processes of knowledge integration hold the characteristics of emerging institutions for resource management: formal and informal rules were developed for the participatory process of the RENMAN project and proposed for future reindeer management; the rules were geographically specified; legitimate participants in the process were clearly defined; and conflict resolution and sanctioning mechanisms were proposed during the project (Hukkinen et al. 2002, 2003a; Heikkinen et al. 2003a).

Ecological Studies, Vol. 184
B.C. Forbes, M. Bölter, L. Müller-Wille, J. Hukkinen, F. Müller, N. Gunslay, Y. Konstatinov (Eds.)
Reindeer Management in Northernmost Europe

While there is plenty of research to advise us on the requirements of social institutions that facilitate sustainable natural resource management (Ostrom 1990; Gunderson et al. 1995; Hanna et al. 1996; Ostrom et al. 1999; Dietz et al. 2003), less work has been done on the cognitive and organizational processes by which such institutions emerge. North (1992) emphasizes the importance of dominant mental models in the shaping of new institutions, but focuses on the macro-level evolution of economic systems. Ostrom (1990) discusses the successes and failures of common pool resource management in different institutional settings, but not the details of the cognitive and organizational processes as new institutions emerge. Hukkinen (1999) does focus on the feedback between cognitive models and institutions, but diagnoses failures in the institutions of modern environmental management. While recent developments in agent-based modeling have increased our understanding of the changing properties of human–environment interaction, institutions in that work have been treated as exogenous factors of individual decisions. The construction of models in which norms, rules, and institutions emerge from group interactions has been identified as an important future challenge (Parker et al. 2002). Since the goal of our empirical work was to develop practical participatory processes of institutional development, we did not strive for a model of individual–institutional dynamics. However, the knowledge-integration process that we examine here provides a structured understanding of the dynamics of cognitive–institutional interaction, which is a necessary first step in the development of decision rules and boundary conditions for more formal models.

The cognition and knowledge of relevant participants in a resource management system are core issues in the development of environmental governance systems. Meaningful participation of relevant stakeholders has been identified as a cornerstone in the development of effective institutions for natural resource governance (Berkes and Folke 1998; Dietz et al. 2003). Participation implies knowledge integration, i.e., ensures that knowledge sets from different cultural, professional, and scientific backgrounds encounter each other and that differences in knowledge be recognized, if not reconciled. Since institutions incorporate shared working rules, some type of knowledge integration must take place for the institution to emerge. In this chapter we describe a participatory research project, during which the processes of knowledge integration themselves became promising emergent institutional structures for sustainable governance of northern natural resources. We will describe the knowledge integration processes in the RENMAN project by tracking the evolution of key results from initial ideas to final policy recommendations. In doing so, we also outline the structure and dynamics of the knowledge networking through which the knowledge integration took place. Finally, we identify characteristics in the knowledge networks that persuade us to call them emerging institutions for sustainable reindeer management in

Finland (for a background on reindeer management in northernmost Fennoscandia, Chaps. 1 and 2).

The overall thrust of the RENMAN project was to "integrate the indigenous people in an integrative study between politics and science", with the specific objectives of "participatory assessment and systems analysis of different reindeer management regimes", the development of "integrative scenarios and management plans for future sustainability", and "enhancing the well-being and quality of life of local populations" (Forbes et al. 2004 pp. 7-9). Based on earlier recorded experiences in "participatory action research" (PAR) (Hall 1979; Hiebert and Swan 1999), our research group, or Workpackage 1, decided to conduct all research activities in this research mode. Researchers and practitioners (reindeer herders) became expert partners in WP1; thus all members of WP1 are authors of this chapter, and six of them are reindeer herders or owners. The research questions set forth by WP1 were formulated in collaboration between researchers and reindeer herders. The contributions to this chapter are based on joint fieldwork, discussions, communications, meetings, writings, translations, and reviews, and thus represent the authors' common points of view. Furthermore, the views are grounded in three participatory workshops, which were organized during the three-year project (2001-2004) and led to final recommendations developed in intensive deliberations among researchers, herders and government officials (Hukkinen et al. 2002, 2003a; Heikkinen et al. 2003a).

During the RENMAN project the PAR approach adopted by WP1 triggered questions among officials and researchers concerning the sources of knowledge in policy design. At the final symposium of the project in Rovaniemi in November 2003, for example, we were requested to distinguish between the preferred recommendations of the herders, the preferred recommendations of the researchers, and the optimal recommendations; all this presumably in order to have an impact on public policy-making. The request was unexpected and contrary to our intentions. At the same time it underlined the framework under which research today is conducted. Our research problems and questions were freely formulated by both the academic community and the "stakeholders" or practitioners (reindeer herders). Yet project implementation functioned under a fairly rigid bureaucratic and politically driven system, which complicates the impact of research results. The request placed us at a crossroads in Rovaniemi. How could we, after our experience, separate between those ideas that came from the professional herders and those that came from the professional researchers, when the entire analytical process was designed from the very beginning to integrate the analysis, the questions, the processes, and the results? Still the request did raise a valuable point about analytically disentangling the processes of participation, partnership, and related knowledge transfer (Chap. 4). Since we felt strongly that harmful obstacles to communication had been brought down and fruitful collaborative linkages had been established during our work, it would indeed be worthwhile to analyze the processes by which this had been achieved. This is what we aim to do in this chapter.

3.2 Knowledge Integration: The Path from an Idea to a Policy Recommendation

Our first analytical task is to trace the evolution of initial observations made during informal exchanges of fieldwork into final policy recommendations for improving the governance of reindeer management in Finland. The analysis of these processes of knowledge integration will follow the three stages of networking mentioned above (Bruun et al. 2002; Langlais et al. 2004). The recommendations are institutional reform proposals (Table 3.1): the first pro-

Table 3.1. RENMAN policy recommendations

Policy recommendation	Specifying statements
Recommendation 1: Develop reindeer management as a social innovation and a symbol.	• Reindeer management is a socioecological innovation and a national symbol of Finland. • As an innovation and a brand, reindeer management is worthy of significant public funding for investment, operation, management, research and development. • Reindeer management institutions, policies and practices must be aimed at safeguarding locally based livelihoods and the linguistic and cultural expressions attached to such livelihoods. • Reindeer management and reindeer products should be labeled as normal or organic.
Recommendation 2: Establish scale sensitive coalitions and strategies in reindeer management with participatory decision making procedures.	• There is a need to develop regional cooperative councils on land use that set the rules for resource management in their respective regions, enhance the viability of reindeer management, and strengthen the position of reindeer herders in land use decisions. • Divergent interests in reindeer management should be brought to permanent roundtable forums with significant decision making powers over mutually accepted management strategies. • Reindeer herders must be accepted as professional managers of a high reliability ecosystem whose expertise needs to be fully and formally acknowledged in reindeer herding and ecosystem management.

Table 3.1. (*Continued*)

Policy recommendation	Specifying statements
	• In Finland, strategic coalitions of reindeer management need to be redefined at different scales that enable the reindeer herders to undertake differentiated, locally and regionally adapted management strategies. • Research strategies based on participatory action research should be emulated in other projects aiming at pragmatic sustainability policies.
Recommendation 3: Monitor and regulate reindeer management with dynamic sustainability bandwidths.	• Regulation of reindeer management should be based on dynamic sustainability bandwidths instead of fixed carrying capacity indicators. • Limit summer access to winter rangelands
Recommendation 4: Resolve tension between planned and market economy in reindeer management.	Incompatibilities between reindeer quotas and meat markets in Finland must be addressed and negotiated urgently.

posal identifies mechanisms for clarifying the image of reindeer management as a modern livelihood, the second articulates rules for improved organization and decision making, the third outlines the principles of a monitoring system and the fourth proposal develops enforcement rules. We will first describe the three modes of networking in general terms and then use the terminology to describe the details of the emergence of the policy recommendations.

3.2.1 Modes of Networking

"Pioneer knowledge networking" is the most informal and least structured mode of knowledge integration, characterized by exploration and discovery (Fig. 3.1 and Table 3.2). There is no coordinator, as such, since networking relies on broad integration of knowledge. In this type of networking, individuals succeed in identifying mutually understandable "boundary objects" of knowledge. These can be repositories of knowledge, standardized forms and methods, objects or models, or maps of boundaries, for example (Star and Griesemer 1989; Bruun et al. 2002). During the RENMAN project, the bound-

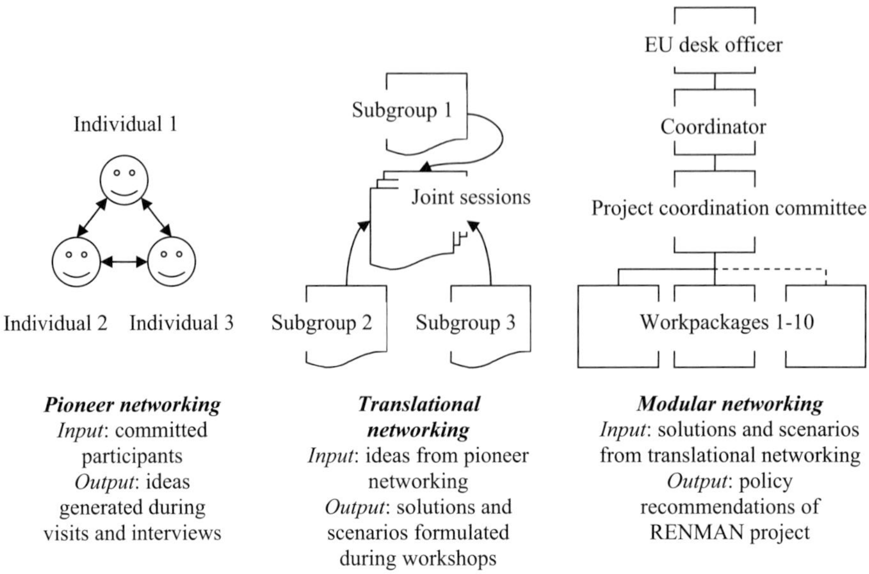

Fig. 3.1. RENMAN knowledge networking

Table 3.2. Characteristics of knowledge networking in the RENMAN project

	Pioneer networking	Translational networking	Modular networking
Processes	Ritualized field visits and interviews	Workshops	Final symposium, PCC meetings
Outcomes	Ideas for workshop agenda items	Problem and solution identification, preferred scenarios	Policy recommendations
Stakeholder groups involved	Government officials, researchers, reindeer herders	Government officials, researchers, reindeer herders	EU desk officer, government officials, researchers, reindeer herders
Number of people involved	2–5 per visit (45 total)	27–32 per workshop (32 total)	80 in symposium

ary objects consisted of "ritualized" field visits to reindeer management sites, field interviews, case studies of local reindeer management, and shared repositories of reports put together during the project. The ritualized field visits can be characterized as regular encounters that took place during the three-year project between the researchers and reindeer herders, all members of WP1. Often these visits were to the same place and followed the same routine of informal exchange of ideas, combined with participation in and observation of practical reindeer management activities. In some cases the field visits were preceded by negotiation encounters requested by both researchers and herders to establish the foundation for and functioning of partnership and participation: items of negotiations dealt with particular local circumstances, logistics, and human resources. Furthermore, the application of anthropological research methodologies were extremely useful in stressing long-term and continuing commitment by researchers to communities and individuals, thus creating an environment of mutual respect, trust and knowledge of local conditions, and providing for feedback between both partners (Pelto and Pelto 1978; Müller-Wille and Hukkinen 1999). In addition, WP1 researchers conducted a total of 45 interviews with reindeer herders and officials, which are not discussed here, but were published elsewhere (Raitio and Heikkinen 2003). Furthermore, the professional reindeer herder members of WP1 wrote three case studies on the recent evolution of reindeer management practices in their own co-operatives during the latter part of the 20th century. WP1 members prepared the reports collaboratively and then distributed them to the larger communities of reindeer herders, officials and researchers (Heikkinen et al. 2003a,b; Hukkinen et al. 2002, 2003a, b).

The ritualized field visits, interviews, case studies, and shared reports were boundary objects, because they evolved into bodies of shared cumulative knowledge. The routines of the field visits and interviews ensured that the participants perceived the occasion as the "normal" way of identifying problems and seeking solutions. The case studies by reindeer herders were conducted and conceived by the herders in their native Finnish but written with assistance from the professional researchers from WP1 to ensure that the message was indeed understandable to the broader research community (Heikkinen et al. 2003b). Finally, the reports published by WP1 ensured that the attained set of knowledge on reindeer management was codified and accumulated in a familiar format. The reports were written in an intensive commenting and redrafting process: drafts of the workshop reports were subjected to comments by all workshop participants and the case study reports were deliberated among the WP1 members (Hukkinen et al. 2003b).

"Translational knowledge networking" has a higher degree of formalization than pioneer networking. The different parties interact with each other in a systematic way in generating and sharing knowledge. Communication is typically organized through some kind of standardized interfacing device, which can be a standardized language, a protocol for action, or a mediating group (Bruun et al. 2002). In the RENMAN project, annual workshops organized with reindeer herders, officials, and researchers performed the role of a standardized interfacing device. First, all three workshops followed more or less standardized organizational procedures that were the underpinnings of an ongoing social process. Each workshop was convened with an invitation letter, which included a detailed description of the themes and procedures of the workshop. WP1 members formulated the themes of each workshop on the basis of the accumulated knowledge of preceding workshops, interviews, and fieldwork. The workshops had identical schedules, beginning midday on the first day with a joint session of introductions and background presentations, continuing on the second day with the deliberation of the workshop themes in smaller groups, and concluding by midday on the third day with a joint session summarizing and synthesizing the subgroup deliberations (Fig. 3.1). Both the joint sessions and the subgroup sessions had chairpersons and note takers, who were members of WP1. The second standardizing feature was that the workshops had a fairly large core group of participants. Of the total of 52 people (including WP1 members) who participated in the workshops, 9 participated in all three workshops, 13 in two workshops, and 30 in just one workshop (Heikkinen et al. 2003a; Hukkinen et al. 2002, 2003a,b; Table 3.2). The core group ensured that the deliberations took place in an informal atmosphere and that the standardized procedures of the workshops became the shared knowledge of all the workshop participants, that is, both the core group and the new participants.

"Modular knowledge networking" is the most formal type of knowledge networking, characterized by a clear division of labor among the network partners. In this mode, the participants make contributions, which are then integrated by a coordinator (Bruun et al. 2002). WP1 itself was organized in a modular fashion. Each subtask – that is, workshop, interview, case study or report – had responsible WP1 members who reported to the WP1 leaders (J. Hukkinen and L. Müller-Wille). But WP1 was also part of the modular RENMAN project. WP1 leaders reported to the RENMAN project coordinator on administrative matters and to the Project Coordination Committee (PCC) consisting of selected lead researchers on scientific matters. Finally, the RENMAN project coordinator reported to the EU desk officer in Brussels (Fig. 3.1 and Table 3.2). The role of modular networking in RENMAN was programmatic, that is, to ensure that what was promised by the funded proposal was done on time and that the tasks carried out somehow spoke to each other.

The three types of knowledge networking can best be understood as simultaneous embedded processes, with knowledge feedback taking place between the different types of networking. Ideas born within the pioneering discussions in the field or during interviews, for example, were translated into an understandable form during the workshop discussions and finally fed into the modular network as potential policy recommendations in the annual PCC meetings. But the PCC discussions were also fed back into the next year's workshop and field discussions for further deliberation (Table 3.2). This feedback will become more understandable by looking, as an example, at the evolution of one policy recommendation from initial ideas that emerged during pioneer networking into the policy recommendation proper that was formulated during modular networking. At this stage in the discussion, we clearly understand the fuzziness projected when recommendations for policies are derived from scientific results; the intertwining of science and policy can be a slippery slope.

Policy Recommendation 1 of the RENMAN project is to recognize reindeer management as a social innovation and a symbol due to the processes of modernization and globalization. We have chosen this recommendation as our case example because it resonates with anthropological findings on reindeer management from elsewhere, for example in the Komi Republic (Habeck 2003) and in Sweden (Chap. 5; Dahlström 2003). Both of these studies illustrate the fact that while the outside public wishes to see the herders as "traditionalists", the herders themselves insist on being engaged in a modern occupation. The implementation details of this policy recommendation were specified in four statements in the final report (Forbes et al. 2004 pp. 61–64):

- Reindeer management is a socioecological innovation and a national symbol of Finland.
- As a social innovation and a symbol, reindeer management is worthy of significant public funding for investment, operation, management, research, and development.
- Reindeer management institutions, policies, and practices must be aimed at safeguarding locally based livelihoods and the linguistic and cultural expressions attached to such livelihoods.
- Reindeer management and reindeer products should be labeled as "normal" or "organic".

These statements are radical departures from the officially accepted wisdom concerning what reindeer management is and should be in Finland (Kemppainen et al. 1997; Raitio and Heikkinen 2003; see also Chap. 2). In contrast to widely held assumptions, the statements imply that reindeer management is not a socially and economically marginal livelihood of the nation's northernmost regions; that support for reindeer management is not subsidy but innovation funding; and that product development, not marketing, is a key part of the solution to the livelihood's current problems. How did this redefinition of accepted wisdom come about during the modes of pioneering and translational knowledge networking in the RENMAN project?

3.2.2 Pioneer Networking

Let us first take a closer look at pioneer networking (Table 3.3). Obviously, there were several pioneering exchanges of knowledge during RENMAN that generated the initial ideas about reindeer management as a social innovation and a symbol. We will focus on just two, the first of which highlights the social innovation characteristics of reindeer management and the second its symbolic characteristics.

Table 3.3. Evolution of a pioneering idea into a policy recommendation

Pioneer networking: idea generated during field visit becomes input to translational networking	Translational networking: idea from pioneer networking is first condensed into a problem and then into a solution/ scenario, which becomes input to modular networking	Modular networking: solution/scenario from translational networking is formulated into a policy recommendation
Government policy encourages few herders with large herds • calf marking and culling no longer the social events they used to be • benefits of scale increase are lost to investments in machinery and wages to helping hands • conflict between current herders and embittered former herders	*Problem*: Institutional and environmental problems in reindeer management cause internal conflicts within RHCs and lead to social and cultural disruption of communities *Solution/scenario*: • Regional differentiation of policy, management and government support • Coherence between regional management differences and RHC boundaries	*Recommendation 1*: Develop reindeer management as a social innovation and a symbol

This is how the social innovation aspect became evident: During the field visits and also in the case studies on the evolution of local reindeer management practices (Heikkinen et al. 2003b), reindeer herders often lamented the diminishing number of reindeer herders and the concurrent increase in herd sizes. The cause of this development was seen as a combination of the need of individual herders to supplement their reduced income from reindeer management with other jobs, and government policies encouraging larger herd sizes by increasing the minimum-herd size requirement for subsidies. As reindeer herder Hannu Magga explains in his case study, diminishing incomes had much to do with dramatic environmental changes: "In the year 1970/1971 the Lappi RHC had 60 full time reindeer herders who got their primary income from reindeer management. When I write this, their number is in my assessment 15.... Pasture losses resulting from environmental changes are part of the reason.... Pasture losses to forest cutting are about 150,000 hectares, to reservoirs 56,100 hectares and to tourism 18,000 hectares. We have lost about 49 % of pastures as a result of environmental changes..." (Heikkinen et al. 2003b pp. 61-63).

Fewer herders with larger herds eroded the livelihood, because the increased income that might have been achieved by increasing the scale of activities was lost to costs of intensified mechanization (machines, gasoline, fences, etc.) and wages to helping hands during calf marking and culling. The symbiotic human–reindeer relationship became looser. In earlier times with more herders, calf marking and culling were social events during which reindeer management tasks were done communally without cash payments or as part of the local subsistence or barter economy. A further repercussion of this development has been social disruption in the local reindeer herding communities, whereby embittered former herders who have been forced to sell their herds are now actively opposing any initiatives that might benefit the remaining herders.

The social significance of reindeer management is articulated succinctly in reindeer herder Outi Jääskö's case study: "Reindeer management is one of the livelihoods of the Sámi community, but it is also the community's social security system, entertainment, art, bodily culture, school, language, calendar, clock, logistical system, map, land use plan, civil register, justice system, and world view" (Heikkinen et al. 2003b p. 126). The realization that reindeer management is actually a key element of the social fabric of local communities in the north gave us confidence to propose to the workshops – or to the translational knowledge networking phase – that the livelihood be reconceived and promoted as an indispensable social innovation.

The idea that reindeer management has brand value in modern Finland came from field visits that two of us (J. Hukkinen and L. Müller-Wille) made to reindeer owners and herders who had also become entrepreneurs with highly developed reindeer meat products. By brand value we mean marketable value over and above that of meat, such as positive image and trustworthiness brought about by reindeer management practices (for a detailed account of branding, see Klein 2000). Hannu Lahtela of the Northern Salla Reindeer Herding Cooperative (RHC), for example, cuts reindeer meat to exact specifications for individual clients. He has also realized that mass marketing is not the issue, since the average annual reindeer meat consumption is only 400 grams per person in Finland. Instead, he has created niche markets for his custom-made meat products by personally visiting restaurant chefs in Helsinki (H. Lahtela, pers. comm., 17 August 2003). As reindeer herder Satu Nevalainen from Kuivasalmi RHC points out in her case study, the much-criticized centralized slaughterhouses that were constructed in the 1990s to comply with hygiene requirements of the European Union had in fact an unexpected positive side effect: "Reindeer meat slaughtered in centralized slaughterhouses and processed in meat processing factories is not good enough for private buyers. Demand for traditionally slaughtered reindeer that is cut specifically for household use is higher than we can produce" (Heikkinen et al. 2003b p. 101). Jaakko Kantola from the Alakylä RHC also processes his own reindeer meat and sells it to a pool of trusted individuals who have purchased meat from him over several years. The location of his operation near the road to one of the major ski resorts in Lapland helps, too (J. Kantola, pers. comm., 15 August 2003). Vieno Länsman of the Galdoaivi RHC also benefits from ideal location at one of the main border crossing points between Finland and Norway (V. Länsman, pers. comm., 21 August 2003). Finally, reindeer meat is not everything, as Jari Heikki Länsman of Galdoaivi RHC pointed out. He plans to generate revenue by turning the annual calf marking and culling events into tourist attractions (J.H. Länsman, pers. comm., 21 August 2003). Together with the fact that reindeer has for long been an emblematic symbol of Finland (Müller-Wille 2001; Heikkinen 2002), these surprisingly modern adaptations of reindeer management gave rise to the idea that reindeer management should be regarded as a national symbol worthy of significant public attention and investment.

3.2.3 Translational Networking

How, then, were these pioneering ideas transformed during the translational knowledge networking that took place in the workshops? This evolution can be traced from the workshop reports, which follow a logical sequence from problem definition in the first workshop, to proposed solutions in the second workshop, and finally to preferred scenarios and management plans in the third workshop and the final RENMAN symposium (Hukkinen et al. 2003b; Forbes et al. 2004). That reindeer management is a key social innovation in the northern communities was reflected in the problem statements made by reindeer herders in the first workshop: "Since reindeer management is difficult to categorize in modern culture, outsiders find it difficult to understand reindeer management and reindeer herders find it difficult to obtain influence in the outside culture", "Institutional and environmental problems in reindeer management cause internal conflicts within RHCs and lead to the social and cultural disruption of communities", and "Young people do not continue as reindeer herders and the elderly are forced out of the livelihood" (Hukkinen et al. 2002). During the second workshop, several solutions were articulated to alleviate the problems: "Employers must educate forestry and tourism professionals about reindeer management", "All livelihoods in a region should be required to participate in the resolution of conflicts over reindeer management", and "Reindeer management must be adapted regionally so that some regions can opt for cooperative arrangements and others for market-based arrangements" (Hukkinen et al. 2003a; Table 3.3).

The emergence of the notion of reindeer management as a brand can also be traced in the problem statements and proposed solutions of the workshops. Many problem statements express fears that the positive brand image of reindeer management is being damaged or stolen: "The press maintains a negative image of reindeer management" and "Outsiders are taking over the program services related to reindeer management" (Hukkinen et al. 2002). To solve these problems, "The Finnish society should promote a new profile for reindeer management, which in addition to reindeer management includes natural resource management, 'predator management,' fishing and hunting" and "In tourism, reindeer herders should be awarded a license to use reindeer, and zoning and taxing should be used to limit the damages of tourism on reindeer management" (Hukkinen et. al. 2003a).

The evolution from problem statements to preferred scenarios took place in a structured workshop process. The workshops served as the standardized interfacing device that enabled knowledge integration not only among the workshop participants but also between the pioneering and modular networking phases. Input to the workshops came from the pioneering networking. WP1 members formulated the agenda points for each workshop on the basis of the ideas generated during pioneer networking. In the workshop subgroups, discussion focused on the agenda points, with the specific task of developing concise statements about problems, solutions, and preferred scenarios for reindeer management. On the basis of the notes taken in each subgroup, WP1 members drafted workshop reports, with intensive commenting from the workshop participants. The final reports then became input to the PCC meetings and the final symposium of the modular networking phase (Fig. 3.1).

3.2.4 Modular Networking

The final phase of knowledge networking was modular networking (Table 3.3). The pioneering ideas that were further specified in the translational phase were condensed into our first policy recommendation at the final RENMAN symposium: "Develop reindeer management as an innovation and a brand." The recommendation was specified with the four sub-statements listed in Table 3.1. Judging by the positive reaction at the symposium and the media coverage thereafter (Lerner 2003; Lessing 2003), the keywords of "socioecological innovation" and "national brand" struck a chord. The other RENMAN workpackages and the rest of the symposium participants reacted positively to this policy recommendation, which then found its way into the final report (Forbes et al. 2004).

In the modular networking phase, the processes of knowledge integration were more structured and hierarchical than in the pioneering and translational networking. The problems, solutions and scenarios of reindeer management that were articulated in the workshop reports became the input to the annual PCC meetings, where they were discussed with the other WP leaders and compared with research results from the other WPs. On the basis of the feedback from the PCC meetings, members of WP1 formulated initial policy recommendations and preferred scenarios for reindeer management, which then became WP1's contribution to the agenda points of the final symposium (Fig. 3.1). On the basis of the deliberations over each WP draft recommendation and scenario, during the final symposium the PCC formulated the RENMAN project's policy recommendations and scenarios, which were then published in the final report to the EU (Forbes et al. 2004).

3.2.5 The Challenges of Knowledge Networking

As always in deliberations over highly contested political issues, there were twists and turns along the path from the first pioneering ideas to final policy recommendations. During the fieldwork, interviews and workshop deliberations, we repeatedly encountered differences of opinion between reindeer herders and government researchers or officials over what reindeer management is all about. For most reindeer herders, reindeer management is not just a livelihood, but rather a northern regional way of life with deep sociocultural underpinnings. In contrast, many government researchers and officials see it as a commercial enterprise in need of modernizing measures aimed at increased efficiency and profitability (Kemppainen et al. 2003; Raitio and Heikkinen 2003). Furthermore, discussions at a research seminar on sociocultural aspects of reindeer management in Finland, organized at the University of Oulu in February 2002, revealed differences within the research community over what constitutes appropriate research methodology for obtaining new knowledge about reindeer management. In the seminar, WP1 researchers emphasized participatory research methodologies, while some others felt that researchers should clearly divorce themselves from the people and policies that their research deals with. Finally, there were internal conflicts within the reindeer herding communities over community representation, ethnic identity, reindeer ownership, regional specifics, and many other issues (Hukkinen et al. 2003b).

The differences in worldviews and opinions did not, however, prevent knowledge integration from taking place. Deliberation over conflicting individual viewpoints on reindeer management took place in the translational networking of the workshops. During the workshops, the various individual perspectives that had been obtained in the interviews and field visits, were deliberated upon and crafted into a format that could be presented to the modular networking phase. Selection of workshop participants was designed to facilitate constructive deliberation despite differences in individual perspective. Drawing from positive experiences in an earlier related research setting (Müller-Wille and Hukkinen 1999), we applied the so-called self-organizing snowball sampling in the selection of workshop participants. On the basis of their extensive experience with reindeer management in Finland, members of WP1 could agree on a set of candidates for each workshop, representing a broad range of stakeholders and dimensions of interest. Once the initial set of candidates was settled (approximately 20 individuals), an uncontrolled social process of acceptances, refusals, and substitutions ensued within the communities of reindeer herders, officials, and researchers. During this process, if some individuals felt reluctant to participate, we asked them to identify other individuals whom they expected might be interested in participating in the workshop. At the same time, we asked them to "keep the ball rolling" by passing on the workshop's ideas, goals, and guidelines, should subsequently requested individuals feel reluctant to participate. In the end, the self-organizing snowball sampling produced a broad representation of interest groups. Direct open conflicts, however, were absent in the workshops, precisely because of the self-organized selection of those individuals who themselves felt they could communicate constructively with each other, regardless of their conflicting viewpoints or affiliation with a particular interest group.

The knowledge networking processes were not just an effective way of generating institutional policy recommendations to facilitate sustainable reindeer management. In themselves, these processes were also the first stage of testing and implementing the institutional recommendations that were generated. It is this emergent characteristic of the proposed institutional reforms that we turn to in the next section.

3.3 Building on Cognitive Potential: Knowledge Networking as an Emergent Institution for Sustainability

We call the processes of knowledge networking during the RENMAN project emergent institutions, because during the networking, divergent social rules evolved into shared social rules. As such, our case study is an analysis of how new institutional arrangements are being born through simultaneous, multi-level processes of knowledge sharing. What emerged during the RENMAN project, and even continues to evolve afterward (see the concluding section of this chapter), is not a set of ad hoc rules for participatory reindeer management developed during a few inspirational workshops. Instead, the results are proven working procedures tested during the recursive processes of informal discussions in the pioneering phase, in participatory workshops in the translational phase, and in formal project meetings in the modular phase. In short, our analysis shows how formal institutions are being shaped in the modular knowledge networking phase simultaneously and in intensive feedback with the informal institutions expressed in pioneer knowledge networking. Furthermore, the institutions that germinated during the RENMAN project also have characteristics that have been shown in multiple earlier studies to support sustainable governance. We will now move on to show, first, that what evolved during RENMAN were indeed the first stages of institutions, and second, that they were institutions likely to facilitate sustainable reindeer management.

To show in detail that the knowledge networking processes were emergent institutions, we will test the characteristics of the rules that were formed during knowledge networking against the characteristics of institutions identified in earlier studies of environmental governance. Summarizing governance literature, Ostrom (1990) defines institutions as working rules that (1) determine eligible decision makers in an arena and (2) prescribe what actions are required, permitted, or forbidden. They are working rules only if they are (3) common knowledge, (4) monitored, and (5) enforced. Let us see how well these criteria are met in the RENMAN case.

First, the knowledge networking processes of the project soon converged on a core group of individuals who became aware of the project and its purpose, their position as key decision makers in the project, and their position as informal representatives of the broader reindeer management community. As was mentioned in the previous section, of the 52 people who participated in the workshops, 22 participated in two or more workshops (Table 3.2). Throughout the networking process, a core group of individuals was therefore maintained, who became aware of the mission of the project and the format of the workshop procedures. This core group also represented the heterogeneous reindeer management community in several dimensions: it included representatives of reindeer owners from the northern, central, and southern regions of the reindeer herding area in northern Finland; officials from central authorities in Rovaniemi and Helsinki and researchers from within and outside RENMAN; Sámi and Finns; men and women.

Second, the knowledge networking generated rules for action. Obviously, given the pioneering nature of the RENMAN project, the rules forbidding, requiring, or permitting actions were often not very specific. But some were. Under Recommendation 1 (reindeer management as an innovation and brand), for example, a specific call was made to label reindeer-related products as normal or organic, and under Recommendation 2 (scale-sensitive coalitions with participatory decision making), there were specific calls for cooperative councils on land use, equal acceptance of reindeer herders' expertise in ecosystem management, and regionally adapted management strategies (Table 3.1). In fact, as the case studies by reindeer herders themselves describe in remarkable detail, modern reindeer management has already had to adapt to the rules of a highly reliable, just-in-time management system: tightly scheduled calf marking and culling with occasional airplane assistance, EU-regulated slaughtering with tightly scheduled veterinary inspections, part-time herders with other jobs to handle, and fewer and fewer herders to manage it all on pastures diminished by intensive hydropower development, forestry, and tourism (Heikkinen et al. 2003a).

Third, the rules that emerged during knowledge networking became common knowledge among the project participants, by the very design of the networking process. The main principle that ensured the rapid awareness among all project participants of the key rules of participatory reindeer policy and management was the simultaneous emergence of formal and informal institutions. The emergent formal institutions were the modular knowledge networking operations that took place in the context of the formal EU project. The emergent informal institutions were the pioneer knowledge networking operations that took place during the field visits and interviews throughout the project. The translational knowledge networking that took place in the annual workshops ensured constant feedback between the formal and informal institutional arrangements (Fig. 3.1 and Table 3.2).

Fourth, the policy recommendations outline a monitoring system for the rules that were formulated during the project. The building blocks of a new type of monitoring system are outlined in Recommendation 3 (monitoring with dynamic sustainability bandwidths; Table 3.1). Bandwidths are the complex set of limits that define the range of solutions available for the different actors involved in the management of a particular ecosystem without exceeding its carrying capacity (van Eeten and Roe 2002). The idea here is that management options should be considered under different carrying capacities that depend on alternative assumptions of what the future desired social and ecological context of reindeer management will be. Typically these assumptions vary from one actor to another. In practice, one alternative scenario of social, economic and ecological sustainability may have a completely different ecological carrying capacity than another alternative sustainability scenario (Hukkinen 2001, 2003a,b; van Eeten and Roe 2002). Ecological studies conducted under RENMAN in the Näkkälä RHC in Finland and adjacent sites in Norway show that while intensive grazing and trampling result in a different vegetation pattern from that of a less intensive reindeer management regime, such intensity does not threaten the carrying capacity of the system as a whole (Chap. 9). In fact, the RENMAN modeling studies indicate that even if reindeer numbers were assumed to be at their socially and economically sustainable maximum in Finnish Lapland, the ecological sustainability (i.e., carrying capacity) of the system would still not be threatened (Chap. 15, 16).

Finally, the policy recommendations also touch upon enforcement issues. As the fourth policy recommendation articulates, the tension between planned and market economy should be resolved by fixing the incompatibilities between reindeer quotas and meat markets (Table 3.1). Reindeer meat markets in Finland currently operate in an informational asymmetry created by the combination of quota regulation by the state and price fixing by a few large meat purchasers. The result of the uneasy combination of planned and market economy is downward price fixing by the meat purchasers (Heikkinen et al. 2003a; Hukkinen et al. 2002, 2003a,b). Several alternatives come to mind to remedy the situation. First, reindeer statistics and quotas could be declared confidential. Second, the state could commit itself to securing an agreed-upon income for reindeer herders regardless of the meat price. Third, reindeer quotas could be turned into marketable commodities. A RHC that slaughters more than is required to maintain the number of reindeer below the permissible maximum would have assets to sell, whereas a RHC that slaughters less would have to purchase additional quotas. Fourth, the quota system could be dropped altogether and replaced with a system of local self-regulation (Hukkinen et al. 2003a).

We will now move on to the second part of our proof, namely, that the emergent institutions are likely to facilitate sustainable reindeer management. Here we will rely on what to our knowledge is the most recent, concise, and comprehensive review of the characteristics of sustainable resource management institutions, recently published in *Science* to commemorate the 35th anniversary of Garret Hardin's "Tragedy of the commons" article (Dietz et al. 2003). According to the review, robust governance of environmental resources is guided by the following principles:

1. Devise rules that are congruent with ecological conditions.
2. Clearly define the boundaries of resources and user groups.
3. Devise accountability mechanisms for monitors.
4. Apply graduated sanctions for violations.
5. Establish low-cost mechanisms for conflict resolution.
6. Involve interested parties in informed discussion of rules.
7. Allocate authority to allow for adaptive governance at multiple levels from local to global.
8. Employ mixtures of institutional types.

The recommendations of RENMAN fulfil these criteria of sustainability (Fig. 3.2).

First, the rules prescribed in the policy recommendations strive for congruence with the ecological conditions, as reflected in the call for labelling reindeer products into normal and organic (under Recommendation 1) and the recommendation to monitor and regulate reindeer management with dynamic sustainability bandwidths (Recommendation 3). The recommendations also aim at the definition of the boundaries of resources and user groups, as in the calls to develop regional cooperative councils and strategic coalitions, both of which would require the formal acceptance of reindeer herders as high level professionals of northern ecosystem management in their own regions (Recommendation 2). There are also recommendations to devise accountability mechanisms for monitors, apply graduated sanctions for violations, and establish conflict resolution mechanisms, as for example in the ideas to establish regionally specified management systems with sustainability bandwidths (Recommendations 2 and 3) and to resolve the tension between planning and markets in reindeer management (Recommendation 4). Finally, the recommendations reflect an effort to involve interested parties in analytic deliberation about the rules of reindeer management, to develop governance at multiple levels of government, and to employ institutional mix-

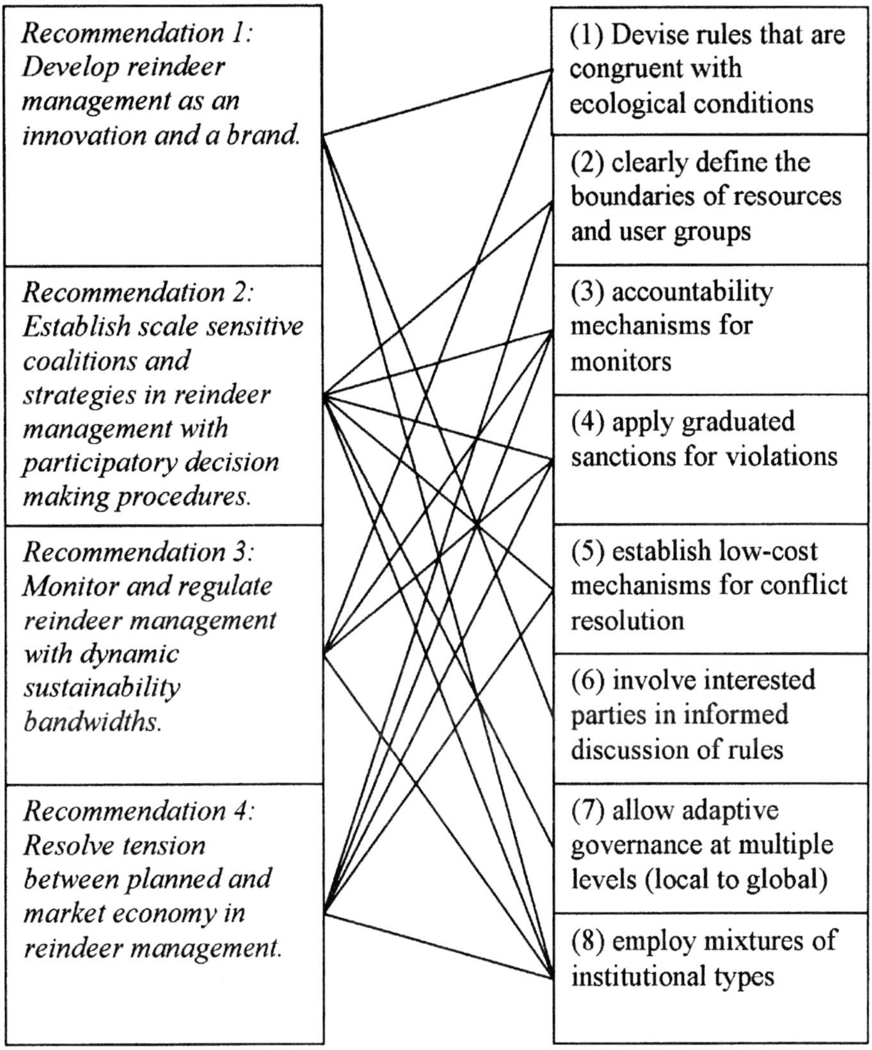

Fig. 3.2. Sustainability characteristics of RENMAN policy recommendations

tures, for example in the recommendations to ground reindeer management policies in local livelihoods and cultures (Recommendation 1) and to establish scale sensitive coalitions and strategies with participatory decision making (Recommendation 2) (Fig 3.2 and Table 3.1).

3.4 Conclusion

Throughout this chapter we have used the term 'emergent institution' to describe the processes that took place during RENMAN WP1. A biological analogy helps us to summarize the idea of institutional emergence. While we are aware of efforts to extend the notions of biological genomics to theorizing about sociocultural evolution (Dawkins 1989; Ehrlich 2000), our analysis does not permit such an extension. The analogy here is strictly an inspirational and illustrative one (von Ghyczy 2003). Modern biology considers the behavior of an organism as an orchestrated interplay between inherited and environmental influences acting on the same substrate, the genome (Robinson 2004; Ehrlich 2000). Now, for the sake of constructing the analogy, think of "institutions" as the "genome" – not too far-fetched, since both follow codified systems of rules.

The emergent institutions resulting from WP1 processes can be understood as the outcome of an orchestrated interplay between the objectives and instructions laid out for the RENMAN project in the EU research contract (the "inherited influences") and the contingencies of the actual research process (the "environmental influences"). The orchestrated pattern that we have in retrospect discerned in the RENMAN process can be described in terms of the three parallel processes of modular, translational, and pioneer knowledge networking. Modular knowledge networking displays strong influences inherited from the EU research contract requirements for project organization and management. In contrast, pioneer knowledge networking can be seen as a way of dealing with the environmental contingencies in a somewhat organized manner through interviews, ritualized field visits, and standardized reporting. Translational knowledge networking in the form of workshops is the interface between the inherited formal EU requirements and the informal environmental contingencies of reindeer management.

The biological analogy can be taken further. The signals that the RENMAN WP1 process sent to its environment as policy recommendations (the "behavior of the organism") reflect the institutional rules (the "genome") that emerged during the process. In other words, the policy recommendations calling for institutional reform, particularly those under Recommendation 2 (Establish scale sensitive coalitions and strategies in reindeer management with participatory decision making procedures), call for the kinds of institutional reform already tested in practice during the WP1 processes. As the behavior of an organism mirrors its genetic makeup, so do the RENMAN policy recommendations mirror the rules put in place during the project.

The biological analogy lends itself to asking questions for the future, as well. The process of emergent institutional development described here is really just a description of "microevolution", which focuses on the motivations and actions of a relatively small community of interviewees, research partners, and workshop participants (Ehrlich 2000). Is there any evidence of "macroevolution", namely, that processes of institutional development similar to those identified in RENMAN would have occurred elsewhere in the society? While our answer is a cautious yes, we do not claim there is a causal relationship between what we did in RENMAN and what took place elsewhere in the Finnish society at the same time. A more credible explanation is that RENMAN is merely one sign of a simultaneous evolution of participatory analysis, planning, policy and management in the society of Finland.

First, during the RENMAN project signs appeared of the participatory research approach being adopted in several other projects relating to reindeer management. For example, during a research seminar entitled "Reindeer Management – Society – Environment", organized by the Thule Institute of the University of Oulu in February 2002 with a focus on the social and cultural aspects of reindeer management in Finland, at least half of the ten research projects included explicit collaboration between reindeer herders and researchers, and occasionally also with officials responsible for reindeer policy. Second, the Reindeer Husbandry Research Programme 2003–2007, published by the governmental Finnish Game and Fisheries Research Institute (GFRI) in 2003, was developed by a group consisting of researchers, officials, reindeer herders, and other experts. Two seminars opened up the development of the research program to an even broader range of stakeholders (Kemppainen et al. 2003). Third, in 2004 GFRI began a series of participatory workshops with the aim of developing reindeer management and planning at the local level. The significance of these actions by GFRI extends beyond research and planning, because of the Institute's governmental status and consequent power in shaping national reindeer policy. Among other things, GFRI plays a key role in determining the maximum allowable reindeer numbers per RHC. Fourth, in 2004 some members of WP1 joined forces with regional reindeer officials to plan a continuation of the workshop series, but this time around with more concrete themes. While modest, we think these developments are nonetheless the first signs of a "macroevolution" of the participatory institutional development principles beyond the sphere of RENMAN.

However well the biological analogy illustrates the emergent stages of an institution, it does not reveal the power relations inherent in the knowledge integration process, both within the RENMAN project and between the project and governmental authorities. Within the RENMAN project, professional researchers wielded considerable power over the structure and functions of the project, despite the underlying principles of participation and partnership. In WP1, for example, it was the professional researchers who allocated project resources, shaped workshop agendas, and finalized project reports. While such power made use of intensive consultation with reindeer herders, regulations from the EU down to the university level placed formal responsibility for such decisions with the project leaders. Had these formal powers been with, say, the herders, the conclusions and their language would probably have looked rather different from what they are now.

But an even more significant power issue has to do with the relationship between the RENMAN project and public authorities. Since the RENMAN project has no official status within the EU or Finnish decision-making system, its credibility could not last over the long run. Intimate involvement of stakeholders in research and planning generates expectations of influence among those involved. In the end, RENMAN could only propose the shape of the improved future institutions and demonstrate how they might function. While the knowledge integration that we observed bears the signs of an institution, it is still just a demonstration project. Because of the structural issues just mentioned, there is a strong possibility that the proposals derived from the workshop may not be adopted by the relevant authorities or governmental bodies. This last point persuades us to make a concluding recommendation about the status of participatory policy research and development more generally.

Modern national and supra-national governments increasingly require stakeholder involvement in policy-related research and development. The requirement tends to be rooted in rather traditional notions of how a representational democracy works, namely, through the balancing of conflicting stakeholder interests. Yet when stakeholders are thoroughly integrated in policy research and development, beginning with the collaborative formulation of relevant policy issues and ending in joint implementation, then the identification of a given stakeholder group with a particular policy recommendation is not only impossible from the analytical point of view – it is counterproductive from the policy point of view. Governmental policy makers would be wise to view the participatory policy process not as democratic testing of competing policy options, but rather as crafting of preferred policies. When successful, a participatory workshop is not a parliament. It is a mission.

Note: An earlier version of this chapter was presented at the Fifth International Congress of Arctic Social Sciences (ICASS V) in the session on *Effective Local Institutions for Collective Action in Arctic Communities*, Fairbanks, Alaska, 19–23 May, 2004 (Hukkinen et al. 2004).

References

Berkes F, Folke C (eds) (1998) Linking social and ecological systems. Cambridge Univ Press, Cambridge

Bruun H, Langlais R, Janasik N (2002) Transepistemic communication and innovation: A conceptual platform. Technology, Society, Environment 3. Helsinki University of Technology Laboratory of Environmental Protection, Espoo, pp 7-52

Dahlström ÅN (2003) Negotiating wilderness in a cultural landscape: predators and Saami Reindeer herding in the Laponian world heritage area. Uppsala University Library, Uppsala

Dawkins R (1989) The selfish gene. Oxford Univ Press, Oxford

Dietz T, Ostrom E, Stern PC (2003) The struggle to govern the commons. Science 302:1907-1912

Ehrlich P (2000) Human natures: genes, cultures, and the human prospect. Island Press, Washington, DC

Forbes BC, Aikio P, Arttijeff T, Beach H, Bölter M, Burkhard B, Furmark O, Grape H, Gunslay N, Heikkilä L, Heikkinen H, Helle T, Horn R, Höller C, Hukkinen J, Idivuoma PG, Jernsletten J-L, Jääskö O, Kankaanpää S, Kantola J, Kemper N, Kitti H, Konstantinov Y, Kumpula T, Kyrö R, Laakso AM, Magga A, Magga H, Magga J, Möller R, Müller F, Müller-Wille L, Nevalainen S, Näkkälä A, Peth S, Pokuri O, Raitio K, Ranta H, Sainmaa M, Soppela P, Thuen T, Turi JM, Turunen M, Uhlig C, Vehkaoja M, Vladimirova V, West N (2004) RENMAN – final report. The challenges of modernity for reindeer management: integration and sustainable development in Europe's subarctic and boreal regions. Arctic Centre Rep 41. University of Lapland, Arctic Centre and EU FP5 QLK5-CT-2000-00745, Rovaniemi, pp 1-100

Gunderson LH, Holling CS, Light SS (eds) (1995) Barriers and bridges to the renewal of ecosystems and institutions. Columbia Univ Press, New York, NY

Habeck JO (2003) What it means to be a herdsman: The practice and image of reindeer husbandry among the Komi of Northern Russia. Doctoral Thesis, Scott Polar Research Institute. University of Cambridge, Cambridge

Hall BL (1979) Participatory research: an approach for change. Convergence 8:24-33

Hanna SS, Folke C, Mäler K-G (eds) (1996) Rights to nature: ecological, economic, cultural, and political principles of institutions for the environment. Island Press, Washington, DC

Heikkinen H (2002) Sopeutumisen mallit. Poronhoidon adaptaatio jälkiteolliseen toimintaympäristöön Suomen läntisellä poronhoitoalueella 1980–2000 (in Finnish; models of adaptation - adaptation of Reindeer Herding to the post-industrial environment in the Western Finnish reindeer herding area 1980–2000). Doctoral dissertation, University of Oulu. Suomalaisen Kirjallisuuden Seura, Helsinki

Heikkinen H, Hukkinen J, Jääskö O, Laakso A, Müller-Wille L, Nevalainen S, Raitio K, West N (eds) (2003a) Poronhoidon tulevaisuus: Raportti EU:n RENMAN-hankkeen Kittilän työpajasta 13.-15.8.2003 (in Finnish; the future of reindeer management: report of the EU RENMAN Project Workshop in Kittilä 13-15 August 2003). Technology, Society, Environment 7. Helsinki University of Technology Laboratory of Environmental Protection, Espoo

Heikkinen H, Magga H, Nevalainen S, Jääskö O (2003b) Kuuluuko sääsken ääni taivaaseen? Poromiesten analyysi poronhoidon murroksista Suomen Lapissa 1900-luvulla (in Finnish; Can the mosquito be heard in heaven? Reindeer herders' analysis of changes in reindeer herding in Finland's Lapland during the 20th century). Technology, Society, Environment 3. Helsinki University of Technology Laboratory of Environmental Protection, Espoo

Hiebert W, Swan D (1999) Positively fit: a case study in community development and the role of participatory action research. Comm Dev J 34:356-364

Hukkinen J (1999) Institutions in environmental management: constructing mental models and sustainability. Routledge, London

Hukkinen J (2001) Eco-efficiency as abandonment of nature. Ecol Econ 38:311-315

Hukkinen J (2003a) From groundless universalism to grounded generalism: improving ecological economic indicators of human-environmental interaction. Ecol Econ 44:11-27

Hukkinen J (2003b) Sustainability indicators for anticipating the fickleness of human-environmental interaction. Clean Technol Environ Policy 5:200–208

Hukkinen J., Jääskö O, Laakso A, Müller-Wille L, Raitio K (eds) (2002) Poromiehet puhuvat: Poronhoidon ongelmat, ratkaisumahdollisuudet ja tutkimustarpeet Suomen Lapissa poromiesten näkökulmasta (in Finnish; Reindeer Herders Speak: Problems, Potential Solutions and Research Needs of Reindeer Herding in Finnish Lapland from the Viewpoint of Reindeer Herders). Technology, Society, Environment 1. Helsinki University of Technology Laboratory of Environmental Protection, Espoo

Hukkinen J, Jääskö O, Laakso A, Müller-Wille L, Nevalainen S, Raitio K (eds) (2003a) Poronhoitokulttuurin arvo Suomessa: Haasteet hallinnolle, ohjaukselle ja valvonnalle (in Finnish; The value of reindeer herding culture in Finland. Challenges for administration, management and control). Technology, Society, Environment 1. Helsinki University of Technology Laboratory of Environmental Protection, Espoo

Hukkinen J, Müller-Wille L, Heikkinen H (eds) (2003b) Development of participatory institutions for reindeer management in Northern Finland: preliminary synthesis and report. Technology, Society, Environment 6. Helsinki University of Technology Laboratory of Environmental Protection, Espoo

Hukkinen J, Müller-Wille L, Aikio P, Heikkinen H, Jääskö O, Laakso A, Magga H, Nevalainen S, Pokuri O, Raitio K, West N (2004) Knowledge integration as participatory institutional development for sustainability: the case of reindeer management in Finland. 5th international congress of Arctic Social Sciences (ICASS V), Abstracts, Fairbanks, Alaska, 19-23 May 2004, pp 46-47

Kemppainen J, Nieminen M, Rekilä V (1997) Poronhoidon kuva (in Finnish; Overview of Reindeer Management). Finnish Game and Fisheries Research Institute, Helsinki

Kemppainen J, Kettunen J, Nieminen M (2003) Porotalouden taloustutkimusohjelma 2003-2007 (in Finnish; Reindeer Husbandry Research Programme 2003-2007). With English abstract. Fish and game reports 281. Finnish Game and Fisheries Research Institute, Helsinki, pp 1-56

Klein N (2000) No logo. Flamingo, London

Langlais R, Janasik N, Bruun H (2004) Managing knowledge network processes in the commercialization of science: two probiotica discovery processes in Finland and Sweden. Sci Stud 17:34-56

Lerner L (2003) Renman-tutkimusprojekti suosittelee parannuksia – maankäyttömuodot poronhoidon keskeisimpiä ongelmia (in Finnish; RENMAN research project recommends improvements – land use practices key problems). Lapin Kansa, 29 Nov 2003

Lessing H (2003) Renman-tutkijat suosittelevat – poro Pohjoismaiden brändiksi (in Finnish; Renman researchers recommend – reindeer should be a Nordic brand). Pohjolan Sanomat, 29 Nov 2003, p 4

Müller-Wille L (2001) From reindeer stew to pizza: the displacement of local food resources in Sápmi, northernmost Europe. Fennia 179:89-96

Müller-Wille L, Hukkinen J (1999) Human environmental interactions in Upper Lapland, Finland: development of participatory research strategies. Acta Borealia 16:43-61

North DC (1992) Institutions, institutional change and economic performance. Cambridge Univ Press, Cambridge, MA

Ostrom E (1990) Governing the commons: the evolution of institutions for collective action. Cambridge Univ Press, Cambridge, MA

Ostrom E, Burger J, Field CB, Norgaard RB, Policansky D (1999) Revisiting the commons: Local lessons, global challenges. Science 284:278-282

Parker DC, Berger T, Manson SM, McConnell WJ (eds) (2002) Agent-based models of land-use and land-cover change: report and review of an international workshop, 4-

7 Oct 2001, Irvine, California, USA. LUCC report series 6. LUCC Focus 1 Office and LUCC International Project Office, Louvain-la-Neuve

Pelto PJ, Pelto GH (1978) Anthropological research. The structure of Inquiry. Cambridge Univ Press, Cambridge

Raitio K, Heikkinen H (2003) Enemmän oma-aloitteisuutta, vähemmän valitusta: Hallinnon näkemyksiä poronhoidon osallistumiseen perustuvien instituutioiden kehittämisestä (in Finnish; More initiative, fewer complaints: the views of Finnish administration on the development of participatory institutions in reindeer management). Technology, Society, Environment 4. Helsinki University of Technology Laboratory of Environmental Protection, Espoo

Robinson GE (2004) Beyond nature and nurture. Science 304:397-399

Star SL, Griesemer JR (1989) Institutional ecology, 'translations,' and boundary objects: Amateurs and professionals in Berkeley's Museum of Vertebrate Zoology, 1907-1939. Soc Stud Sci 19:387-420

Van Eeten MJG, Roe E (2002) Ecology, engineering, and management: reconciling ecosystem rehabilitation and service reliability. Oxford Univ Press, Oxford

Von Ghyczy T (2003) The fruitful flaws of strategy metaphors. Harvard Business Rev 81:86-94

4 The Comparison of Indigenous and Scientific Perceptions of Reindeer Management

L. Heikkilä

4.1 Introduction: Research Questions, Approach, and Material

The objective of this chapter is to compare indigenous and scientific perceptions of reindeer herding and environmental management. More precisely I refer here to Sámi reindeer herders' knowledge, and natural scientific knowledge (biology and environmental sciences) related to managing the use and protection of nature in the mountainous, forested and coastal regions of Northern Fennoscandia. In tourist presentations this region is often referred to as the last wilderness areas of Europe. It is also known as the home region of the Sámi. In this chapter I will present the founding principles of indigenous practice related to concepts and expertise of reindeer management, particularly concerning pasture and pasture:animal ratio assessment, and relate them to scientific knowledge (Chaps. 2, 3, 5, 6). The purpose is to investigate the major discrepancies between the two systems and evaluate the chances of communication between them. Finally, I will assess and compare the impact that the two systems of knowledge have on the encompassing political and administrative structures relevant to reindeer herding both as a means of livelihood and as the material foundation of Sámi culture.

There are two focally important choices that specify the theoretical approach. First, indigenous and scientific knowledge are examined in the context of the actual land use and environmental management processes within the defined area. The potential discrepancies and communicative factors are referential to this particular context, which I will draw in brief outlines. Second, environmental management is approached as *a process of inter-cultural communication*, and indigenous knowledge (reindeer herding) and natural scientific knowledge (biology and environmental sciences) are examined as the *interpretative resources* of the two social categories in question, the herders and the scientists, rather than closed and compact realms

Ecological Studies, Vol. 184
B.C. Forbes, M. Bölter, L. Müller-Wille, J. Hukkinen, F. Müller, N. Gunslay, Y. Konstatinov (Eds.)
Reindeer Management in Northernmost Europe

of knowledge. The accounts and statements of these sources are perceived as culturally constructed interpretations of reality. This starting point implies or enables the existence of several parallel realities, which in some respects are only partly shared. Consequently, a kind of relativity is provided for the starting point, in contrast to a monocultural approach operating with only one reality.

Practically speaking, the context of this study is divided between the borderline area of northern Finland and northern Norway (Fig. 4.1). Consequently, the area is under the jurisdiction of administrative agencies that belong to Finland's and Norway's separate governmental structures. Included in the management programs are: (1) wilderness (Pöyrisjärvi Wilderness Area in Enontekiö, Finland); (2) national parks (Övre Anárjohka National Park in Norway); (3) Land Use Management Areas set by municipalities (designated as "agriculture", "nature conservation", and "open air recreation" in Norway); and (4) nature or remote areas (in Norwegian "utmark" meaning literally outlying areas). Inclusively, environmental management consists of the following: land use management, natural resource management, and management of biodiversity (nature conservation).

The responsible environmental management agencies in Finland are: Forest and Park Service (Metsähallitus) and its Upper Lapland Management District, and the Wilderness Planning Office. In Norway they are: Forest and Park Service (Statsskog), its Land Sales Office (Jordsalgskontor), and its Mountain

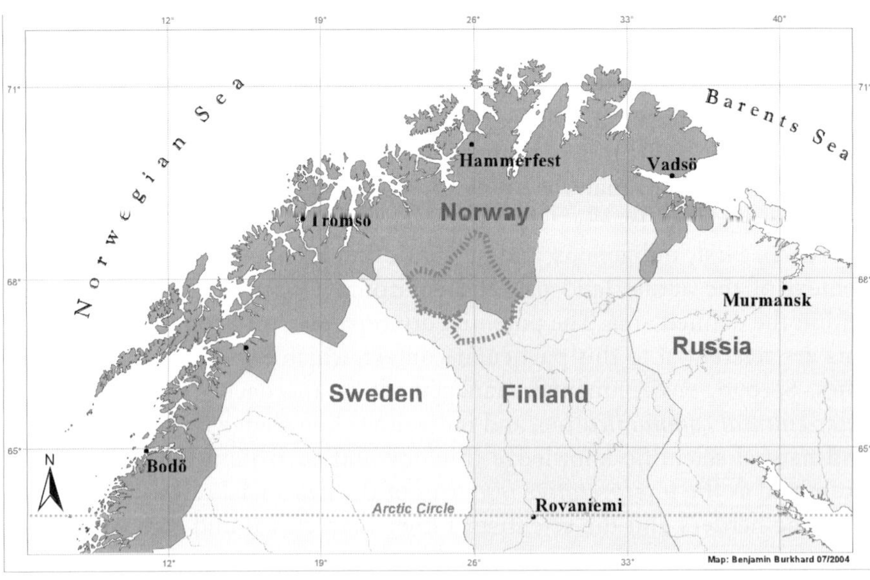

Fig. 4.1. Research area at the border between northern Finland and northern Norway

Service (Fjelltjenst); and regional and municipal planning authorities, namely, the Finnmark provincial administration dealing with the management of national parks and the municipal agencies of Guovdageaidnu responsible for land use planning.

The study is based primarily on naturally occurring data (Wetherell and Potter 1992; Potter 1997) where information is collected from two main sources, Sámi reindeer herders and public servants employed by the reindeer herding and environmental agencies in both countries. In addition, I examined a large number of public reports, policy propositions, and statements, as well as media presentations of herding and pasture issues. To grasp the contents of these accounts and their contexts I listened to conversations among herders as well as between herders and scientists and between herders and administrators. To get a better picture of how local or indigenous knowledge helps herders to assess the environmental observations they make and transform them to information on which they can base their decisions about herding strategies, I initiated a number of open-ended interviews and conversations with them, thus inviting them to put into words much of what is conventionally kept as unwritten information. The total number of informants – depending largely on the definition – was 14 with seven herders on each side of the border. In addition, on the Norwegian and the Finnish sides I interviewed seven civil servants to get information on case procedures as well as their own considerations of problems and challenges related to the reindeer herding industry. Through this combined project design I intended to cover, within the practical frames set by time and mobility, some major aspects of the field of reindeer herding.

4.2 Theoretical Background

Environmental management is often defined as concentrated societal acts for regulating and directing human conduct, in order to meet the exigencies of human life and maintain the regeneration capacities of the ecosystem (Haila 1995). Integration of scientific and indigenous knowledge, in this context, is pursued with the intention of increasing the effectiveness of management measures. This conception generally refers to the integration of types of knowledge such as ecological, biological, resource management, etc., with knowledge gained through interviewing indigenous people and interpreting the results by researchers. In practice it means *finding linking categories* for data collections within the frame of scientific knowledge for the purpose of set environmental management targets.

In my approach, I wish to accentuate the communicative aspects of environmental management procedures, and compare them to a negotiation. At the same time, I seek to generate a larger comprehension of the various dimensions of environmental management and understanding of indigenous knowledge, and its role in the integration process. With this approach I would also like to emphasize that, in the environmental management context, one actually talks about adapted knowledge to make the premises that determine and condition the decision-making visible.

The theoretical starting point of this approach is in social constructionism (Berger and Luckmann 1967), according to which reality (and science for that matter) is *contextual and conventional* – it is a result of conventions, based on rationality and certain culturally approved presumptions, values, and conduct(Harding 1992; Hall 1997). These conventions guide the transformation from observed facts to interpretation of reality, including assumptions of cause/effect connections and predictions of future developments. At the outset, assumptions based on indigenous knowledge may be as correct and realistic as those based on data from scientific investigation, as long as what has been observed corresponds with reality. On one hand this is an important theoretical commitment that bears on the research results. On the other hand it contains a research-ethical statement having reference to indigenous peoples' position in research – as research objects or equal partners – and to the consequent treatment of indigenous knowledge as the property of indigenous people.

Adapting Strauss' conceptions (1978), the term *negotiation* is used in a special sense of the word, where the participants bring forth their individual, joint, or institutional land use claims, drawing from and appealing to certain interpretative resources. "Negotiation" should not be taken literally to mean the interaction between parties in a concrete competition over influence on a decision, but more generally and indirectly to communication of varying descriptions, categorizations, connotations, etc., presented as valid information about reindeer herding. The statements, claims, and comments of the participants are approached as *accounts*. I introduced the term *reindeer talk* (Heikkilä 2003), which refers to the accounts concerning reindeer herding, delivered basically by the environmental management and reindeer herding management authorities, researchers, reindeer herders themselves, their organized interest groups, Sámi organizations and Parliaments, and representatives of non-governmental organizations in environmental negotiations.

These accounts convert into tangible discourses, which are understood partly as shared, partly as separate signification systems, as ways of making sense of the reality relating to the subject (Hall 1997). Although these discourses mostly contain an interest claim, they should not be reduced only to interest talk, as Grove-White (1993 pp. 22-23) notes. Discourses should not be expected to be homogeneous, internally consistent, or coherent (Milton 1993). It is more like a polyphonic communication within a topical area, where an array of subdiscourses or parallel discourses is activated depending in each occasion on the social position, mutual relation, and history of the participants. While one often refers to a field of communication as a discourse, such a field may also comprise several diverse or competing discourses (Milton 1993). In other words, one is talking about a discursive practice or a process that is alive, being reinforced, transformed, and rearticulated in everyday communication practices.

The term negotiation is congruent with the present administrative practice of participatory planning derived from Agenda 21 (Loikkanen et al. 1999). As a result of this practice, a large number of new "stakeholders" have appeared in the environmental negotiations, some of them have a concrete and immediate relation, while the others have a more remote and ideological relation to the negotiated object. In general, this has resulted in more extensive and heavy planning procedures (Marcussen 2003) and has accentuated the significance of the argumentation skills. Since a considerable proportion of argumentation today is performed through or by mass media, negotiating skills and arguing capabilities sometimes play a decisive role.

As an example of the intricate power plays that the negotiations involve, I refer to the controversy between reindeer herders and the other stakeholders concerning the recreational use of wilderness and reindeer pastures (Chap. 2). In this issue the views and interests of the reindeer herders diverge considerably from those of the majority of the participants in the negotiations and from the public interest. The majority of the stakeholders propose or support the protection of an unlimited recreational use based on the Everybody's Right (a well-known expression in Nordic countries), on the grounds that it is a firmly established custom and cultural value in the Nordic societies. In turn, the reindeer herders, referring to the already-caused or potential disturbances and encroachments (cabins, roads, routes, bridges, holiday resorts, etc) call for a restricted access to nature, in order to protect the narrowing natural facilities for reindeer herding. In their official statements and public accounts, reindeer herders and their official representatives usually appeal to reindeer herders' own expertise and judgment of the situation, and to the old customary right established by immemorial usage of reindeer herders, the indigenous people (Chap. 5). By the decision of the management authorities, based on the prevalent legal praxis of the nation-state, the juridical issues are, however, marked off the negotiated agenda. The situation is complicated further by the fact that recreational use and nature tourism have certain shared interests: argument in favor of one ends up easily in support of the other. In Finland, one of the business units of the conglomerate of *Metsähallitus Group* (Forest and Park Service) is involved in nature tourism. It is obvious that the wilderness management authorities have to look after its prerequisites and interests in their management decisions.

As can be noted from this example, the involved discourses do not have an equal position in environmental negotiations. Environmental management agencies have the authority to regulate the negotiation context and therefore to control other discourses and accounts. It other words, theirs is the hegemonic discourse, into which the other account deliverers have to adapt their argumentation and rationale, if they wish to make an effective and "valid" case. Environmental management agencies are also in a dominant position since they are responsible for adopting into practice the manifold claims presented by the account deliverers. This is particularly significant in the case of conflicting claims and scarce land resource.

4.3 The Environmental Management System and its Recognition of Tasks

Sámi tundra reindeer herding has been divided by the two nation states since 1852, and more specifically in 1952 with the establishment of a fence on the borderline. Since then, two diverging administrative systems have evolved (Chap. 2). In Finland there is a more concentrated environmental management procedure, where Forest and Park Service runs the planning duties in the wilderness areas designed according to the Wilderness Act of 1990. In Norway a wilderness act does not exist as such and there is a more fragmented system; environmental management is divided among Forest and Park Service, provincial administrations, and municipalities. Furthermore, in Norway there is a corporate administrative culture of hierarchical echelons with advisory committees at each echelon. Looking back in history, environmental management is a newcomer to the region, emerging mostly from the period after the Second World War, but partly from only the 1990s. Relating to the old system, where reindeer herding was a system of environmental management in itself, establishing the modern planning bodies and procedures for environmental management has signified transfer of authority over pastures from reindeer herders to administrative agencies. It is evident that the legitimacy of the management authorities is not fully established in relation to local population groups, including reindeer herders, a matter causing constant friction (Chap. 3).

Regardless of the different ways of organizing environmental management, it is surprising how similar the results are, with regard to both the substantial management problems and the position of reindeer herding in management procedures. The first point to understand is that management of reindeer herding on a national scale in each country is separate from management of the environment and placed under different administrative bodies. It means that in regard to their most important resources – pastures and reindeer – reindeer herders are dependent on the decisions made by external authorities. Most importantly this concerns the number of animals permitted within each herding district, the time limit allowed for each district to graze their animals within the boundaries of a specific herding district, and the right to exclude or restrict competing forms of land use (Chap. 5).

The right to herd reindeer is governed by law in designated areas. In Finland it comprises all permanent residents; in Norway it is an exclusive right of Sámi (with a few minor exceptions), but legal authority is relatively weak with regard to land use decisions. The dominant ideology in Fennoscandian environmental management is the *diversified use* of land/nature. This concept, which was initially designed for managing forests, has been transplanted, with pertinent modifications, to the wilderness management context. Diversified use covers traditional use such as reindeer herding, hunting, fishing, and others, as well as small-scale industrial use meaning primarily forestry and commercial nature tourism. These uses have been integrated with the original idea of wilderness management, which consisted of protecting the biodiversity and providing or encouraging the facilities for recreation, as it was defined in North America (Saarinen 1998, 2002; Naskali 2002). Diversified use has become the major operative idea of the Forest and Park Service both in Finland and in Norway, regardless of the formal structure of environmental management.

In wider perspective, environmental management practices in the two countries are rooted in the ideological frame of reference of environmentalism (or environmentalist discourse). It is a global appreciation of the world, with an incorporated regard for the definition of the problems and the well-reasoned solutions for them. Fischer and Hajer (1999) emphasize that it is not the environmental perspective per se that labels the new era, but a particular approach to environmental questions. In this interpretation, the concept of sustainable development has played a central role (Chap. 16). It has become the major operative tool for understanding the reality, the actual problems, and the viable solutions (Hajer 1995; Fischer and Hajer 1999; Haila and Jokinen 2001).

In analytical terms, "sustainable development" is the dominant story line of the environmentalist discourse. The articulation of ecological concern with economic growth under this concept becomes the overarching, guiding principle for modern societies. "Sustainable development" has in fact proven to be a functional concept, bringing to life a common language for talking not only about environmental problems (Fischer and Hajer 1999), but addressing socioeconomic issues, too. It represents, however, a narrow interpretation of the social aspects. The main message of sustainable development is one of urgency to adjust existing economic and institutional practices to the ecological realities and still maintain economic growth (Chaps. 15, 16). Fischer and Hajer suggest that sustainable development is, in fact, *a political strategy* with a stress on the programmatic nature of environmentalism. They regard environmentalism as a way of channeling social dissension, thus as a political discourse (Hajer 1995; Fischer and Hajer 1999).

Major premises for the administrative conceptualization of environmental management tasks are based on perspectives such as this. Authorization of the framework of sustainable development has brought along the idea of production rationality into environmental management practice. It means, among other things, that nature is approached as a resource, and competing land use forms are evaluated in proportion to the profit they yield. This trend is intensified by business principles, which the Forest and Park Service in Finland (*Metsähallitus*) and in Norway (*Statsskog*) adopted during the 1990s. Accompanying this approach the idea of scientific management has been introduced (Emmelin 1997; Luke 1999; Naskali 2002) with the emphasis on planning as the central management tool.

In accordance with the dual ideology of sustainable development, the major interpretative resources in environmental negotiations are ecology and economy. Ecology has become the key discipline for designing and supplying the theoretical framework and conceptual equipment for environmental management, whereas economy (profit, regional economic impact, acceptable level of household income) seems to be the uncontested argument that legitimizes environmental negotiations. This has meant, among other things, the rise of the status and demand for expert knowledge, as well as the introduction of an entirely new language. Proficiency in using the scientific language, theories, and categories, has become a prerequisite of successful environmental negotiation.

4.4 Major Challenges in Management

Analyzing the various accounts of *reindeer talk* establishes on the whole that there are two divergent interpretations of the most acute management challenges among reindeer herders and environmental management agents. For reindeer herders the major management concerns are *encroachments* and *disturbance in the diminished total range of pastures*. For management agencies, the condition of pastures and the corresponding *carrying capacity* are the most acute questions (Chaps. 7, 9). These two conceptions are not necessarily conflicting, but they do reflect the diverging concerns of researchers and administrators on one side, and herders on the other.

As an illustrative example, here is a quotation from the statement of the reindeer herders, presented by their official representative agents in the negotiations (author's translation):

"...In the general introduction volume [of the wilderness management plan] there is a tendency to defame reindeer herding in accordance with the current trend of our time. It contains among other things statements of overgrazing, illegal hunting of wolverines by snow-scooters, and damages to terrain caused by the reindeer herders using motorized vehicles. However, in the area [in question], there are no signs of overgrazing caused by reindeer. Instead there are traces caused by the lack of a functioning system of pasture rotation and by the disturbed rotation due to external activities, encroachments, and competing land use forms." (Statement by the Näkkälä Reindeer Herding District about the Wilderness Management Plan for Pöyrisjärvi, May 11, 1997).

Behind the concern of management agencies are the biological studies of pasture inventories and lichen cover, which have shown evidence of a significant degradation of grazing lands occurring over a period of time in some parts of the vast reindeer herding area (Chaps. 7, 8, 15; also see Lyftningsmo 1965; Helle and Aspi 1983; Käyhkö and Pellikka 1993; Prestbakmo 1994; Kumpula et al. 1996, 1997; Gaare 1998; Johansen and Karlsen 1998; Helle 2002; Kosmo and Ims 2002). In this connection, a category called *overgrazing* has been introduced. Although scientifically there is not full unanimity as to the content and accompanying determinants of overgrazing, in an environmental-management context it has gained a dominant position. In the analytical terms of discourse, it has become a major metaphor or key interpretative resource in the current management problem in the boreal, subarctic, and tundra ecosystem (Chaps. 2–3).

The following quotes highlight these issues:

"The major problem of reindeer breeding is to balance the number of animals against the available pasture resources. The continuing increase in number of animals has damaged the industry. The internal competition engendered by the growth is detrimental to the industry, biologically as well as socially. At the same time the wear and tear of the pastures has created negative attitudes in the population at large." (Kosmo 1991 p. 24).

"This study is an attempt to explore a contemporary resource problem; semi-domesticated reindeer overgrazing of vulnerable lichen pastures in the heartland of the subarctic tundra of Sápmi on the territory of Finnmark, Norway. The overgrazing has taken place in the wake of a technological revolution, and an area of social reform, having its core period about 1960–1990…" (Riseth 2000 p. 1).

Biological pasture research usually starts from certain unquestioned theoretical presumptions. The Tragedy of the Commons (Hardin 1968), an example of a theoretical model leading to a situation embodied in a game called the Prisoner's Dilemma (Dawes 1973), is frequently cited. In this model the pastures are approached as a common pool resource for all reindeer herders (Chap. 15). The complex reality of reindeer herders' activities is simplified into a competition over a finite pasture for maximizing one's herd size. Overgrazing is then seen as a result of the fiery and often reckless race among the herders to increase their private property and income at the cost of others, as according to this premise there are no natural or internal checks to such behavior (e.g., Brox 1989; Kosmo 1991; Skonhoft 1998; Helle 2002). This approach is shared by some anthropologists such as Ingold (1980) according to whom overgrazing and the resulting ecological crisis are part of the built-in mechanism of pastoralism (Chap. 5).

According to the Sámi point of view this not only an oversimplification of the matter, but the premises are contentious (Turi 2001). To start with, a wider array of factors is involved. Several social, cultural, societal and climatic issues influence the growth of the herd size and tend to control and reduce it. For instance, the available labor force, the social institutions for sharing pastures, sanctions, poaching, loss of control over large herds, converting surplus into alternative forms of capital, and effects of the climatic variations, as well as the geographical (topographical) fragmentation of pastures function as stock-reducing factors (Sara 1993; Paine 1994). Some biological examinations have confirmed the effect of single influences such as climatic variations (Tyler 1997).

Furthermore, from the Sámi perspective economic factors, such as the price of reindeer meat and the general level of prices, have a strong impact on herding decisions and should correspondingly be noted as regulating forces. It should also be understood, as Oskal (1995) showed, that maximizing herd size is never the only value and aim in Sámi reindeer herding communities. It appears that although the Prisoner's Dilemma, in some limited respect, may describe the decision-making situation of an individual herder or herding household (Ulvevadet 2000), it is neither an adequate explanation for it nor an exhaustive theoretical framework for approaching the matter.

Finally, the concept of the pastures as a common pool resource is not exactly correct. Reindeer pastures are divided and identified through traditional distribution systems based on inheritance, *siida* membership, and interaction or use (Sara 1996). The fact that this custom is not always respected today, and that there may appear confusion in some parts of the reindeer herding area, does not invalidate this state of affairs altogether. It indicates that there are some serious problems and some fractures in the functioning of the traditional land use institution. The reasons behind it are mostly very complex, being connected with the existence of dual land use patterns (the formal and the informal), side effects of market economy and modernization, not excluding the private ambitions of individual reindeer herders. According to Paine (1994), certain forms of government intervention, which were designed to counter the major problems in reindeer herding, including competition for pastures, have in fact worked to maintain or even aggravate the problems. In all appearances, greed is hardly either a sufficient explanation for the situation or the sole motivation of the herders. In fact, land seems to be a common pool resource between the competing overall land use interests and strategies – the way the situation is presented today through the idea of diversified use by the environmental management agencies.

The heated discussion around the issue of overgrazing shows that despite the emphasis on rational decision-making based on scientific knowledge, the argumentation in environmental management is not free from emotion. Following the normative quality of the international agreements, protection of the environment has become a moral issue, and those opposing it or questioning it for whatever reason are signified as immoral. This underlines the point that negotiating environmental management is not only about ecological or economic issues, but also about many aspects of social relations, including value systems and power issues.

If the dominant knowledge system, embodied by the management institutions, defines the relevant ways of signifying the reality, it also cuts out alternative ways and contexts. This has produced an asymmetric negotiation situation, in which large parts of the argumentation – in this case by reindeer herders – are singled out on the grounds of not being scientifically competent, reliable, or relevant (Sipilä et al. 2000). Their assertions are often reduced to opinion, attitude, or fear, not perceived as factual statements. In terms of discourse analysis, it means that reindeer herders are not regarded as experts in their own field. Their knowledge is only validated through scientific mediation, meaning an external, scientific interpretation of it, for instance in the form of environmental impact analyses or other external expert statements in environmental planning.

Furthermore, the disputation of the reindeer herders is often accused by environmental management agencies of being expediently directed to political goals, and therefore not reliable. According to my observation of discourses (talk and text) and participatory planning meetings of wilderness and natural resource management, one of the central qualities of indigenous management concepts (as reindeer herders' knowledge represents it) is exactly the aptitude or suitability for putting into practice. A solution not adaptable to practice is not a solution. In this sense it is adapted knowledge, containing an evaluation based on a practical management perspective, in contrast to the ecological or biological knowledge that, although it also derives its base from observation, has a theoretical more than an applied perspective.

The following examples illustrate the mutual distrust between reindeer herders and environmental management authorities.

"... Reindeer herders [and their representatives] do not need to negotiate, because they can rely on appealing to High Court on the grounds of Indigenous Peoples' Rights..." (Informal statement by an environmental management agent, E 4)

"... Their [the representatives of the reindeer herders] disapproval of the management plan came to us as a total surprise, because during the negotiations, they would not oppose publicly..." (Informal statement by an environmental management agent, E 2)

"... It is no use participating in any negotiations. At the end Forest and Park Service will always do like they wish..." (Informal statement by a reindeer herder, R 3)

"... [T]hey [the environmental management agents] are maybe hearing, but not listening to us..." (Informal statement by the reindeer herder, R 5)

All in all, the accounts and argumentation of *reindeer talk* reflect explicitly the societal position and rationales of the account deliverers. The administrators, as representatives of the public (majority) community, start from the general principle of common use, and seem to be inclined to accept encroachments when they are considered to meet public needs. The herders in turn, in order to sustain the economic viability of their family units, cannot afford to reduce the size of their herds. Overgrazing is then the logical outcome of reduced pasture areas. From this point of view, monitoring the pasture:animal ratio, which is the main management tool, is not producing any data that might help herders deal with dilemmas they have to cope with or assist in forming strategies to adopt.

Undoubtedly biological pasture research has served to make the decision makers and the reindeer herders aware of pasture problems. However, I am concerned that overgrazing has been conceived of as the only authorized interpretation of the condition of pastures, without discussing its occurrence, its complex causes , or alternative interpretations presented by the various parties in the debate, even though biological research results have been criticized due to errors in the methods (Chap. 3; Sipilä et al. 2000). Moreover, there is an apparent shortage of research on other management aspects and concerns although the need for it is gradually recognized (Moen and Danell 2003). On the whole, the RENMAN project has, in an important way, contributed to widening the scope for further research by providing a deeper understanding of the rich variety of factors involved, and of their mutual interdependence (Chaps. 1, 17).

4.5 Monitoring vs. Observing for Practical Purposes: Contrasting Discourses

Regardless of the divergent points of departure, supervising the quality of pastures is in the interest of reindeer herders as well as researchers and environmental managers, different though the methods, purposes, analyses, interpretations, and conclusions may be. Natural scientific methods include primarily sampled and repetitive field experiments and surveys by satellite imagery, occasionally supplemented with herders' observations. The purpose here is to obtain tested data on biomass and growth (regeneration) of selected species in classified vegetative zones. The aim is to generalize theoretical and exact knowledge that contributes to the overall understanding of the regulating mechanisms of the ecosystem and of the resilience of biotopes and species. The conclusions are utilized by the official reindeer herding administration as directives and guidelines for regulating the number of reindeer, which in turn has direct consequences for the economy of reindeer herding households (Chaps. 3, 16).

The reindeer herders' objective in observing the pastures is primarily to follow the reindeer's access to nutriment and the factors influencing grazing conditions. Unlike scientific research, the purpose is to increase one's knowledge repertoire, not to produce tested facts and generalizations of the reality, or elaborate scientific theories. Reindeer herding knowledge is attained in direct contact with nature, through continuous practice, and in communication with other herders, primarily of the herding unit, clan, or family. Herding knowledge is typically practical knowledge or applicable to a practical context, but it is not necessarily experimental in the sense that it involves direct personal participation (Chaps. 7, 13). Herding knowledge is the primary cultural capital of the herder, for it corresponds to his or her competence and prestige in making herding decisions and managing practical operations. The observations cannot usually be separated from a management context although the connection between the two may not always be perceptible (Bielawski 1996). Reindeer herding knowledge is tested, complemented, and stored in the course of active oral communication between the members of the *siida*, the Sámi reindeer herding unit or community (Chap. 2, 5).

Comparing these two knowledge systems or ways of knowing, there appears to be, on one hand, a methodological question: how monitoring and observing differ as methods, how knowledge produced in these two contexts differs, and how it is transferable into another knowledge system? On the other hand, it is a question of interpreting research results and adapting them into practice.

Monitoring certain categorized features and processes of nature has become the internationally validated way of demonstrating concern for the environment. Accordingly, a series of indicators are designed with the intention of monitoring nature and its changes. This system is best suited for monitoring certain clearly outlined physical phenomena such as the components of climate and degree of pollution. However, when designing indicators for societal factors, the inaccuracy of such gauges and their openness to ambiguous interpretation is quickly apparent. Transforming multi-dimensional social and cultural phenomena into quantitative values often implies, or leads to, considerable oversimplification with a high degree of error, and the benefits are marginal.

Monitoring consists primarily of measuring, examining, and supervising through scientific, quantified methods and criteria. In connection with tundra or the forest/tundra ecosystem, which are the vital (winter) pastures for reindeer, monitoring is directed predominantly at the vegetation – above all at the lichen cover, since lichen is the key plant of significance for reindeer of the biotopes involved (Chap. 9, 11). In practice, the biomass of lichen (or predefined warmer-season fodder) is measured, or animal weight is monitored as an indicator of the state of the lichen cover, or of the major fodder at the summer pastures (Ahti 1961, 1977; Lyftingsmo 1965; Helle 1966; Klein 1968; Johansen et al. 1995). Besides the ecological factors some economic factors are also monitored according to cost/benefit logic. For instance, the productivity of reindeer herding is evaluated by the annual turnover (Porotaloustilastot 2002; Totalregnskap 2002).

Monitoring land use patterns with the help of district reindeer herding maps has been introduced as an attempt to produce a planning tool for more efficient use of land (Reindriftslov 1996). However, for the present, the results are incomplete for several reasons. First of all, the type of formal planning that this method represents is not in accordance with the cultural and social customs and conventions of the reindeer herding community. The siida units are used to negotiate the allocation of pastures in a flexible manner, although in later years conflicts over pastures have become more acute, due to the diminished total range of pastures and thus a less fortunate animal:pasture ratio. Second, because of a variety of factors, which are partly irregular or unpredictable, drafting simplified plans is fraught with difficulties and may even not be viable. Third, and perhaps most important, as the reindeer herders repeatedly emphasize, reindeer herding needs a *variety of alternative ranges* in order to maintain the flexibility and sustainability of the livelihood. Because of this, it is not possible or advisable to run tight planning schemes in the name of efficient land use.

In general, observation seems to have a special status in the reindeer herding Sámi community. It is a culturally validated skill that is tightly interwoven into the texture of cultural intercourse and practice. Reading nature and the herds' behavior, the movements of predators and of people, as well as registering the spread and impact of permanent changes such as roads, routes, cabins, bridges, fences, and other facilities in the grazing land, migration routes, round-up fences, etc., in their immediate vicinity, is an essential part of Sámi reindeer herding environmental management. It contributes to the herding decisions and the overall herding practice. Sámi language provides an accurate and abundant vocabulary (Eira 1994) for registering the nuances involved and conveying the information to siida members (Chap. 7). This is a living tradition and a conspicuous feature in my field of daily life where I reside and participate in the reindeer herding community. It has proved to be an adaptive and still viable practice after the many changes brought about by the modernization process.

"Reading nature" consists of observing and evaluating pastures and weather, snow and ice, and the sequence of changes involved, which determine access to nutrients and the behavior of reindeer. In the winter, perhaps the most often used term used in North Sámi is *goaivvesguohtun* or *goaivvis*, which means the digging circumstances under which reindeer reach the vegetation under the snow cover; Sara 1994, 1996). More than just a term, this is a comprehensive category that combines the various aspects mentioned before, connecting them with those of *guohtun ráfi* (grazing peace) – another major management term/category. Threats to grazing peace cover the appearance of predators, movements in the landscape (particularly by outsiders on motorized vehicles or with dogs), and various encroachments. These aspects form the core environment management concerns of Sámi reindeer herders and are negotiated daily between members of the siida. These negotiations are called *ságat,* "(reindeer) news" or "(reindeer) recitals". They form the basis of the communication culture and are an integral part of the social life of the Sámi reindeer herding communities (Paine 1994).

In comparison with scientific monitoring, the reindeer herder's observation may appear sporadic, unsystematic, and subjective. It is dependent on the movements of the herds and the herders in the landscape, follows the seasonally determined behavior of the animals and the herding activities. Furthermore, the scope is usually long term, linking the experiences, reflections and interpretations of the earlier generations with the current situation. It is also more broad-based, not directed at predefined indicators. Not being connected to certain theoretical presumptions and categories, it may not seem decontextualized, but instead possibly more creative in the sense of constructing new approaches to environmental problems. I suggest that indigenous knowledge, as represented here by reindeer herding knowledge, should not be approached as a set of information about nature as an isolated topic, or as a formal iteration of an old custom or model, but more aptly as a capacity for observing each situation and finding means of adjustment. As a consequence, the contribution of indigenous or traditional knowledge to environmental management may be more valuable for administrative decision-making than for supplementing the databank.

4.6 The Integrative Challenges of Indigenous and Scientific Conceptions

No matter how we interpret indigenous knowledge, integrating it with scientific knowledge is not an easy task. We are faced with several intellectual, technical, political, and ideological challenges. Not intending to romanticize or idealize reindeer herding knowledge, I will briefly discuss some practical aspects of integrating Sámi reindeer herding management perceptions and scientific perceptions in environmental management. Examining Sámi tundra reindeer herding as a system of environmental management, it is obvious that the cornerstone is siida unity (Bjørklund 1990; Paine 1994; Sara 1996). In Sámi society the siida represents a small, local, grass-root, worker-oriented management model in contrast to the centralized, external, administration-based, system-oriented management with general environmental, sociopolitical, and national economic goals of Nordic societies (Chap. 2). By its character, siida management is more flexible, resembling not a formal pact but what might be called a Consent formed during a long-term interaction and customary use patterns between the users and user groups, and updated continuously. In contrast to this, regulation by modern environmental management can be compared to contracts drafted by the management authorities acting as the representatives of landowners or land governors after having consulted designated stakeholders. The internal management system of reindeer herding (siida) represents a parallel system of environmental management with a special reference to the community and territory in question. This should be considered when discussing the integration of the two types of knowledge in the context of national environmental management.

In this perspective the Sámi perceptions seem to bring into focus, first, the applicability of relatively powerful planning tools that, with their use of both science and field research, yield management decisions of unquestioned authority. Second, the interlinkage between authority and responsibility is recognized. It means that the people using nature (environment) daily should not be separated from genuine decision-making authority. It also reminds us that management issues should not be separated from land ownership matters in the way it is largely done today. Third, the interlinkage between the wide societal factors and environmental management is emphasized. In other words, environmental management should not be approached only from ecological and economic viewpoints. Fourth and finally, the communicative aspect of environmental management is highlighted. Environmental management is as much an exchange of information regarding pastures, as enforcing and rearranging the social relations of the partners involved.

Up to what extent the Sámi management model is functioning today, and whether it contains features that are applicable, or could be integrated into the complex environmental management situation of modern society, remains still open. There are many comprehensive and ethically oriented questions to be answered related to regulatory mechanisms, distribution of authority, and so forth. However, certain things are evident if one considers the obvious challenges Sámi reindeer herders' perceptions pose to the current management system. For one thing, a re-evaluation of the principle of diversified use is called for. Prioritizing is needed between current user groups (reindeer herders, tourist agencies, and others), and compensation arranged for disturbances and damages to those negatively affected. For another, enlargement of the applied knowledge base in environmental management is required. This implies the validation of experience or practice-oriented knowledge, along with the methods and knowledge of academic disciplines as resources in environmental management. In this context, the economic, ecological, social, and cultural dimensions must be considered as inseparably intertwined, and efforts should be directed at finding the connecting links, and the interdependence and correspondence between these different aspects. All in all, modern environmental management in the tundra may entail, and profit from, developing multiple and variegated system management and the use of common sense and experience, which reindeer-herders' management conceptions, in a manner, represent.

In regard to monitoring, as the prevalent environmental management method, reindeer herding knowledge suggests two new indicators. According to reindeer herders' perceptions, assessing the adequacy of pastures should comprise, in addition to the biomass of selected species, also the extent of fallow, peaceful wilderness areas. Connected with this, a bucolic, undisturbed "grazing peace" should be acknowledged as an important feature of pastures. As presented earlier in the text, this involves the movements of other land users and land use forms, predators, etc. These two measures of grazing adequacy, (1) plant biomass and (2) a peaceful extent of pastures to include an unbroken range of diversified pastoral space and undisturbed grazing conditions, should be recognized as vital prerequisites of sustainable reindeer herding. Management measures should be taken for establishing and protecting them. Implicitly it is a step toward recognizing truly the fundamental dependence on land of reindeer herding.

4.7 Conclusions: Needed Dialogue Between Indigenous and Scientific Perceptions

In summarizing this vast and complex subject and the questions arising from the issue of comparing indigenous reindeer herding knowledge with scientific knowledge and administrative parameters, one should first be reminded that the trend of paying more attention to grass-root or local perceptions (including reindeer herding) is most welcome. The problem remains to be solved how, as a practical matter, the two types of knowledge are to be integrated. Scientific knowledge contributes to our understanding of interrelated phenomena, but it is not possible to devise stringent rules for human conduct from it (Haila and Levins 1992). On the other hand, it is not reasonable to expect that rough insertion of indigenous categories into biological or ecological knowledge would make it more applicable. The adaptation of management knowledge into practice is a complex process that involves political decisions on the various aspects of management, including the principal premises, targets, aims, means. In this process indigenous conceptions may provide alternative angles and approaches. In short, there is an obvious need for a dialogue between indigenous and scientific perception before any sort of integration of the two can take place. On the whole, Sámi reindeer herders question some of the starting points that modern environmental management consider self-evident. The views of these indigenous herders call for reassessment and novel perception. In this sense indigenous knowledge can contribute to the self-reflection of modern society.

Acknowledgements. I thank Trond Thuen (Department of Social Anthropology, University of Tromsø, Norway) for supervising my research work and substantially contributing to the argumentation and content of this article. I am also thankful to Ludger Müller-Wille (McGill University, Montréal, Canada) and Felix Müller (Christian-Albrecht University, Kiel, Germany) for their constructive comments and remarks on my text.

References

Ahti T (1961) Poron ravinnosta ja laitumista. (in Finnish; On food and pastures of reindeer). Lapin Tutkimusseuran Vuosikirja 2:18–28

Ahti T (1977) Lichens of the boreal coniferous Zone. In: Seaward MRD (ed) Lichen ecology. Academic Press, London, pp 145–181

Berger PL, Luckmann T (1967) The social construction of reality. A treatise in the sociology of knowledge. Doubleday, New York

Bielawski E (1996) Inuit indigenous knowledge and science in the Arctic. In: Nader L (ed) Naked science. Anthropological inquiry into boundaries, power and knowledge. Routledge, New York, pp 216–227

Bjørklund I (1990) Sámi reindeer pastoralism as an indigenous resource management system in Northern Norway: a contribution to the common property debate. Dev Change 21:75–86

Brox O (1989) Kan bygdnæringene bli lønsomme? (In Norwegian. Can rural industries be profitable?) Gyldendal Norsk forlag, Oslo

Dawes RM (1973) The commons dilemma game: an n-person mixed-motive game with a dominating strategy for defection. Org Res Bull 13:1–12

Eira NI (1994) Bohccuid luhtte. Gulahallat ja ollášuhttit siidadoalu. (In North Sámi. By the reindeer. Discussing and carrying out communal reindeer herding). DAT. Tangen Grafiske, Dramme

Emmelin L (1997) Föreställningar om rationalistisk planering (in Swedish. Conceptions of rational planning). In: Hilding-Rydevik T (ed) Samspel mark – vatten – miljö. Fysisk planering för att nå samhälleliga mål (The interplay land – water – environment. Physical planning for societal purposes). Forskningsrådnämnden Rapport 97/1, Stockholm, pp 50–68

Fischer F, Hajer M (1999) Beyond global discourse: The rediscovery of culture in environmental politics. In: Fischer F, Hajer M (eds) Living with nature. Environmental politics as cultural discourse. Oxford Univ Press, New York, pp 1–20

Gaare E (1998) Kan vi beregne hvor mange rein beitene tåler? (In Norwegian. Is it possible to assess how many reindeer the pastures can sustain?) Rangifer Rep 3:103–109

Grove-White R (1993) Environmentalism. A new moral discourse for technological society? In: Milton K (ed) Environmentalism. The view from anthropology. Routledge, London, pp 18–30

Hall S (1997) Introduction. In: Hall S (ed) Representation: cultural representation and signifying practices. Sage, London, pp 1–59

Haila Y (1995) Kestävän kehityksen luontoperusta: Mitä päättäjien tulee tietää ekologiasta? (In Finnish. The natural foundation of sustainable development: what the decision-makers should know about ecology?) Suomen Kuntaliitto, Helsinki

Haila Y, Jokinen P (2001) Ympäristöpolitiikka. Mikä ympäristö, kenen politiikka. (In Finnish. Environmental politics. Which environment, whose politics). Vastapaino, Tampere

Haila Y, Levins R (1992) Humanity and nature. Ecology, science and society. Pluto Press, London
Hajer M (1995) The politics of environmental discourse. Ecological modernization and the policy process. Clarendon Press, Oxford
Hardin S (1968) The tragedy of the commons. Science 162:1243–1248
Harding S (1992) After the neutrality ideal: science, politics, and "strong objectivity". Soc Res 59(3):567–599
Heikkilä L (2003) Ympäristöuhka vai perustuslaillinen oikeus? – Poronhoitoa koskeva puhe ympäristönhallinnan kontekstissa. (In Finnish. Environmental threat or constitutional right? – Reindeer talk in environmental management context.) In: Suopajärvi L, Valkonen J (eds) Pohjoinen luontosuhde. Elämäntapa ja luonnon politisoituminen (in Finnish; Northern nature relationship. Lifestyle and politicisation of nature). Lapin yliopiston yhteiskuntatieteellisiä julkaisuja B 43:115–143
Helle R (1966) An investigation of reindeer husbandry in Finland. Acta Lappon Fenn 5:1–66
Helle T (2002) Luonnollista vai ihmistoimintaa: poron laidunnusvaikutukset luonnonsuojelu- ja erämaa-alueilla (in Finnish; Natural or human activity: the impact of reindeer grazing on protected and wilderness areas). In: Saarinen J (ed) Erämaapolitiikka: pohjoiset erämaat arjen, hallinnan ja tutkimuksen kohteena. (In Finnish. Wilderness politics: the northern wilderness areas as objects of everyday activities, management and research.) Metsäntutkimuslaitoksen Tiedonantoja 827:57–65
Helle T, Aspi J (1983) Effects of winter grazing by reindeer on vegetation. Oikos 40:337–343
Ingold T (1980) Hunters, pastoralists and ranchers. Reindeer economics and their transformations. Cambridge Univ Press, Cambridge
Johansen BE, Karlsen SR (1998) Endringer i lavdekket på Finnmarksvidda 1987–1996 basert på Landsat 5/TM data. (In Norwegian. Changes in the lichen cover on the Finnmarksvidda 1987–96 based on Landsat 5/TM data.) NORUT, Tromsø
Johansen BE, Johansen ME, Karlsen SR (1995) Vegetasjons- og beitekartlegging i Finnmark og Nord-Troms (in Norwegian; Vegetation and pasture mapping in Finnmark and North Troms). NORUT, Tromsø
Käyhkö J, Pellikka P (1993) Porolaidunnuksen vaikutus kasvipeitteeseen Enontekiön Näkkälän paliskunnassa – SPOT XS satelliittiaineiston digitaalinen tulkinta (in Finnish; Remote sensing of the reindeer grazing impact on the vegetation in the Näkkälä reindeer herding co-operative, Enontekiö municipality, NW Finland, using SPOT-XS data). Terra 105(2):83–97
Klein DR (1968) The introduction, increase and crash of reindeer on St. Matthews Island. J Wildlife Manage 32:350–367
Kosmo A (1991) Hvorfor har vi for mange rein? (In Norwegian. Why there are too many reindeer?) Reindriftsnytt 2:24–26
Kosmo A, Ims AA (2001) Høyeste reintall for distriktene i Vest-Finnmark (in Norwegian; Maximum number of reindeer for the districts in West Finnmark). Høringsdokument. Reindriftsforvaltningen, Alta
Kumpula J, Colpaert A, Kumpula T, Nieminen M (1996) Poronhoitoalueen pohjoisosan porolaidunten inventointi. (In Finnish. Inventory of the pasture resources of the northern part of the reindeer management area.) Riista- ja kalatalouden tutkimuslaitos ja Oulun yliopisto, Maantieteen laitos, Oulu
Kumpula J, Colpaert A, Kumpula T, Nieminen M (1997) Suomen poronhoitoalueen talvilaidunvarat (in Finnish; The winter pasture resources of the Finnish reindeer management area). Riistan- ja kalantutkimuslaitos, Kala- ja riistaraportteja 93:1–42
Loikkanen T, Simojoki T, Wallenius P (1999) Participatory approach to natural resource management. A guide book. Metsähallitus. Suomen Graafiset Palvelut Oy, Kuopio

Luke T (1999) Eco-Managerialism: Environmental studies as a power/knowledge formation. In: Fischer F, Hajer M (eds) Living with nature. Environmental politics as cultural discourse. Oxford Univ Press, New York, pp 103–120

Lyftingsmo E (1965) Norske Fjellbeite – Oversyn over fjellbeite i Finnmark (in Norwegian. Norwegian mountain pastures – overview of the pastures in Finnmark), vol XV. Selskapet for Norges Vel, Mosjøen

Marcussen HS (2003) National environmental planning in the Third World: sustaining the myths? J Transdicipl Stud 2:1–15 (http://www.journal-tes.dk/; accessed 25 Sept 2004)

Milton K (1993) Introduction. In: Milton K (ed) Environmentalism. The view from anthropology. Routledge, London, pp 1–17

Moen J, Danell Ö (2003) Reindeer in the Swedish mountains: an assessment of grazing impacts. Ambio 32(6):397–402

Naskali A (2002) "The Great Wilderness Debate" – Erämaa-ajatuksesta villiyden korostamiseen (in Finnish; From the idea of wilderness to the emphasis on wildness). In: Saarinen J (ed) Erämaapolitiikka: pohjoiset erämaat arjen, hallinnan ja tutkimuksen kohteena (Wilderness politics: the northern wilderness areas as objects of everyday activities, management and research). Metsäntutkimuslaitoksen Tiedonantoja 827:87–119

Oskal N (1995) Det rette, det gode og reinlykke. (In Norwegian. The right, the good, and reindeer fortune.) Doctoral dissertation. Institutt for samfunnsvitenskap, Filosofiseksjonen, Universitet i Tromsø

Paine R (1994) Herds of the tundra: a portrait of Sámi reindeer pastoralism. Smithsonian Institution Press, Washington

Porotaloustilastot (2002) (in Finnish; Statistics on Reindeer Economy) Paliskuntain yhdistys. Rovaniemi

Potter J (1997) Discourse analysis as a way of analysing naturally occurring talk. In: Silverman D (ed) Qualitative research. Theory, method and practice. Sage Publications, London, pp 144–160

Prestbakmo H (1994) Lavbeitene på Finnmarksvidda. Endringer fra 1960 til 1990. (In Norwegian. Lichen pastures on the Finnmarksvidda.) Småskrift 1. Reindriftsadministration, Alta

Riseth JÅ (2000) Sámi reindeer management under technological change 1960–1990: Implications for common-pool resource use under various natural and institutional conditions. A comparative analysis of regional development in West Finnmark, North Trøndelag, and South Trøndelag/Hedmark, Norway, vol 1. Doctoral dissertation. Department of Economics and Social Sciences, Agricultural University of Norway, Ås

Reindriftslov (1996) (In Norwegian; Reindeer Herding Act). Storting, Oslo

Saarinen J (1998) Wilderness, tourism development, and sustainability: Wilderness attitudes and place ethics. USDA For Serv Proc RMRS-P 4:29–34

Saarinen J (2002) Erämaan muuttuvat merkitykset: pohjoisen luonnon traditionaalinen käyttö, moderni suojelu ja turistinen tulevaisuus. (In Finnish. The transforming significance of wilderness: the traditional use of northern nature, modern conservation and touristic future.) Alue ja Ympäristö 31(2):25–36

Sara MN (1993) Reindriftnæringens tilpassning og reindriftspolitikk. (In Norwegian. The reindeer industry's adaptation and reindeer breeding policies.) Dieđut 2:1–188

Sara MN (1994) Eatnamat (The landscape). In: Eira NI (ed) Bohccuid luhtte. Gulahallat ja ollášuhttit siidadoalu. (In North Sámi. By the reindeer. Discussing and carrying out communal reindeer herding). DAT, Tangen Grafiske, Drammen, pp 26–53

Sara MN (1996) Boazu lea biekka buorri. Studie av reindriftens tilpasningsformer i østse del av

Kautokeino reinsogn på 1950-tallet. (In North Sámi and Norwegian; Reindeer is faster than the wind. A study of the adaptative forms of reindeer herding in the eastern part of the Kautokeino reindeer herding zone in the 1950s.) Master's Thesis. Institute for Samfunnsvitenskap, Universitet i Tromsø

Sipilä P, Magga H, Aikio P (2000) Luppoa etsimässä. Lapin paliskunnan alueen luppolaidunten inventointi 1999–2000. (In Finnish. Searching for arboreal lichen. Inventory of grazing grounds with arboreal lichen within the Lappi reindeer herding district in 1999–2000). Lapin paliskunta, Oulun Painotuote Oy, Oulu

Skonhoft A (1998) Bioeconomic modeling of Sámi reindeer pastoralism. In: Jentoft S (ed) Commons in a cold climate. Coastal fisheries and reindeer pastoralism in North Norway: The co-management approach. Man and the biosphere series, vol 22 UNESCO. Parthenon, New York pp 41–56

Strauss A (1978) Negotiations. Varieties, contexts, processes and social order. Jossey-Baas, San Francisco

Totalregnskap for reindriftsnæringen (2002) (in Norwegian. Total accounts for the reindeer breeding industry 2002). Økonomisk utvalg for reindriftsnæringen. Reindriftsforvaltning, Alta

Turi JM (2001) Opening speech. The 2nd World Reindeer Herders' Congress. Arctic Centre Rep 36:10–14

Tyler N (1997) Årlig variasjon i vekst hos rein: er "bæreevne" et gyldig begrep? (In Norwegian. Annual variations in reindeer growth: is "sustainability" a valid concept?) Reindriftens Fagråd, Rapport fra fagseminar i reindriften Kautokeino, November 1997. Reindriftsforvaltningen, Alta, pp 141–147

Ulvevadet B (2000) Penger teller, kultur avgør. Analyse av Statens virkemiddlerbruk for å oppnå en bærekraftig reindriftsnæring i Vest-Finnmark, og reindriftsamenes reaksjoner på denne politik (in Norwegian; Money counts, culture decides. Analysis of the State's use of measures to achieve a sustainable reindeer herding economy in West Finnmark). Master's thesis. Institutt for sosiologi og samfunnsgeografi, Universitetet i Oslo

Wetherell M, Potter J (1992) Mapping the language of racism. Discourse and the legitimation of exploitation. Columbia Univ Press, New York

5 The Challenges and Dilemmas of Concession Reindeer Management in Sweden

J.-L. JERNSLETTEN and H. BEACH

5.1 Introduction

From a circumpolar perspective, reindeer herding in the Fennoscandian region, despite many affinities with herding in Eurasia and North America, demonstrates a number of distinctive characteristics, not least through its relations to Sámi indigenous issues (Chap. 2). While the legal structures for herding in Fennoscandia, designed by the nation states of Norway, Sweden, and Finland, present a wide array of regulations, these can also be recognized as having common denominators occasioned by historical circumstance which at times have even bound these different countries in unions with common legal codes. Moreover the Fennoscandian countries, all being modern so-called First Nations with highly developed policies of social welfare and commitment to human rights, have confronted the same dilemmas of resource equity for all citizens versus resource protectionism for indigenous minorities. While their dilemmas are similar, the solutions devised in the Fennoscandian countries have been quite variable (Chaps. 3, 4). We do not wish to argue that there is necessarily an optimal formula of resource openness or protectionism applicable to all countries and indigenous peoples. Quite the contrary, we argue that reasonable and moral policies must vary over time and with context (Chap. 6).

We also maintain that one of the most important elements in any assessment of context for devising resource regulations pertains to the state of the environment and our growing understanding of ecological principles (Chap. 15). This said, it is equally important to guard indigenous interests against what Beach has termed "eco-colonialism" (1997) and "vulgar ecology" (1998) in which the unavoidable political nature of ecological resource management is masked by appeals to what is presented as pure science (Chap. 4).

This chapter will focus upon a unique form of reindeer resource management in northern Sweden – small in territorial extent, yet highly significant in

Ecological Studies, Vol. 184
B.C. Forbes, M. Bölter, L. Müller-Wille, J. Hukkinen,
F. Müller, N. Gunslay, Y. Konstatinov (Eds.)
Reindeer Management in Northernmost Europe

principle (Jernsletten and Beach 2004). It is unique in its position of mixed openness for local residents and protectionism for indigenous Sámi reindeer management. This form of herding, "concession herding", offers its own positive solutions to the ecological problems born of private ownership of animals on common grazing lands, and those born of conflict between different competing land-based industries. Naturally, the same regulations that secure their strengths also generate their own set of problems. This chapter is therefore dedicated to elucidating the major points of principle characterizing the regulations specific to concession herding and seeing how these impact upon the project of sustainable herding, sustainable settlement, and the maintenance of traditional livelihood in the North.

There are many aspects on which reindeer management forms might be compared (Beach 1990). However, of utmost significance to this study are the social relations structuring work and ownership as they impact on ecological sustainability. Hence, we are particularly interested in regulations defining access to herding work and reindeer ownership. Within this Fennoscandian context, besides the importance of distinguishing those of Sámi ancestry who all hold herding rights from those without such ancestry and without the right to herd , it is essential to distinguish between the legal right to herd reindeer and the legal right to own them. It is also important to realize that those people who fit into the category "herder" as a legal construct do not necessarily exhaust the category of those people who actually do herding work.

With respect to the social relations targeted here, we find that the reindeer industry in Fennoscandia is closely connected to the Sámi and the Sámi society. Both Norway and Sweden have made reindeer management a protected industry, meaning that only Sámi can be employed full time by it and have this as their main income. The situation in Finland differs considerably from that in Sweden and Norway, since in Finland the industry is not protected and therefore is in principle open to everyone living in the reindeer herding area (Chaps. 2–4). When it comes to reindeer ownership, however, the situation becomes immediately more complex. In the "tame reindeer district" (in Norwegian *tamreinslag*) in the southern part of Norway, the majority of the reindeer owners are Norwegian. With the "contract reindeer system" (in Swedish *skötesrensystemet*) in Sweden and notably in this system as expressed and elaborated regionally in the "concession reindeer management area" in the Kalix and Torne Valleys (Tornedalen; *konsessionrennäringen*) both Sámi and non-Sámi can own reindeer without holding legal status as herders. Moreover, some of these non-herders can do a good deal of actual herding. We shall proceed to portray the social relations of reindeer management in the concession area from the aspects of ethnicity, legal herder categorization, reindeer ownership, and actual herding work.

The reindeer management system in the concession area differs from the Sámi reindeer management system in the rest of the Swedish herding area on several vital points. One important difference is that the year-around reindeer pastures are "below", that is south or east of the curving *Lappmarksgränsen*[1] and thereby on what in the rest of Sweden comprises traditional winter pastures (Fig. 5.1). Another important difference is the strong position held by

[1] Lappmarksgränsen is translated by the Swedish state as "Sámi Territory Border" (SOU 2001:101, page 65), and was a line drawn on the map by the Crown in the 1700s to implement policy decisions. It was never an accurate demarcation of original Sámi territorial residency. While Sámi territorial rights have thereby come to be restricted by this border, we have choosen to use the original term to avoid giving the impression that original Sámi occupancy was properly described by it.

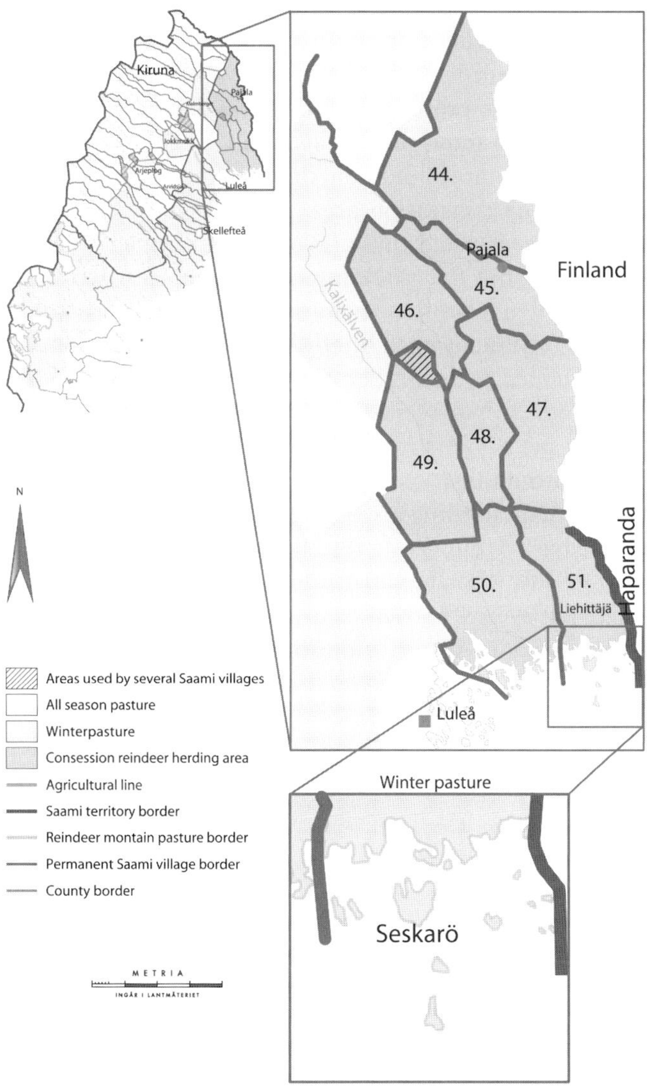

Fig. 5.1. The borders of the concession management area in Sweden

the contract reindeer owners (Sect. 5.4.1). And finally, a concession from the local and national authorities is needed to practice this form of reindeer management (SOU 2001 p. 101).

Knowledge about the reindeer industry from the majority population and the local authorities is surprisingly low throughout Fennoscandia, and the industry is almost "invisible" in every respect. This is a situation not only connected to Fennoscandia, but found throughout the circumpolar region, i.e., in Norway, Sweden, Finland, Russia, Alaska, and Canada (Jernsletten and Klokov 2002). The concession reindeer industry in Tornedalen is not only quite "invisible" to the local population, but it is also invisible in the academic mindscape.

The main literature connected to this topic consists of various official documents from the Swedish government or from some concrete investigation connected to a specific problem in the concession area (e.g., Rönnbäck and Kvist 1990). Literature about the Forest Sámi Villages or Forest Sámi reindeer management is much more common; this herding occurs not in, but immediately adjacent to the concession area. *Skogslapparna i Sverige* (The Forest Sámi in Sweden) by Ernst Manker (1968), for example, describes the Forest Sámi area in detail, but the adjoining concession area (with practical forest reindeer management traditions not unlike those of their neighboring Forest Sámi Villages) is just mentioned briefly at the end of the book. Why has academic interest stopped at the border of the Sámi territory? Could it be as simple as that reindeer management is conceived as a secondary source of income, even as just a hobby, or does it reflect a lack of "indigenous appeal" to the scientific communities?

In the larger Sámi herding area in Sweden, it could be easy to explain this lack of knowledge by connecting it to the protection of this industry; that is, since it is only Sámi who are allowed to herd, the majority population is little concerned. But even if partially true, this explanation cannot be fully adequate, for the industry is still quite invisible in the concession area, too, where the concession model opens the industry much more to local residents.

5.2 The Unexpected Field

The above-mentioned academic invisibility becomes even more evident when one starts to look empirically into the field. With no literature describing or analyzing the concession reindeer management system, the field is wide open for descriptions based on a stereotypic misunderstanding.

Fieldwork was conducted for an eight-month period in 2003 and provided surprising revelations. Some of the most unexpected findings during the fieldwork were:

1. A low level of integration of the reindeer industry into the local community exists even though this is an industry that includes the local population to a higher degree than it does in Sámi Management Systems.
2. The contract reindeer owners are closely connected to the industry.
3. The contract reindeer owners hold a high level of knowledge of the animal and the earmark (notches cut in the animal's ear as a sign of ownership).
4. The contract reindeer owners have a close relationship with their animals and are able to find their own animals in the herd.
5. The reindeer industry has developed its own management system and is much less an imitation of the Sámi reindeer system than expected.

The surprises mentioned here are good examples of (mis-)understandings (also shared originally by both authors of this chapter) of the concession system occasioned by a far more prevalent knowledge of Sámi-based reindeer management.

During the fieldwork in Tornedalen it became more and more obvious that the concession reindeer system represented a model for "coexistence" between different industries and an example of different ways of solving some of the challenges the reindeer industry is facing with regard to the majority society.

5.3 The Reindeer System in Sweden

There are 51 Sámi Villages in Sweden. The northernmost ones are located in the county of Norrbotten and the southernmost Sámi Village is situated in the county of Jämtland (Fig. 5.1). According to the "Swedish Reindeer Herding Act" (SFS 1971 p. 437), a "Sámi Village" is defined as the organization that manages the reindeer herding in a designated geographical area. A "Sámi Village" is also an economic business firm; legally it is similar to an economic corporation with a chairman and an assembly. All Sámi rights concerning hunting, fishing, and use of the forest are connected to membership in a Sámi Village, and it presumes active participation of its members in reindeer herding. All activity connected to reindeer management is (supposedly) organized by the Sámi Villages and supervised by the County Administrative Board (Labba and Jernsletten 2004).

The reindeer industry in Sweden, including concession areas, involves approximately 950 private family business units distributed over the 51 Sámi Villages. According to the government, a management unit can be defined as an economic enterprise managed by a *responsible* reindeer herder (host herder or Sw. *husbonde*) and his/her household/family (SJVFS 1993 p. 162). A management unit can be legally regarded as an enterprise with either one or several family members who own reindeer. The activity within a management unit can be viewed as cooperation between (1) the responsible reindeer herder (*husbonde*), and (2) the Sámi Village (Labba and Jernsletten 2004).

5.4 Early Legislation and Regulation

Arguments for maintaining the possibility of non-Sámi ownership are invoked from the earliest Reindeer Acts of 1886 and 1898. Both of these Acts upheld the rights of the Sámi people of Sweden to continue the nomadic life connected to reindeer management (although this legislation promotes the concept that these rights are in fact only privileges bestowed by the State). On the other hand, reindeer gave the Swedish farmers in the north an important supplement of meat, which could mean the difference between success and failure. The Reindeer Acts, therefore, encoded the already extant and important system of social relations between the Sámi reindeer families and the local population, the system that we today know as the "contract reindeer system", a plan that gave the local population the opportunity to own reindeer as part of a Sámi reindeer family herd. The contract reindeer system was confirmed as a supplement to the local people's income. "...[T]he term 'contract reindeer' [to] indicate those reindeer owned by non-herders, Sámi or Swedes. These owners must arrange for their reindeer to be cared for by a herder. They must establish some kind of contract with a host herder." (Beach 1981 p. 110)

Of course by "herder" and "host herder" in this legal concept is meant not simply someone who might herd in the physical sense a person who has the legal right to control reindeer and thereby carries responsibility for them, one who by Swedish terminology is defind as "husbonde" for the reindeer. When wishing to draw attention to someone possessing rights according to this legal concept of herding, we suggest use of term "host herder".

As indicated above, the contract reindeer system embodies a strong relationship between herders and owners of the contract reindeer. The local people were not supposed to make large earnings on the reindeer – animals that the government wanted to be an important source of income and a way out of poverty for the Sámi pastoralists. This contract reindeer relationship between the Sámi reindeer management system and local people changed dramatically in the 1930s, and today the interdependence has almost ceased to exist (Nordin 2002). This is not the case in the concession reindeer herding area. Here this relationship is still very strong.

5.5 Current Legislation and Regulation

5.5.1 The Unique Tornedalen Case

Concession reindeer herding as we know it today was introduced by the government in the 1928 Reindeer Grazing Act. However, the concept "concession" was first introduced in the 1971 Reindeer Herding Act. Many events have conspired to compose the background for the ensuing regulations of reindeer ownership and herding between the Sámi and the local population. However, the events back in 1912 and 1913 in Arjeplog and Arvidsjaur, where local Swedish businessmen owned big herds and exploited the Sámi as herders, and the repercussions in the Swedish Parliament (*Svenska Riksdagen*) are among the strongest policy determinants. According to the recorded discussion in the Parliament, its members were concerned about this development and wanted to stop it. One of the solutions discussed was the possibility of make ownership of reindeer, not only their herding, exclusively Sámi.

The reactions to make both exclusively Sámi were many, and several members of the parliament were afraid that the farmers of Norrbotten who were dependent on reindeer meat would not be able to continue farming. But there came also an especially strong reaction from Tornedalen in 1918. This was from Victor Wikström, priest (*kominister*) of the Muonionalusta kappelförsamling (parish congregation), who emphasised the importance of local ownership of reindeer and that this was a tradition in Tornedalen with deep historical roots. The arguments were that the people of Tornedalen had always owned reindeer, that they used reindeer instead of horses in their forestry work during the winter, that reindeer transportation was very important in winter, that reindeer were important as a meat resource, and that the reindeer connected the local population in a positive way with the Sámi families in the area.

Space here does not permit us to look deeply into this story, but it began in 1917 when the government decided that all the reindeer in Tornedalen should be slaughtered because herding them was illegal. Then between 1917 and 1928 there came an about-face in the Swedish government policies concerning reindeer – from slaughter in 1917 to incorporation in 1928 into Reindeer Grazing Act. Suffice it to say that while the contract ownership system became heavily regulated everywhere in the herding area, a special arrangement was devised for the Tornedalen area and embedded in what became designated the "concession herding system".

While reindeer management in Tornedalen was regulated for the first time by law in 1928, but the wording and content of the act has changed since. The official presentation of the reindeer management system in the concession area today is as follows: "Concession reindeer management is conducted below the Sámi territory border in the Kalix and Torne river valleys. Reindeer management may only be conducted all the year round once the county administrative board has given one or several Sámi permission (a concession) to have both their own reindeer and those belonging to the landowner in the area (managed reindeer) within a particular area. As well as the holder of the concession, the owner of the managed (i.e., contract) reindeer also participates in the reindeer management to some extent. It is the owner of the managed reindeer who, together with the holder of the concession, makes the decision concerning how the reindeer management is to be conducted." (SOU 2001:101, p. 65)

An important concept is mentioned in the first sentence, namely, "the Sámi territory border". The Sámi Territory Border is an administrative boundary created back in 1752 (Hultblad 1968; Cramér and Prawitz 1970) and is viewed by the government as the limit of the Sámi reindeer pasture area. According to the Swedish Reindeer Herding Act the Sámi hold rights for winter pastures below this border, but not for summer pastures, or "year-around" pastures. It is not difficult to understand that this border was created in the first place to regulate disagreements between the newly arrived settlers along the coast and the Sámi reindeer families.

The fact that it is only Sámi that could hold a concession for *year-around* reindeer management below the Sámi territory border on *private owned land* brings up an essential question: is it indeed a concession for the Sámi to utilize grazing for their reindeer year-around below the Sámi territory border where landed farmers hold sway? Or is it rather the reverse: a concession for the landed settled contract reindeer owners to exercise the Sámi right to herd? One must also consider what the effects are of the introduction of the concession management system into Kalix and Tornedalen. And what is the future prospect of this system?

As will be shown below, there is a great deal of vagueness and confusion connected to the legal framework of concession herding and the relationship between the concession holder and the contract reindeer owners.

5.5.2 Concession Reindeer Herding in Tornedalen Today

The concession reindeer herding area is found in Kalix and Tornedalen in the northeast corner of Sweden, bordering Finland. In 2003 the area covered 16,800 square kilometers and consisted of eight concession Sámi Villages with a total number of 16,000 reindeer. The eight concession Sámi Villages are (from north to south): Muonio, Sattajärvi, Tärendö, Kurjo, Ängeså, Pirttijärvi, Kalix, and Liehittäjä (Svensk rennäring 1999).

A concession Sámi Village is at first appearance not very different from a Sámi Village elsewhere in Sweden. But as one looks more closely into the concession system, several important differences will appear. First, the Swedish Reindeer Herding Act states that the leader of the herding activities must be Sámi. The chairman of the board of the concession Sámi Village does not need to be Sámi. The leader of the herding activities who must be Sámi must also be the person who holds the concession from the national reindeer authorities. Some of the villages have one concession holder, the necessary minimum, and some of the villages have several.

Liehittäjä is one of the eight concession Sámi Villages and will serve as an example of how this system is put together. The members of the village consist of the concession holder (in this case only one) and persons with a registered farm inside the village grazing area, a total of 96 persons with as many farms in 2003. Each farm has one registered earmark in addition to those of the concession holder family's earmarks, and one earmark for the village (to be used when an individual owner cannot be specified). Each village member's right to an earmark is connected to the registration number of his or her farm, but the earmarks for the concession holder and his family are personally owned. This means that a member who, for example, chooses to relocate outside the concession Sámi Village area, or to sell his or her farm for a regular apartment or house, will lose the right to an earmark.

The concession held by the Sámi herding leader states the maximum number of reindeer for the concession Sámi Village and, as a part of this total, the maximum number of reindeer to be owned by him, the concession holder. The difference between the village maximum number and that of the concession holder is the number of reindeer to be taken as "contract reindeer" (Beach 1981). In addition, the Reindeer Act restricts the number of reindeer that can be owned by each contract reindeer owner to 30. A normal division is 20–80 between the concession holder's reindeer and the contract reindeer in the concession village. The maximum number of reindeer permitted in Liehittäjä by the reindeer authorities is 1,200 animals. This means that all 96 members cannot have 30 reindeer each, since this would make 2,880 contract reindeer in addition to the reindeer owned by the concession holder.

According to the presentation above about the "framework" of the concession system, it is clear that this system in principle does not differ a lot from the contract reindeer system found in the Sámi reindeer management system in the rest of the country's herding regions. But two main differences have to be mentioned. The first is the distribution of power between the concession holder and the contract reindeer owners, and the second is the participation of the contract reindeer owners in the herding.

5.5.3 Distribution of Power

The voting system in a concession Sámi Village differs from the voting system found in a Mountain or Forest Sámi Village, a construct that makes a significant difference when it comes to the distribution of power in a village. In a Mountain and Forest Sámi Village the numbers of Sámi reindeer owners outnumber by far the contract reindeer owners; contract owners have no vote at all in Sámi Village affairs anyway; and it is just in special cases that the contract reindeer herders have an influence on the voting power.

"Which herder is accounted as caring for an inactive member's reindeer can be quite significant for a group's total voting power. If there are two active, class-1 brothers, one having 291 head and the other 280 head and their sister, a class-3 member, has only 10 head, counting her reindeer with her first brother's herd will give him an extra vote, while counting them with her second brother's will not add a vote." (Beach 1981 p. 387)

The crucial point here is that in the Mountain and Forest Sámi Villages the contract reindeer can be counted in the calculation of their individual contract herder's voting power, but it is the *host herder* responsible for the care of these deer and not the contract *owner* who casts the vote, whatever its strength, as he pleases.

By contrast, in the concession area each contract reindeer owner holds one vote independently of the number of animals owned. The concession holder, on the other hand, gets one vote for every newly started 20 head of reindeer, meaning that the concession holder wields between 10 and 30 votes depending on the number of animals owned. With a voting system like this, the contract reindeer owners' collective voting power can dominate a concession Sámi Village, a condition that is not possible in the Mountain and Forest Sámi Villages where voting blocks, if any, are not drawn along the host herder/owner distinction at all. In Liehittäjä concession Sámi Village, as one example, there is one concession holder and 96 contract reindeer owners – a situation in which the contract reindeer owners easily can control any voting situation. In Liehittäjä, however, the graduated voting system has never been implemented. Trust in the decisions of the concession holder is high, ethno-politics has not invaded herding practicalities, and consensus is strong. This is not necessarily the situation in other concession villages.

5.5.4 Concession for What?

The above heading is a question posed by one of the contract reindeer owners in Liehittäjä concession Sámi Village. As we have seen previously, this question presented itself unavoidably as we attempted to outline the parameters of the concession system. It is significant that it is also posed by a concession Sámi Village member, and the answer to this question could contribute to clarifying a complex legal framework. We noted earlier two basic alternative answers to this question: 1) The concession is directed toward the Sámi and gives them a concession to practice year-around reindeer management below the Sámi Territory Border, or 2) the concession is directed toward the local population, (the land-owners) permitting them to practice herding activities inside a concession Sámi Village with powers greater than those permitted non-Sámi elsewhere.

According to the content of such a concession (the actual concession letter to the concession holder), and documents from the County Administrative Board of Norrbotten (Rönnbäck and Kvist 1990), the intention of the reindeer herding authorities is identical to the first explanation above. This intention is also supported in the legal framework, since the Sámi concession holder does not have any Sámi fishing and hunting rights inside the Sámi Village area (as Sámi herders do in the Mountain and Forest Sámi Villages) and has limited rights connected to the building of fences, cabins, etc. (SFS 1971:437, §88)

Even if we can surmise the official understanding of what the concessions are given for, the problem connected to the legal status of the local landowners remains. According to the Reindeer Herding Act of 1971, it is only Sámi who hold the right to herd reindeer and to enjoy full membership rights in the Sámi Village. The reindeer herding right in Sweden is explained by a Swedish court judge in the following way: "The reindeer herding right means that a person who is a Sámi has the right to utilize grazing land and water for support of himself and his reindeer. This means that the reindeer herding right is a right of use connected to real estate property. Real estate property is land and water. A right of use is a right to utilize land and water without having ownership" (Geijer et al. 2003; authors' translation).

The right to herd reindeer is closely connected to the Sámi people of Sweden, and this is of course the reason, given by the government, that only Sámi can be the concession holders. To become a full member of a Mountain or Forest Sámi Village the requirement is that the person holds the right to herd reindeer (in other words, is "of Sámi heritage"). Looking into the legal framework of the concession system, one finds this legal principle is upheld with respect to the concession holder, but contradicted with respect to general membership. It states: "(§86). For the practice of reindeer herding within the concession area there shall be a Sámi Village. A member in such a Sámi Village is the owner of the concession, his or her spouse and at-home-living children, those with herding eligibility who assist the concession holder in herding work and have no other major source of salaried employment, as well as owners of contract reindeer within the concession area" (SFS 1971 p. 437).

The regulations regarding the Sámi Village and its administration apply in appropriate parts to matters concerning the Sámi Village for concession herding with the following deviations: "The owner of contract reindeer is to be considered like a reindeer herding member" (SFS 1971:437, §86; authors' translation).

The government has here introduced a remarkable hybrid construction whereby the contract reindeer owners are to be "like" reindeer herding members (full Sámi Village members) of the concession Sámi Village, but without Sámi reindeer herding rights! (Note that this is the same kind of hybrid construct reinvented for the Fishing Sámi by the Swedish Sámi Rights Commission in the government Proposition 1992/93:32 (in Beach 1995). In the unique concession *Sameby* (Sámi Village) construction, we must distinguish between the inalienable Sámi reindeer herding right (for the concession holder) and the right to herd reindeer according to §86 – linked on this issue also to §13 – accruing to contract owner membership. According to §13, "By reindeer herding member in a Sameby is meant a member who...engages in reindeer herding with his own reindeer within the Village's grazing area." Significant here is that in the Sámi Villages with concession, contract owners are *reindeer herding* members. The question arises therefore: How does the Sámi reindeer herding right of the concession holder differ (if at all) from the contract owners' right to herd as reindeer herding members?

The right to herd of the contract owner is alienable. That is, should a contract owner/herder ever lose his connection to the land (homestead) on which his contract ownership right is based, he loses the right to herd. Someone of Sámi heritage maintains the Sámi herding right no matter what, everywhere in Sweden (although a Sámi must gain Sameby membership somewhere to *practice* herding). Neither the concession holder nor the contract owners/herders have Sámi rights to hunt and fish on concession village lands. A contract owner or concession holder may hunt and fish on his own private lands, but this is based on his ownership right and not on his herding right.

5.6 A Small Scale System of Coexistence?

The experiences during fieldwork in Liehittäjä concession Sámi Village brought up interesting questions. One question mentioned earlier was the active status of the contract reindeer owners and the relationship between the concession holder and the contract reindeer owners. This relationship and the way it is worked out by the members and the concession Sámi Village with their surroundings illustrates what can be called a small-scale system of coexistence.

This coexistence is in many ways a requirement, or at least an important factor, for the reindeer authorities in their decision-making process before a concession is granted. This requirement occurs in the text of the legal framework: "A concession can be assigned only if the continuation of the reindeer industry is of predominant importance for the local community, and only if the applicant for the concession can be assumed to pursue this reindeer herding in a suitable way." (SFS 1971:437 §85; authors' translation)

The formulation that the "reindeer industry should be of predominant importance for the local community" is difficult to understand and calls for some unknown criteria for an evaluation. On the other hand, this paragraph emphasizes the importance of coexistence between the local community and the reindeer industry.

In the text following we will supply some empirical examples of how such a coexistence is evident in Liehittäjä concession Sámi Village. The yearly migration with the herd from the winter pasture to the summer pasture will illustrate the complex interplay between the concession holder and the contract reindeer owners, an interplay that is rarely found between host herders and contract owners in the Sámi reindeer management system.

5.6.1 Winter Pasture Activities

The winter pasture is situated in the south along the coast of Haparanda and in the archipelago of the Gulf of Bothnia (*Bottniska viken*). The reindeer cross the sea ice in the beginning of December out to Seskarö, the main island of winter pasture (Fig. 5.2).

Winter pastures among the islands of the archipelago are subject to a high risk of icing. The temperature can drop to –30 °C, rise to +5 °C and fall back to –15 °C over just a couple of days. A temperature shift like this will create a thick layer of ice and snow over the pastures with no access for the reindeer to the lichen. In order to avoid starvation of the animals, two fodder stations have been established on Seskarö. These two stations and the reindeer enclosure constitute the very "heart" of the winter pasture, around which all activities are organized.

Fig. 5.2. Moving animals over the sea ice

Liehittäjä concession Sámi Village has only one concession holder, and he has been working for the Sámi Village for many years – first as an assistant and now as the concession holder with the whole responsibility for the herding activities. Approximately 20 % of the total herd belongs to him; the remaining 80 % are contract reindeer.

In the last week of March the gathering of the winter herd starts. The animals are scattered around among 10–12 different islands in the archipelago. The concession holder and two contract reindeer owners drive by snowmobile around the whole pasture area to collect them to take back to Seskarö where they are fenced in. This operation normally takes a week.

5.6.2 Spring Migration

Another example of the coexistence of concession management with the larger society and the integration of work between the Sámi concession holder and the contract reindeer owners is evident in the spring, too. Depending on the state of the sea ice, the reindeer are moved from the winter pasture to the summer pasture in one herd.

Several contract owners are involved in this migration process. The concession holder has the opportunity to select different migration routes depending on snow conditions, disturbances and accessibility. The selection of the migration route is solely the concession holder's responsibility. When the route is selected, the preparation work starts. The concession Sámi Village must come to an agreement with the police for the crossing of highway E4, the main road in this area. Because the highway has wildlife fencing on both sides of the road, the reindeer must be taken across in a well-controlled herd cluster in a short time. The reindeer are, therefore, initially enclosed. Then the highway is closed, the wildlife fencing is opened on both sides, and the herd is ushered across.

The contract reindeer owners put up all the temporary fences for the night stops beforehand. They also bring fodder for the temporary night enclosures, food for the herders, and fuel for the snowmobiles. After the herd and herders start off each morning, all equipment and fences are collected by other contract reindeer owners not actively engaged with the animals and brought back to the storehouse in the concession Sámi Village.

The actual travel with the herd proceeds in more or less the same way in all the reindeer herding systems. The animals are prodded along in one big herd with herders pushing in the back and "closing" on both sides.

5.6.3 Summer Pasture and Calf Marking

The reindeer management system in Liehittäjä is partly based on traditional forest Sámi reindeer management and partly on a local reindeer herding system adapted to the environment and the pastures inside the border of the Sámi Village. The summer pasture is situated in the forest area of the northern Haparanda municipality on the border of the Övre Torne municipality. The pasture consists mainly of pine and birch forest with a grass constituting the main summer diet.

One of the main events in a concession Sámi Village is calf marking (Fig. 5.3). In late June, when the temperature rises and mosquitoes enter the scene, it is time to collect the animals and bring them into a marking corral. It is long and hard work in the dense forest to find, round-up, and bring the animals into these enclosures. Because the animals are scattered around over the whole summer pasture, they are built at different locations, the animals to be brought to the nearest one. Rounding up reindeer in dense forest is difficult and requires a substantial knowledge of the landscape. The reindeer know the area and its landscape well, and use the same "roads" every year. It is essential for a herder to know their pattern of movement to have a successful roundup.

Fig. 5.3. Catching calves in the fence

Not all the contract reindeer owners possess this kind of expertise, but those who do actually participate in this work, develop an excellent knowledge of landscape and animal movement.

While the concession holder and some of the contract reindeer owners are busy rounding up the herd, other members prepared the enclosure and the equipment for the marking. When everything is set for marking the calves, all members are informed by telephone. The marking starts usually at 1800 (6 PM) and continues until early morning, depending on how many calves there are in the corral. The work inside the fence is followed from the sidelines by many of the contract reindeer owners. Some of the owners do not participate in the actual marking but are responsible for the equipment, some are making coffee, and others are just checking if they have calves present and that they get marked. The calf marking is a big social event for the members, and it arouses huge interest. This is the place where one meets other members, where stories are told, where people get the opportunity to catch up with old friends. Depending on the weather – temperature and wind – the marking period could continue for two to three weeks before all the calves are marked.

These examples illustrate the close connection between the contract reindeer owners and the industry and the contract reindeer owners' high level of knowledge of the reindeer and of the practical work with them.

5.7 Old System, New Challenges: The Future of the Concession System

The concession reindeer management system is of unique interest, because it offers an opportunity to look into a management system that, compared to other reindeer management systems, is different in history, in legal framework, in management practise, and thereby in its approach to the pursuit of sustainability.

The Sámi reindeer management system in the west of Fennoscandia (Norway, Sweden and part of Finland) is characterized by a strong Sámi role, where contract reindeer owners have minimal impact on the Sámi Village/industry. In the east (Finland) the situation is quit the opposite, with the reindeer industry open to everybody living inside a reindeer district (in Finnish *paliskunta*) (Chap. 3). The concession area (Kalix and Torne Valley) is situated between these two reindeer management systems, and represents a third option: controlled openness permitting a majority of (usually) non-Sámi reindeer owners with voting powers in the Village Board, accompanied by a small, uniform, maximal herd size limit for each contract owner/homesteader. It is a system designed far more to supplement the many-faceted household economy of the northern settler than to support the legal rights and culture of an indigenous people.

These three different management systems illustrate the effects and issues connected to the questions of protective versus non-protective reindeer management systems, and, despite other influential variables, provide also an opportunity to seek correlations between these different management regimes and sustainability.

In Sweden, the reindeer industry is protected. The local population is not important within the industry itself (even if highly important to its infrastructure, e.g., operation of slaughterhouses). This creates classical conflicts over land use, and a confrontational situation between Sámi and the local population. The industry is designed (in its raw, legal structure) to be a rational, market-oriented money machine for relatively few persons, although the practitioners' own ideals of cultural continuity and group solidarity cause it to distribute its diluted economic benefits to a far greater number.

In Finland, the reindeer industry is not protected (Chaps. 2, 3). The only limitations found in the Finnish reindeer industry are on the total number of reindeer inside a district and the number of animals (quota) for each member of the district. In the southern part of the reindeer herding region, the maximum number is 300 reindeer per member, while the corresponding number is 500 reindeer per each district member in the northern area (Jernsletten and Klokov 2002). This creates conflicts in the districts between the goals of ecological sustainability and individual livelihood, because, while there are set individual quotas, the number of reindeer owners can swell. However, although no upper limit of herder numbers is assigned, the number is not without constraint. In both Finland and Sweden it is the existing membership of the herding units (*paliskunta* and Sámi Villages) that decide to accept or reject new applicants. Open access supposedly exists (for those who are eligible members) to what in practice is usually a rather closed shop (Beach 1995). Nonetheless, in Sweden, for example, the closed shop system cannot bar the membership of children born within the Village, and membership can still swell. By herder increase and/or by reindeer decrease, the situation can occur in which herders come to have so few animals each that active herding is not worth the effort (Beach 1981). Should the herd enter a phase of rapid growth, it can explode its ecologically sustainable limits rapidly without triggering responsible controls by its own all-too-uninvolved (and possibly also all-too-numerous) owners.

Moreover, because of the concern for the welfare of individual families (to provide each with enough head to secure a decent living standard from the herding livelihood rather than as a smaller subsistence supplement), individual herder quotas on reindeer numbers are not mandatory. In fact, to limit individual herd sizes so that the total would not meet the so-called rational herd limit allowed for, and encouraged for, a Village would be a decidedly non-rational management set-up. In effect it would mean passing up available sustainable profits (Beach 2000). Instead, there is no formalized feedback between available ecological "space" in a Swedish Sámi Village and the number of herders in it. The competition of all herders in the Village striving to attain herding as a sustainable decent livelihood (as opposed to supplementary subsistence) commonly pushes for reindeer numbers in excess of the rational limit. Another reason for herding authorities to allow unrestricted individual reindeer numbers is that a sizeable herd will promote a herder's responsible herding activity. In effect both of these well-intentioned goals (decent living standard and responsible herding) will tend to push Village herd size toward the brink of ecological sustainability–and soon beyond it. Concession herding sidesteps these structural problems.

In Tornedalen, the reindeer industry is semi-protected. Only Sámi have the right to herd reindeer (when given concession) as a main livelihood, but the local population is participatory and important. Most of the reindeer are owned by non-Sámi who enjoy voting rights in the Village. There is both a total Village limit and a quota for the number of animals per individual owner (usually a higher quota for the Sámi concession holder). Under this system, the concession holder (for whom herding is a full-time job and constitutes the main income) secures enough animals to obtain a decent market-based economy. For the contract owners, the lesser quota satisfies their cultural continuity and subsistence needs (reindeer meat on the table, but not sufficient market-based income from reindeer sales). Since the reindeer owners are also the owners of the private grazing lands utilized by their reindeer, potential for conflicts is greatly reduced. Importantly, owner numbers are automatically regulated within bounds through relation to the landed "homesteads" within the concession area. There can be no more contract owners than there are homesteads in the concession area. If a new owner is included as a member in a concession Village, so is his homestead and its pastures. Hence reindeer number "consumers" are also "providers" of resources for reindeer. Herein lies an important link of ecological principle (although it was most likely unintended by the legislators of the time).

To be sure, the sustainability controls gained through the concession model does not come without a price tag. Herding in Tornedalen supports few Sámi "livelihood (host) herders"; it is mainly a limited subsistence support for local landowners. However, the goal of supporting cultural survival must not be regarded as justifying the support of Sámi "livelihood herding" alone. "Supplementary subsistence" herding has its own deep cultural roots in Tornedalen, but also throughout the Sámi herding area, past and present. In fact, while there may be no formal structural feedback processes guaranteeing a comfortably sustainable fit between herder numbers and reindeer numbers in a Swedish Sámi Village, there certainly is the "school of hard knocks" forcing herders from the field and causing their children to seek other prospects in life. Insufficient pastures, weather catastrophes, and causal combinations of these, predatory pressures on the reindeer, land conflicts, and internal herder competition can all conspire to place herders at a herd size level well below "livelihood" level. Of course one should not forget the "carrot" aspects of a non-herding life either. Herding is no easy life. A large proportion, maybe over one third, of the Sámi Village herders today outside of the concession area have a total herd size at or below the maximum stipulated for supplementary subsistence of the individual contract owners in the concessions, i.e., 30 head.

The maintenance of the supplementary hunting and fishing rights which accompany the herding right might play a role in the decision of a small-scale "hobby herder" to stay in the game, although this incentive has been short-circuited by the termination of exclusive Sámi small game hunting rights by the State in the early 1990s. That is, individuals can hunt and fish without being members of a Sámi Village (Beach 1981, 1994).Beyond such economic strategies, however, the deciding factor is generally the desire to maintain cultural continuity and herding *enskilment*; that is, to gain recognition for the "apprenticed learning" of knowledge and the ability to piece together a life in the towns and camps of one's home landscape. Toward this end, supplementary subsistence herding and market-oriented "livelihood herding" have much in common, and many individual herders may slide along a continuum between them. They cannot be regarded as forming isolated poles, even if legal language distinguishes such categories.

This reindeer industry is important for the local communities in Tornedalen and it works as an example of alternative ways of integrating the reindeer industry into a small-scale system of coexistence in the local community. The reindeer industry creates employment opportunities in the local communities besides concession holding, reindeer herding, to wit, work in slaughterhouses, transportation, tourism, and tannery, in addition to other adjoining industries that benefit from the side effects of this industry. As mentioned previously, an indication of the low level of conflict generated by this industry and its importance in the local community is the support given by local politicians. According to the law (SFS 1971:437, 85§), local politicians must submit an opinion regarding the prolongation of a concession before the final decision is taken by the reindeer authorities. To date, their opinion has always been in favor of continuing the Liehittäjä concession.

5.7 Conclusions: ... but Some Conflicts Exist ...

Even though the reindeer industry in Tornedalen could be described as important for the local communities and integrated within the local population, conflicts do exist. It is not difficult to find "classic" conflicts between the reindeer industry and a fast growing infrastructure, for instance conflict with a planned new railroad and new highway. Such a development will further complicate the move between the summer and winter pastures for Liehittäjä concession Sámi Village, or even cut the pastures into separate parts. Liehittäjä concession Sámi Village is entitled to comment on these plans, and they have been invited to several meetings with the Swedish railroad authorities. In such a circumstance, both Sámi Villages and concession Sámi Villages are regulated by the same law.

On the other hand, when it comes to the challenges connected to forestry, the situation differs in the concession areas. Forestry and the reindeer industry have been locked into one of the most persistent resource conflicts in Sweden, as recognized in many reports (Beach 1981; Gustavson 1989; Jernsletten and Klokov 2002). Both industries are dependent on the same areas. The concession model, with farmers as contract reindeer owners and at the same time full members of the Sámi Village, often means that contract reindeer owners are big land and forest owners. This results in these members often finding themselves in a dual role, on both sides of the table, with contradictory interests – but most importantly with the will and practical position to reach a good settlement.

As mentioned in the Introduction, reindeer herding (as opposed to mere owning) is a protected livelihood for the Sámi, and therefore closely connected to issues of Sámi indigenous rights and ethno-mobility. However, in Tornedalen the situation is somewhat different, as on-the-ground control of the reindeer management system is not exclusively in Sámi hands. As one might imagine, therefore, the members of the concession villages are inclined to hold a negative attitude toward the victories of indigenous rights legislation and are apprehensive about Sweden's possible ratification of International Labor Organization's Convention No. 169 on the rights of indigenous populations. For the majority of contract owners in the concession villages at least, problems related to Sámi versus non-Sámi ethnicity and rights derive from the rest of Sweden. In fact, from their perspective, it can be aggressive Sámi organizational efforts to undermine the concession system in favor of granting grazing rights below the Sámi Territory Border to regular Forest Sámi Villages that forms a major threat. From the concession vantage point, the ethnicity of those working with the reindeer is a problem only to the extent the law and Sámi rights lobbyists make it so. The essence of this contrast rests in the variable foundations of herding conceived in the different regions: is it to be a small-scale form of supplementary subsistence for a wide number of people to help them maintain their homesteads? Or should it be designed to maintain distinct nomadic cultural traditions as the main livelihood for a smaller handful?

The concession herding system embodies points of principle that command our attention to a far greater degree than its economic impact would otherwise justify. While situated on the easternmost periphery of the Swedish herding area, from the larger Fennoscandian perspective, it hinges the Swedish and the Finnish herding systems. On a higher level, it shares rather than contrasts with the aspirations of those favoring a strictly Sámi-protected herding form for all of Sweden. The concession villages, just like the regular Sámi Villages, strive to enable their members to continue living where they were born. They are proud to be Tornedalians, with a history, land rights, cultural rights, cultural pride, and reindeer herding of their own.

Despite its small number of animals and a rather invisible position in the academic mindscape, the concession reindeer management system is of importance politically and as to the principles involved. This system is highly significant as an analytical model, and it offers not only problems but also solutions to many of the challenges the industry faces today.

References

Beach H (1981) Reindeer-herd management in transition: the case of Tuorpon Saameby in Northern Sweden. Almqvist and Wiksell International, Uppsala

Beach H (1990) Comparative systems of reindeer herding. In: Galaty J, Johnson D (eds) The world of pastoralism: herding systems in comparative perspective. Guilford Publ, New York, pp 255–298

Beach H (1994) Shots heard round the world. In: Arnesson-Westerdahl A (ed) Beslutet om Småviltjakten - en studie i myndighetsutövning. (In Swedish. The decision about small game hunting – a study in the exercise of public authority.) Sametinget, Kiruna, pp 44–69

Beach H (1995) The new Swedish Sámi policy - a dismal failure: concerning the Swedish government's proposition 1992/93:32, *Samerna och samisk kultur m.m.* (Bill) (The Sámi and Sámi culture etc.). In: Gayim E, Myntti K (eds) Indigenous and tribal peoples rights 1993 and after. Juridica Lapponica 11:109–129

Beach H (1997) Negotiating nature in Swedish Lapland: ecology and economics of Sámi reindeer management. In: Smith EA (ed) Contested arctic: indigenous peoples, industrial states, and the circumpolar environment. University of Washington Press, Seattle, pp 122–149

Beach H (1998) Vulgärekologi? (Vulgar ecology?) In: Anttila S, Vikman P-Å (eds) Forskning och förvaltning för hållbar utveckling i landets fjällområden. Rapport från 1998 års fjällkforskningskonferens. (In Swedish. Research and management for sustainable development in the country's mountain areas. Report by the 1998 Mountain Research Conference.) Östersund, Fjällforskningsinstitutet, Mitthögskolan

Beach H (2000) Reindeer-pastoralism politics in Sweden: protecting the environment and designing the herder. In: Hornborg A, Pálsson G (eds) Negotiating nature. Lund Stud Hum Ecol 2:179–211

Cramér T, Prawitz G (1970) Studier i renbeteslagstiftning. (Studies in reindeer grazing legislation.) Stockholm, Norstedt

Geijer U, Hannu U, Andersson L (2003) Renskötselrätt. (In Swedish. Reindeer herding law.) Umeå, Sámiid Riikkasearvi/SSR

Gustavsson K (1989) Rennäringen: en presentation för skogsfolk (in Swedish; Reindeer herding: a presentation for people working forestry). Arlöv, Berlings tryckeri

Hultblad F (1968) Övergång från nomadism till agrar bosättning i Jokkmokks socken. (In Swedish. Transition from nomadism to agrarian settlement in Jokkmokk paris.) With an English summary. Almqvist and Wiksell, Stockholm

Jernsletten J-L, Beach H (2004) Cultural perceptions and implications of ecological management. In: Forbes B et al (eds) The challenges of modernity for reindeer manage-

ment: integration and sustainable development in Europe's subarctic and boreal regions. Arctic Centre Rep 41:16
Jernsletten J-L, Klokov K (2002) Sustainable reindeer husbandry. Tromsø, Universitetet i Tromsø, Senter for samiske studier
Labba N, Jernsletten J-L (2004) Sweden. In: Ulvevadet B, Klokov K (eds) Family based reindeer herding and hunting economies and the status and management of wild reindeer/caribou populations. Arctic Council, Tromsø, pp 131–146
Manker E (1968) Skogslapparna i Sverige. (In Swedish. The forest Sámi in Sweden.) Almquist and Wiksell, Stockholm
Nordin Å (2002) Relationer i ett samiskt samhälle : en studie av skötesrensystemet i Gällivare socken under första hälften av 1900-talet. (In Swedish. Relations in a Sámi society: a study of the contract reindeer system in Gällivare parish during the first half of the 1900s.) Umeå universitet, Umeå
Rönnbäck J-I, Kvist W (1990) Utredning om förhållandena i Muonio Sameby. (In Swedish. Investigation about the conditins in Muonio Sámi village.) Länsstyrelsen i Norrbottens Län, Luleå
Svensk rennäring (in Swedish. Swedish reindeer herding) (1999) Data om rennäringen (Data concerning reindeer herding). Statistiska centralbyrån, Stockholm
SFS (1971) Svenska Författningssamling. (In Swedish. The Swedish statute book.) Rennäringslagen 1971/437 (The Reindeer Herding Act). Riksdagen, Stockholm. Latest update according to SFS 2001:439
SJVFS (1993) Statens jordbruksverks författningssamling. (In Swedish. The Swedish Department of Agriculture Statute Book.) Statens jordbruksverks föreskrifter om renräkning, renlängd och företagsregister för rennäringen 1993:162 (The Swedish Department of Agriculture regulations with respect to reindeer counts, reindeer registry, and company registry in reindeer herding), Stockholm
SOU (2001) Statens offentliga utredningar. (In Swedish. The state's public investigations.) En ny rennäringspolitik: öppna saembyar och samverkan med andra markanvändande (A new reindeer herding policy: open Sámi villages and cooperation with other market applications). SOU 2001:1001. Fritzes, Stockholm

6 Changes in Property Regimes and Reindeer Herding Management in Post-Soviet Herding Collectives: The Case of the Municipality of Lovozero (Murmansk Region, Northwest Russia)

Y. Konstantinov and V. Vladimirova

6.1 Introduction: The Setting

Along with a fieldwork focus on the reindeer herding part of Murmansk Region, this chapter discusses the current dynamics of three major developments affecting reindeer property and management in the subarctic regions of the Russian Federation. The first of these is the new agrarian legislation passed since 1990, the second is the management practiced by the heirs of former state farms (*sovkhoz* in Russian; pl. *sovkhozi*), and the third is comprised of the coping strategies as practised by rank-and-file members of such farms, now referred to as "limited liability companies" (*TOO*; *tovarishchestvo ogranichennoi otvestvenosti*), or "agricultural production cooperatives" (*SKhPK; sel'skokhoziaistvennaia proizvoditel'naia kooperaciia*). We describe these developments and try to show how a local reality is shaped by the major protagonists: distant legislators in Moscow; local administrators and cooperative managers; and, finally, workers and their dependants on such cooperative farms – which in our case is a small reindeer herding community in the center of the Kola Peninsula in Northwest Russia.

We first present a brief historical survey of land reforms in Russia in their relevance for what today is Murmansk Region, and the way in which land issues were settled in the period of collectivization (1929–1937). We turn then to the more prominent documents from the post-1990 reform-oriented legislation concerning the restructuring ("privatization") of former state farms (*sovkhozi*). The specificity of their application in the reindeer herding sector of Murmansk Region is treated next; and an anti-reformist ideological tendency as applied to practice is described, which we call for working purposes *sovkhoism* (Chap. 2). Here an analogy is sought with other "-isms": for instance the imagined reality of communism, originally denoting, in terms of everyday life, one ideally lived in an egalitarian commune. By the same token sovkhoism again attempts to describe a desirable life, but this time retrospec-

Ecological Studies, Vol. 184
B.C. Forbes, M. Bölter, L. Müller-Wille, J. Hukkinen, F. Müller, N. Gunslay, Y. Konstatinov (Eds.)
Reindeer Management in Northernmost Europe

tively – one that was ideally lived in the protected social reality of the state farm. The state farm is thus seen, as it was by Humphrey (1983, 1998), as a "total social institution" – and at present a fatally lost world, to be somehow regained.

Such a retrospective disposition, often intensely emotional and nationalistic in a pro-Soviet Union sense, can be attributed to a plane of cultural ideals and ideologies (von Benda-Beckmann and von Benda-Beckmann 1999). At the same time, a critical economic situation requires effective coping strategies on the level of the individual domestic economy, and for this reason we turn next to a pivotal coping mechanism for the local herding community. This is a "private-within-the-collective" coping device realized by having privately owned reindeer mixed with the cooperative herd. In its relation to land, this system continues an age-long tradition of usufruct rights over a given set of resource sites (*ugod'ye*, pl. *ugod'ya*), transformed in Soviet times into reindeer herding team territories (*brigadnie teritorii*). In this way, in terms of retained continuities, relations to land do not seem to have been changed much by the Soviet period, as discussed below.

We conclude that in such a context "transition" (from command socialism to market-based capitalism, whatever the virtues and vices of such a process) is not in evidence, as implied by the meaning of reforms, if the latter are taken on the level of officially proclaimed goals. Seen from the grassroots, what we observe looks much more like an acceptance of events, tragically but inevitably imposed from above, and accompanied by resistance to them by overt public demonstrations, or by covert, informal, and loosely concerted means in the realm of the domestic economy. Thus, at the base of the social pyramid, sovkhoism appears as a contradictory bundle of vernacular ideological dispositions and applied strategies, reinterpreting previous forms of processing and surviving of upheavals, or, to paraphrase slightly Creed's felicitous definition - of "domesticating revolutions" (Creed 1998).

6.2 Brief Historical Survey of Land Reforms

The overall attitude to land reforms in Murmansk Region after 1991 and 1992 should be seen against the background of responses to previous agrarian reforms. These have a long history. In a retrospective unwinding of crucial events, the following may be seen to have the greatest relevance for the formation of attitudes and consequent present responses: (1) the manner in which land was allotted to newly created collective farms *kolkhozi* and *sovkhozi* in 1929–1937; (2) the passing of the decree "On Land" (*O zemle*) in 1917; (3) the abolition of serfdom (1861); and (4) the Edict (*Ukaz*) of 1766 prohibiting the use of land as private property by the "state peasant" (*gosudarstvennie krestiane*).

6.2.1 "State Peasants"

This formulation is very important for understanding ownership of land in a historical perspective. It is a corollary of the "engines of expansion" of the South to this part of the North. The colonization of what became Russian Lapland, and is today Murmansk Region, was the earliest Russian expansion to the North during the 11th century. It was driven by the search for precious furs on the part of Novgorod hunters and sailors, who eventually settled in the southern part of the Kola Peninsula. Today we know them as Pomors (*pomori*). A subsequent motivating force several centuries later was the desire of the Russian State, unified since 1471, to block expansion from the West, especially the spread of Lutheranism. This led to the inclusion into the fold of Russian Orthodoxy of the indigenous Sámi population during the 16th century, a campaign carried out by the great monasteries on the peninsula. These monasteries eventually became the biggest landowners – until 1766 when their possessions were severely reduced by the state. Thus, it can be said that until the 1760s the land belonged to the monasteries and the state as the biggest feudal owners, whereas after that the state remained the only owner who gave the right of usufruct to its peasants, Pomors and Sámi.

It thus happened that when the radical reforms of 1861 occurred and the Russian serfs were given the opportunity to acquire or rent land from the landowners (*pomeshchiki*), the reform was hardly noticed in the North because here pomeshchiki had never existed. Since the 18th century there had been only one pomeshchik, the state.

In this way, by the time of the revolutionary fiat "On Land" of 1917, local relation to land had been for centuries one of officially sanctioned (and taxed) traditional usufruct rights to resource sites (*ugod'ia*). These rights existed in an extended version prior to 1766, in the sense of saleability of usufruct shares (sing. *dolia, luk*). They were restricted afterwards, in the sense of the state prohibiting to its state peasants using their resource sites (*ugod'ia*) – i.e., forests, lakes, rivers, land – as private property, i.e., for selling, mortgaging, giving them as dowry, etc. (Ushakov 1997).

From this vantage point, the reforms of 1861 were indeed not very relevant to the local situation, as the Pomors and Sámi continued using their traditional resource sites as before. Concerning the Sámi population, a factor of real – one might say, fateful – relevance was not any change of property regime connected with the abolition of serfdom in 1861, but the coming of the Komi reindeer herding settlers in 1883 and 1884 (Chap. 2; Konakov 1993), and the subsequent rearrangement of use of pastures and of the reindeer resource itself. As for the relation to land, however, things remained much as before, and even the revolutionary decree of 1917 seems not to have produced a significant impact on this relationship.

6.2.2 Kolkhoz and Sovkhoz Building

The process of collectivization in Murmansk Region began in the late twenties and can be considered as fully developed about ten years later (Kiselev and Kiseleva 1987; Ushakov and Dashchinskiy 1988). In the overall scheme of developments this is considered late, and the reasons can be attributed to geographical distance, but most of all to the fact that serious thinking about specifically reindeer herding collectives began fairly late in the Soviet state in comparison to collectivization of other, mainly southern, agricultural specializations. Despite the fact that discussions began in 1919 about the great potential that reindeer herding held for Soviet agriculture in the North (Kertselli 1919), it was not until the 7th Plenum of the Northern Committee (April 23–30, 1930) that more detailed guidelines for reindeer herding collectives were provided (Northern Committee 1931).

Specifically about how to deal with land, the Committee's resolution was naturally predicated on the Decree "On Land" (1917), and its subsequent amendments (Krassov 2000). As we can see from the documents of the period, an issue here was not who was the previous owner, or traditional user of a resource site, but how the range was to be divided in the best interests of efficient and "scientific" reindeer herding.

An important feature of the situation is the fact that the resource sites became allotted to the herding units of the respective reindeer herding kolkhoz or sovkhoz. These units, referred to at first as *zvena* (teams), and subsequently as *brigadi* (brigades), were allotted range according to the average size of a herding unit, i.e., herds of approximately 3,000 head after harvest. The range is estimated by the formula of 0.1 km^2/head × 3,000 head × 10 years (time needed to regenerate a grazed range). Thus a small farm with five units would need a territory roughly 150 by 100 km in size (Kudriavtsev 1930).

While demarcating land in this way and allotting it to the various units (zvena, brigadi) a parallel – and it seemed a spontaneous, but actually an officially approved process – was going on: herding units (brigadi) were organized from herders who were family or close kin, and who used to graze their previously private herd in the same territory, now been designated as sovkhoz brigade territory (*brigadnaia teritoriia*). The relationship between family/kin and land was reinterpreted as a relationship between a new collective farm team (brigada) and more or less the same piece of land. In this way, more by default than by design, continuity was retained. The relationship to land, despite all the revolutionary innovations and what might be called "scientific shamanism", remained essentially the same as had pertained at least since the mid-18th century, i.e., the same as between "state peasants" and traditional resource sites on the basis of usufruct rights.

6.2.3 Agrarian Reforms: Work History vs. Restitution

The first stage of the Agrarian Reforms in the Russian Federation, carried out in 1990 and1991, affected the former state agrarian enterprises (*sovkhozi*) in two principal ways. In the first place, the Law for Private Property (*Zakon o sobstvennosti*) of 1990 postulated that part of the net profit of an enterprise had to be shared among employees. The resulting sum, calculated per each person, formed a share (*pai*). On the basis of ownership of such shares, dividends to members were to be paid out periodically. We do not have data about other cooperatives, but insofar as the one we have chosen as an ethnographic focus is concerned, i.e., SKhPK "Tundra" of Lovozero, we have enough data to say that dividends have never been paid. On the other hand, should a member of the same cooperative decide to leave, the value of a share to be paid out is symbolic (in 2004 30.00 roubles, equal to1.00 American dollar). It is thus the case that there is little meaning in the concept of shares from the vantage point of the rank-and-file members of the cooperative. Hence the predominant relevance of coping strategies in which the relation to ownership is relevant only insofar as it concerns private deer – an issue to which we turn further down.

The Law of December 15, 1990 "On Change and Supplementation of the Constitution of the Russian Soviet Federative Socialist Republic (RSFSR)" (*Ob izmeneniiakh i dopolneniiakh Konstitucii RSFSR*), recognized state farms (#) as the owners of the property, bound to them. One of the implications of this law seems to be that shares of employees' property are formed on the basis of work history and not on property owned by them or their parents or grandparents prior to the collectivization drives of the late 1920s and 1930s – naturally so in relation to land, as we have seen from the previous section. As such ownership had, in fact, never existed, or at least since the land reforms of 1766, subsequent sharing out was based on how long a person had worked on a territory. The same principle was now to apply also to property in respect of which there had existed ownership – the most important being ownership of reindeer.

The principle of sovkhoz transformation of property is, thus, not primarily one of restitution – as for instance in the Central European countries of the former Soviet Bloc – but on work history. A sense of final departure from pre-Soviet property regimes and practices seems to be implied by federal reform legislation and its various regional interpretations and applications, insofar as previous property rights over deer and other types of property was concerned. But as regards the relation to land, there is, conversely, a deep historical continuity. State peasants or members of collective farms had never had ownership rights to land, but usufruct rights on resource sites. The sole owner of the land, ever since 1766, had been the state, and prior to that again primarily the state and the monasteries.

While in other parts of the Russian North the non-restitution tone of reforms is being attacked by problematizing the issue of indigenous land use and occupancy, Murmansk Region, at least in its agricultural sector, seems to remain indifferent to such implied meanings and their practical outcomes. In other words, local ideological concerns and parallel coping applications, are not in favor of a resurrection of pre-Soviet management regimes, but of an ongoing reinterpretation of such, for which the state farm, or sovkhoz, stands as a paradigmatical core concept.

The working term offered above, sovkhoism, attempts to capture such a worldview. It is seen here as an ideology in practice not officially proclaimed, but effected to a greater or lesser degree by all principal players. It is important to note, however, that such a view is not one-dimensional in its practical expressions – and this adds a quite bewildering complexity for the outsider. While public rhetoric of a pro-Soviet, antireformist kind is predicated on lofty ideals and sentiments with a strong bias for collectivism, social order, care, and protection, coping practices are of an intensely individualistic pragmatic nature, looking for loopholes in the expanded informality of the public economic domain.

Of the principal players in this indeterminate, contradictory, and increasingly open-to-local-interpretation atmosphere of "reforms", the most relevant for us are regional (*oblast*), and municipal (*raion*) administrators – those of Murmansk Region and the Municipality of Lovozero (*Lovozerskii Raion*). The domain occupied by reforms is highly indeterminate and contradictory precisely because a lot of lip-service is paid to anti-statist and anti-bureaucratic rhetoric, development of small businesses, sustainable development, etc., but this seems to be mainly for foreign or other (preferably credulous) consumption, and serves to legitimate high-orbit political establishments, again mainly for export purposes. When it comes to local political realities the "communist (anti-reformist) vs. democratic (pro-reformist)" debate is of main concern and here the anti-reformists are certainly more popular on both the regional and municipal levels.

Thus, in sum, reforms proclaiming privatization may have indeed been passed, but to what an extent both the principal actors and the population at large believe in them and implement them in their original sense is quite a different matter. In some cases leanings for restoration of the old regime have become radical practices, as in Belarus, in others – as in our case in Murmansk Region – they exist in a realm of tense contradiction and indeterminacy. This space is inhabited by "export-oriented slogans" coming from the center and locally intended for foreign investors and aid industries, side-by-side with nostalgic sentiments for former security and relative affluence, hatred for those who "destroyed the great country", and an equally passionate hatred for the newly rich and for private entrepreneurs in general (*komersanti*). The whole complex of political orientations is living alongside day-to-day actual practices, whose effective potential for the home economy is mostly based on available openings into the informal ("gray") one. In this latter sense parallelism with the state socialist period is clear and is in harmony with such generalizing metaphors as "solitude of collectivism" (Kideckel 1993), or the already-mentioned "domestication of the revolution" (Creed 1998). Of the popular generalizations of those who've lived the regimes, a telling one in the home economic sense is "mine is my own, that of the others is common" (in Bulgarian, *moeto si e moe, chuzhdoto e obshto*).

6.2.4 Herding Specifics

The socialist-time saying just cited gives us a vernacular lead into the commons, i.e., common property, debate. In the world of herding, as in any pastoralist society, the commons dilemma exists in a very graphic form almost by definition, since grazing ranges are rarely exclusive private properties. Especially with extensive herding such as it is on the Kola Peninsula, the situation is problematic irrespective of "capitalist", "socialist", or "post-socialist" trappings. Here both synchronic and diachronic similarities appear. Synchronic similarities on the theme "tragedy of the commons" (Chaps. 3, 15; Hardin 1968; Paine 1992) can unite such diverse terrains as those of the Russian Far North and Fennoscandia in the sense that problems suggested by the Bulgarian adage occur in both regions. Diachronic ones can be seen in terms of pre-Soviet, Soviet, and post-Soviet continuities. In all the variety of cases that may postulate such a comparison, the main driving force is inevitably connected with extracting the most private security possible from within, and often at the expense of, a publicly owned and collectively managed resource (Chaps. 3–5).

While the state farm, which is our case study focus ("Tundra" of Lovozero), was still in existence Hugh Beach observantly noticed that in comparison to Fennoscandic herders, sovkhoz herders in Russia enjoy "a basic income security independent of their 'reindeer luck'" (Beach 1992). Both in this and in an earlier article (Beach 1986) he notes the importance of private deer for the herders' well being. Most importantly, his informants, when asked about the future of reindeer herding in the area, told him that "there would probably be a state farm as today" (1992). It was a premonition of what we here call sovkhoism; this type of statement was registered in Russia at the very beginning of reforms in 1990. Subsequently, in the experience of our own research team, it was repeated in various ways over the years with many different reindeer herding brigades on the Kola Peninsula.

It is beyond dispute that privatization reforms have removed a substantial part of previous state support – which was, after all, their very aim. Since the farms were run like state-owned factories, the state took care of them in many ways, but most prominently by paying relatively high salaries and premiums, and also by ensuring the marketing of production. Of this little is left today. As one herder from Brigade 8 ("Tundra") put it: "Only we and the sailors received the highest salaries. But the sailor is at sea, while we walk on dry land and can see our families". Beach echoes retrospectively this statement from 17 years before when reporting the situation in the Tomponski Sovkhoz (Yakutia) back in 1984: "The State seeks to stimulate herding enlistment by offering herders the highest salary of almost any Soviet employee along with other bonuses such as free work clothes, food delivery, cultural activities, housing and free vacation travel anywhere in the country" (Beach 1986).

Today salaries have gone dramatically down in their buying potential; prices of just about everything have rocketed. The state has also withdrawn from taking care of the production and it is mainly due to a Swedish company, based in Lovozero, that there is a buyer at all – a rather fortunate situation in comparison to many other herding areas (Chap. 2). In this rather depressing context one may well ask what is it that still makes the skeletal remains of the former state farms or aging industrial facilities such overwhelmingly attractive options?

The answer is to be sought not in the direction of economic effectiveness, but in collective social security with opportunities opened for informal gains above what the official salary allows. Sovkhoism (to be recalled as our working term) attempts to capture the yearning for state-ensured social security – "institutionalised theft" (Creed 1998 pp. 197-200) – that tolerates a degree of private informalism. The existence of the private under the umbrella of the collective seems to have significant attractive power, and Hann (2000) is certainly right in saying that a simplistic private vs. collective property dichotomy would not do when attempting to explain current post-socialist developments. Not "private vs. collective", but "private-in-the-collective" is the preferred option, and its recent history seems to have found a new vigor.

A re-imposition of pre-collectivization principles of ownership and management practices had only very recently begun to be sought in this region. Here, mainly due to a strong Fennoscandic influence and consequent attention to Sámi traditional rights, some interest has recently been inspired concerning the issue of indigenous land use and occupancy in their connection with native rights (Chap. 2). On the whole, however, the region has stood apart both from Fennoscandia to the west, and from many parts of the Russian North and Siberia to the east, in a general reluctance to pursue "ethnic forms" of land use (Sillanpää 1999).

A further level of complexity is introduced by the fact that the cooperative is a child of the reformist legislation that "privatized" former state property. In reality, however, those who continued to be members of the former sovkhoz in its new guise – and this is close to a hundred per cent of the herders – have little sense of becoming private owners united into a cooperative. As mentioned above, no dividends from shares have ever been paid, and while in principle one can reclaim one's private share of the property, in practice the situation is very discouraging, not to say prohibitive. It appears that should a member of the cooperative decide to leave, all he or she can get is the equivalent of one American dollar per share of the property. The majority of the members have total value of shares not exceeding 100 American dollars. The reindeer herd itself is a constitutive part of the Investment Fund of the cooperative and according to its Statute this fund is not divisible into individual shares. In other words, in case one decides to leave, he or she could get no animals from the collective herd. All they can do is take their private deer from the herd, or sell them for slaughter during the winter corralling. Not surprisingly, only a few have tried to become private reindeer herders. We do not know, in fact, of more than five such attempts, which by now have all failed according to local reports. These failures have been attributed primarily to poachers from urban centers and military bases. Other reasons are connected to a lack of investment for building up transport infrastructure, refrigerating facilities, finding market outlets, and others – e.g., a failure to appear of instruments that the reforms were supposed to bring in to help build up private businesses.

We can thus see that words like "private" and "privatization" tend to have polysemantic meanings, with shifting and indeterminate contextual nuances, often effected as equally indeterminate practical acts and states of property. Striking examples may be cited, similar to ones described by Verdery when she talks about "elasticity of land" or of "fuzzy property" (Verdery 1994, 1999), or in descriptions in Bulgarian material of "real vs. nominal restitution of property" (Konstantinov 2002). Or, again, as in our case in the Raion, of the indeterminate status of the cooperative (SKhPK) between that of a state enterprise, which it ostensibly supplants, and a cooperative of private owners, which it hardly is. It is in this indeterminacy that proclaimed reforms grind to a halt.

6.3 Coping Strategies

6.3.1 Agrarian Reforms as Power Tools

Before describing grassroots coping strategies we turn to those formal applications of the reforms that we have observed at this point in the research. So far, the only application we have registered that seems to make operational sense, and is not simply media rhetoric, is when a reform is used as a power tool by the higher orbit local actors. To describe how the mechanism works we need to go into a brief sketch of this part of reform dynamics.

Of local relevance is the law mentioned at the beginning ("On Change and Supplementation of the Constitution of the Soviet Russian Federation") that recognized state farms (*sovkhozi*) as the owners of the property, bound to them. Indications, however, of how this law was to be applied did not appear until two years after it was published. Signs of some real transformation became visible only with Russian Federation presidential decree of 1992, titled "Towards Urgent Measures for Accomplishing the Land Reform in the RSFSR" (*O neotloznikh merkakh po osushchestvleniiu zemel'oi reformi v RSFSR*). The document prescribed that all kolkhozi and sovkhozi should be re-registered in accordance with the new liberal reform legislation.

Together with this procedure, the new legislative documents demanded the formation of a specialized committee in each kolkhoz or sovkhoz. The activities of these committees were specified within the "Regulation for the Committee for Privatization of the Land and the Reorganization of the Kolkhoz/Sovkhoz" (*Polozhenie o komisii po privatizacii zemli i reorganizacii kolkhoza/sovkhoza*; Bogoliubov and Minina 2000 p. 13). The Regulation is the first document to define who had the right to be given plots of land and shares of property, the principles for determining the size of these plots and shares, and the rights of the new owners with respect to their property. Data from the former sovkhoz "Tundra" of Lovozero suggests that what assumed most relevance on the basis of this document was not the land or other property, but the issue of shares – more precisely, the possibility of gaining power in the post-sovkhoz cooperative by redistributing them. Some more detailed description shall be offered below to illustrate how this structural feature works.

6.3.2 The "Tundra" Collective of Lovozero

"Tundra" is one of the two reindeer collectives in Murmansk Region, the second being SKhPK "Olenevod" of Krasnoshchelye (Chap. 2). The latter has been discussed in earlier publications (Konstantinov 1997, 2000) and structurally presents much the same situation. The combined herd at present of both cooperatives counts roughly fifty thousand after-harvest head, grazing on about one third of the entire territory of the region (Seppänen 1995 p. 27; Rybkin 1999).

During the initial period of reforms "Tundra" realized the new directives by changing its organizational form in 1992 from sovkhoz to "limited liability company" (TOO; Sect. 2.5). In 1998 the TOO was changed to SKhPK, a transformation to which we shall return. At the same time the property of the former sovkhoz was divided among its employees according to the Regulation as a number of shares (*pai*), allotted to each member. The main principle employed seems to have been, again, the length of service in the sovkhoz (i.e., work history) and not property owned before the collectivization of the late 1920s and 1930s (i.e., restitution). In other words, the number of reindeer an owner had prior to collectivization seems to have no bearing on the number of shares in the cooperative a descendant owns at the moment. The veracity of this statement needs, however, to find more extensive confirmation in the future, but so far there has been no evidence of restitution-based practices either in the legislative documents we have worked with or in oral accounts.

6.3.3 Land Issues

When it comes to the land used by the heir of the former sovkhoz, the situation is even less clear, especially when it comes to the reindeer grazing range that is of primary concern to us. Of major importance is the issue of native legal rights of the Sámi who are listed in the Federal Law about the Guarantees of the Rights of the Small Indigenous Peoples in the Russian Federation (Federal Law 1999). Despite the various provisions there, allegedly assuring right of occupancy in ancestral lands, the land situation has remained pretty much the same as before. It is similar to what Osherenko (1995) writes about Yamal, i.e., that in the system of land ownership land is at one and the same time public, common, and private. While the land is officially owned by the state, it is mainly managed by the local authorities, which give it for use to the sovkhoz (Osherenko 1995 p. 1091). The same applies to post-sovkhoz cooperatives, i.e., the SKhPKs that interest us here. On the Kola Peninsula, all that is really considered important to know is that the regional authorities have been allowing preferential tax rates for lands used as reindeer pastures.

6.3.4 SKhPK Structure

In the mid 1990s new changes concerning farming organizations were introduced. The two new laws prescribed the mechanisms for the formation and function of cooperatives. According to them the cooperative is an organization consisting of people whose aim is to perform a common activity, yet not pool common capital. That is why, with no appreciable monetary incentive, participation in the work of the cooperative is compulsory for each member. Members are paid small salaries according to work done, although no dividends in accordance with their shares. The net profit is meant to take care of the needs of the cooperative. At the same time every member bears a subsidiary responsibility for the debts of the cooperative (Bogoliubov and Minina 2000 pp. 135-137). Regarding "Tundra", it has to be noted again that no one takes seriously the issue of dividends, for the obvious reason noted above that they have never been paid.

The major decision-making event in the SKhPK type of organization, which "Tundra" became in 1998, is the general meeting of all the members of the cooperative. Each one of them has one vote in the meeting. With no right to vote are the associate members of the cooperative, which are in a special category of shareholders, consisting usually of old age pensioners or previous employees of the sovkhoz who do not participate directly in the work of the cooperative.

A number of decisions concerning the management of the cooperative depend exclusively on the competence of the participants of the general meeting. Among the actions of the meeting is election of a board of managers and a board of observers, both of which bodies serve for a period of five years. In this system the position that was known formerly as "the director of the sovkhoz" has become "the head of the board of managers" and subject to the provisions given above. It is telling, however, that the "head" is still popularly called "director", and the SKhPK itself "the sovkhoz".

These persisting appellations reflect two facts. One is connected with the general disposition for regaining the sovkhoz as a way of life, which we called sovkhoizm. Embedded in it is the actual practice of the so-called "head of the board of managers" to behave exactly like former directors, in fact, as the current instance shows, even more authoritatively. He makes the majority of decisions without consulting the board of managers and thus their decision-making power is mostly symbolic.

A brief story shall illustrate the point. A recent "director" (in office until October 2003) was short-tempered and quick to fire employees when, as he used to put it, "they made him angry". In a typical case, which occurred with herders from a brigade we are working with, he met two senior herders from the brigade in the streets of Lovozero when they were actually supposed to be at the intermediate tundra base of their brigade, 150 km away. (The truth was they had deserted the brigade temporarily to ferry a group of geologists to the village on a private track vehicle belonging to one of the herders, and thus make some extra money on the sly.) The director told them there and then that they were fired. After a couple of months, however, he changed his mind in the face of the general shortage of hands in brigade teams, and told the herders to go back to their brigade. "But you fired us", one of them said. "Ah, is that so?" replied the director, "I have forgotten. You must have made me very angry then." As the incident illustrates, critical decisions such as firing of herders, in the face of catastrophically understaffed brigades, can be made (and often are made) temperamentally on the spot, with no meetings or other such ceremonies. This type of relationship between management and herders is a far cry from the management of a cooperative and its members as private owners. The directors and the workers are here revealed not as cooperating, but as conflicting elements, engaged in outsmarting each other.

6.3.5 Administrative Management

In theory the primary executive body of the cooperative is the board of managers. Its work is controlled by the board of observers, usually consisting of three members of the general meeting. Within the competence of this latter body is arrangement of meetings, cessation of the five-year mandate of the board of managers, and assumption of managerial functions until the summoning of a general meeting.

Despite such mechanisms, prescribed by the legislative documents, decisive remnants of the previous system have remained intact, with only slight surface changes, mainly terminological, to be in tune with the reforms. One of these is the influence of local municipal and regional authorities over decisions of the general meeting when it comes to electing managers, and decidedly the choice of the head of managers (alias the director). The opportunity for such practices to occur is not averted by the current legislation - which only recommends that the federal and the local authorities do not interfere directly in the management and functioning of organizations other than federal or municipal entities. The role of such authorities is to create favorable conditions for their existence and performance by "indirect economic or psychological influence" (Bogoliubov and Minina 2000). How this appears in actual practice may be seen by an account of the procedure of removing the old, and electing a new head of the cooperative in 1998. At the meeting of the shareholders in that year, among other representatives of the Regional and Raion administrations, the vice-governor of the Region was present as the highest-ranking guest. Commenting on the need to remove the old director he said: "We [sic!] have received many letters describing reindeer herding as being neglected. We have seen that there is complete lack of management in the cooperative. But it is you that have elected your head and we do not have any right whatsoever to fire him. It is your decision."

Following this comment, the Regional head of agricultural management (*nachal'nik upravleniia sel'skogo khoziaistva*) said: "In actual fact, your director has turned out to be incompetent. Before we used to take her side, but we see now - she does not deserve our confidence." After that the head veterinary inspector of the Region took the floor and spoke in much the same vein, then other Regional and Raion dignitaries followed suit, until finally "the shareholders decided to elect a new head, and that happened on 19 May" (LP 1998). Evidently this is what is meant by "indirect psychological influence". The process does not seem to differ in but one detail from former Soviet practices. Such moves and decisions were then made not at shareholders' meetings, but at open Party meetings.

6.3.6 Private Reindeer (*lichnye olen'i*)

This is the backbone of the primary informal activity - let's call it "gray herding" - practiced as a coping strategy by the herders. "Gray herding" has always existed in sovkhoz practice, but its present very high proportions have been reached due to a decision of the SKhPK head removed in that fateful meeting of May 1998. According to this decision, all people working in the cooperative acquired the right to own an unlimited number of private animals grazing with the cooperative brigade herds without paying any fees, i.e., any rent for using cooperative facilities, pastures, and labor force. Previous to this decision, up to 50 head of private deer per herder or a herder's heir could be grazed in the collective herd with a small annual fee paid for each animal; this was a limit and general arrangement common in northern reindeer herding.

With the liberation of this regime at the time of the directorship of 1994–1998, accomplished in the name of improving recruitment and incentive, the number of private deer began to grow. At the time of writing (June 2004) the number of private and collective reindeer is close to becoming equal. Of a total herd of approximately 23,000 after-harvest head, nearly 10,000 are already private. We shall not go into the details of the mechanism that turns collective into private deer, (having done it elsewhere under the general heading of sovkhoist crypto-entrepreneurship; Konstantinov and Vladimirova 2002).

The possibility of increasing the number of private deer largely at the expense of the collective herd ("dilemma of the commons" in a herd, as differing from the classical range variety) is to a critical extent based on the almost total separation of the worlds of urban administration (in the settlement of Lovozero), on one hand, and that of the herders out in the tundra, on the other. On this point we note Habeck's felicitous distinction between the urban world of the administration (*kontora*), and that of the tundra reindeer herding teams (*brigada*; Habeck 2003). Unless they meet accidentally in the streets, as in the story of the director and the truant herders, the two sides do not see much of each other.

The members of the administration work in their modern offices in Lovozero where they are under the vigilant eye of the "director", and are entirely dependent on the salaries attached to their positions. The workers in the tundra, the reindeer herders, have very marginal positions vis-à-vis the urbanite administration and, in fact, have no role in the management of the cooperative. They work in very severe conditions, but, on the other hand, are mostly out of the control of the administrators while, at the same time, have access to the basic tundra resources, reindeer and fish. They are, after all, "the custodians of the tundra". On this basis, their main response to what they perceive as negative "policy" (the herders' word) of the administration, directed against them, is expressed in various types of informal action.

These activities form a two-pronged movement. One aims at total disposal of risk. This is achieved by assigning all losses to the cooperative while private property almost never sustains them – a device popularly known as "immortality of private deer".

The second prong is maximization of private profit at the expense of the cooperative by a variety of means. The main one is to build up private herds at the expense of the collective, as mentioned previously. At the same time, a great variety of other opportunities exist, taking off from work, using cooperative transport for private purposes, using its fuel and spare parts, and, in a word, anything that can be taken. A main parallel activity supported in this way is freshwater subsistence associated with commercial fishing. Other forms of wild resource extraction and gathering activities can be added to this list.

6.4 Conclusions: An Apocalyptic Future for Reindeer Management?

All the practices mentioned above work against the interests of the cooperative as a whole. At the moment the erosion is working on three different levels: that of legislative directives and the use made of them in upper-orbit power struggles; of administrative practices; and of informal coping strategies. Caught between such conflicting forces, the SKhPK, which is otherwise a potentially profitable enterprise, is balancing on the edge of economic existence, and the prevention of complete collapse is primarily due to the presence of a foreign buyer. Were that buyer to withdraw for some reason, it is difficult to see how the SKhPK could survive. In fact, when foreign buyers abandoned the mining concern "Sevredmet", the biggest employer and taxpayer in the Municipality of Lovozero, it quickly reached bankruptcy – and that has placed the Raion in an extremely tense position. It should be mentioned also, that a third reindeer herding entity on the Kola Peninsula, the former Reindeer Herding Experimental Station in Loparskoe near Murmansk (MOOS; *Murmanskaiia opitnaia olenevodcheskaiia stantsiia*, subsequently called Cooperative "Voskhod"), was declared bankrupt a few years ago, and all its assets were sold.

Against such a background, the general desire for the SKhPK to remain as the only conceivable support base, within the framework of a sovkhoist ideology and practice, can be easily understood. What may seem puzzling is perhaps the fact that notwithstanding such a desire (i.e., for the sovkhoz to stay and possibly regain its former stability), this very sovkhoz, in its present reformed SKhPK state, is being eroded by every informal means at hand, both from the top and from the bottom. While the higher administrative elites have been allegedly using SKhPK funds for private commercial operations (at least until October 2003 when a change of leadership occurred) the herders rely on the niches open to them, mostly in the direction of what has here been called "gray herding" as a form of crypto-entrepreneurship.

Against this background it is surprising how reindeer herding in the region has survived the reforms at all. An interesting question in fact is what have been the factors that have averted a total failure, besides the mentioned opportune presence of a foreign buyer. Apart from monitoring the work of the erosive forces outlined above, a main task of future research has to be the definition of such preventive factors and, in an applied sense, the search for possible mechanisms for immediate strengthening of their action (assuming they exist).

The local community is incessantly circulating predictions of apocalyptic outcome of current practices, unless, as a herder said recently during an interview, "...the present politics is changed and life becomes good once again" (*esli politika ne peremenitsia i opiat' stalo khorosho*). Upon being asked who is to change the present politics, the herder looked upward and said: "I don't know...up there, the bosses...." (*ne znaiu ... tam na verkhu ... nachal'niki*). The statement is indicative of the general mood and does not indicate anything remotely like transition from one type of system (command socialism) to another (market capitalism) significantly taking place in local worldviews. At the same time, the challenging questions remain. What are to be the inner long term defensive strategies of the community? How are they to be discussed, formulated, and articulated?

References

Beach H (1986) Visit to a Siberian Sovkhoz. Antropologiska Studier 38/39:73–79

Beach H (1992) Reindeer herding on the Kola Peninsula – report of a visit with Saami herders of sovkhoz tundra. In: Kvist R (ed) Readings in Saami history, culture and language III. University of Umeå, Umeå, pp 113–142

Bogoliubov SA, Minina EL (eds) (2000) Agrarnoe pravo. (In Russian. Agrarian Law.) Norma, Moskva

Creed GW (1998) Domesticating revolution. From socialist reform to ambivalent transition in a Bulgarian Village. Pennsylvania State University Press, University Park, Pennsylvania

Federal Law (1999) Federal'nii zakon "O garantiiakh prav korennykh malochislennykh narodov Rossiiskoj Federatsii. (In Russian. Federal Law 'On guarantees of the rights of the indigenous small peoples of the Russian Federation.) Approved by the Federation Council, 22 April 1999, Russian Federation, Moscow

Habeck JO (2003) What it means to be a herdsman. The practice and image of reindeer husbandry among the Komi of northern Russia. PhD Thesis, Scott Polar Research Institute, University of Cambridge

Hann CM (2000) The tragedy of the privates? Postsocialist property relations in anthropological perspective. Max Planck Institute for Social Anthropology Working Papers 2:1–22

Hardin G (1968) The tragedy of the commons. Science 162:1243–1248

Humphrey C (1983) Karl Marx Collective. Economy, society, and religion in a Siberian collective farm. Cambridge Univ Press, Cambridge

Humphrey C (1998) The domestic mode of production in post-Soviet Siberia? Anthropology Today 14:2–8

Kertselli S (1919) Znachenie olenevodstva i meri dlya ego razvitiya. (In Russian. Significance of reindeer herding and measures for its development.) In: Kertselli S, Houdadov V (eds) Olenevodstvo. (Reindeer husbandry.) Izdatel'stvo Narodnogo Komissariata Zemledeliya, Moscow, pp 3–8

Kideckel DA (1993) The solitude of collectivism: Romanian villagers to the revolution and beyond. Cornell Univ Press, Ithaca, New York

Kiselev AA, Kiseleva TA (1987) Sovetskie Saami. Istoriya, ekonomika, kultura. (In Russian. The Soviet Sami: history, economy, culture.) Murmanskoe knizhnoe izdatel'stvo, Murmansk

Konakov ND (1993) Ecological adaptations of Komi resettled groups. Arctic Anthropology 30:92–102

Konstantinov Y (1997) Memory of Lenin Ltd. Reindeer-Herding Brigades on the Kola Peninsula. Anthropology Today 13:14–19

Konstantinov Y (2000) Pre-Soviet pasts of reindeer herding collectives: ethnographies of transition in Murmansk Region. Acta Borealia 2:49–64

Konstantinov Y (2002) Soviet and Post-Soviet reindeer-herding collectives: transitional slogans in Murmansk region. In: Kasten E (ed) People and the land: Pathways to reform in Post-Soviet Siberia. Reimer, Berlin, pp 171–188

Konstantinov Y, Vladimirova V (2002) Ambiguous transition: agrarian reforms, management, and coping practices in Murmansk Region reindeer herding. Max Planck Institute for Social Anthropology Working Paper 35:1–29

Krassov OI (2000) Zemel'noe pravo. (In Russian. Land jurisdiction.) Yurist, Moscow

Kudriavtsev Y (1930) K organizatsii sovkhozov na Severe. (In Russian. Towards organizing state farms in the north.) Sovetskij Sever 4:48–65

LP (Lovozerskaya Pravda) (1998) "Tundra" – vremya obnovleniya. (In Russian. "Tundra" – time for renewal.) Lovozero Raion, Murmansk Oblast, 29 May, pp 1–6

Northern Committee (1931) O stroitel'stve olennykh sovkhozov na Severe (in Russian; On building reindeer herding state farms in the North). Postanovlenie VII Plenuma Komiteta Severa. (Resolution of the 7th plenum of the Northern Committee.) Komitet Severa, Moscow, pp 186–197

Osherenko G (1995) Property rights and transformation in Russia: institutional change in the far North. Europe Asia Studies 47:1077–1108

Paine R (1992) Social construction of the 'Tragedy of the Commons' and Saami reindeer pastoralism. Acta Borealia 2:3–14

Rybkin A (1999) Olenevodstvo v Murmanskoj oblasti. (In Russian. Reindeer breeding in Murmansk Oblast.) Association of World Reindeer Herders, Tromsø, pp 23–30

Seppänen S (1995) The Barents region: an emerging market. International Business Statistics, Helsinki

Sillanpää L (1999) Small nations of the Russian North and their rights as indigenous peoples: some observations and preliminary hypothesis. In: Von Benda-Beckmann K, von Benda-Beckmann F, Finkler HW (eds) Law and legal pluralism: societies in transformation. Papers of the XIth international congress, 18–22 Aug 1997. Nijmegen, Commission on Folk Law and Legal Pluralism, pp 109–120

Ushakov IF (1997) Izbrannye proizvedeniya (in Russian. Selected works), vols 1–3. Murmanskoe knizhnoe izdatel'stvo, Murmansk

Ushakov IF, Dashchinskiy SN (1988) Lovozero. Goroda i raioni Murmanskoi oblasti (in Russian; Lovozero. Towns and municipalities in Murmansk Region). Murmanskoe knizhnoe izdatel'stvo, Murmansk

Verdery K (1994) The elasticity of land. Problems of property restitution in Transylvania. Slavic Rev 53:1071–1109

Verdery K (1999) Fuzzy property: Rights, power and identity in Transylvania's decollectivization. In: Burawoy M, Verdery K (eds) Uncertain transition. Ethnographies of change in the postsocialist world. Rowman and Littlefield, Lanham, pp 53–81

Von Benda-Beckmann F, von Benda-Beckmann K (1999) A functional analysis of property. In: van Meijl T, von Benda-Beckmann F (eds) Property rights and economic development. Land and natural resources in southeast Asia and Oceania. Kegan Paul, London, pp 15–56

Part II

Reindeer Herding – Effects on Soils, Soil Biota and Vegetation

Photo B. Burkhard

Preface

This section focuses primarily on the patterns and processes that characterize the biophysical environments of the intensive study sites in Finland, Norway, and Sweden. The study areas were located in three distinct physiographic regions with different vegetation and soil types of winter and summer pastures on the one hand, and contrasting herding histories, strategies, and contemporary land use regimes on the other.

This part begins with a transition from the socioeconomic issues (Part I) as it treats a matter right at the nexus of the natural and social sciences – how best to characterize pasture "quality" (Chap. 7). Assessments of forage quantity and quality, derived primarily from low-resolution satellite surveys coupled with ground truthing, have provided the basis for pasture inventories and policies concerning population regulation in recent years. Yet, herders in Sweden and Finland have their own sets of qualitative indicators that they rank at least as high as vegetation cover and abundance in terms of importance. These criteria vary in space and time, deriving from diverse sources – biological, geographical, socioeconomic, management, and climate-related – and so bear heavily on their actual patterns of pasture use. Decisions taken based on these factors may change from hour to hour, day to day, and year to year, a level of complexity that cannot be captured by even the highest resolution satellite imagery. The conclusion is that pasture quality may be characterized more comprehensively by combining the best available scientific data with up-to-date information from practitioners themselves.

We go on in this part to present the data from our studies of vegetation, soils, and soil microbes. We cannot cover all aspects of these topics, but limit ourselves by our project goal, which is to elucidate changes on vegetation and soil ecology due to land use and its management. A special topic is the tundra area in Jauristunturit near the Sámi village of Näkkälä (Chap. 1), where recent changes are visible. It is an area in which changes have become evident from actual processes, from effects of recovery stages and long lasting changes.

Knowing well the recently published sets of papers on similar investigations by Kumpula (2001), Grellmann (2001), Eilertsen (2002), Stark (2002), our approach aims to describe the processes from belowground physical parameters to effects on plant nutrients by UV loads. This depiction provides data for further models and shows effects on both soil structure and soil biology.

Chapters 8 and 9 contrast vegetation cover and structure patterns on either side of the Finnish–Norwegian border, where strongly divergent management systems have been in place for nearly 50 years. Given the prominent role of remote sensing in contemporary reindeer management, it is important to understand the potential for state-of-the-art assessments of pastures. Very-high-resolution satellite imagery is useful for developing timelines of visible patterns and changes in the land at scales that are meaningful for herders. Results from an analysis of IKONOS-2 imagery (Chap. 8) allow us to distinguish 12 classes of treeless tundra, in addition to five different classes associated with varying intensities of reindeer grazing/trampling. *Salix lapponum* can be regarded as an indicator for riparian habitats (Chap. 9). The measured reductions in biomass and height of the "downy willow" in Finland appear to be directly dependent on summer grazing pressure. But wetlands subject to long-term grazing in summer appear rather resistant to change.

Chapter 10 investigates the quality of summer plants under ultraviolet radiation. It focuses on common forage species in boreal wetlands and reveals their individualistic short-term responses to UV-B radiation, an important component of global change. Differences in phenolic contents in different plant species and at different times during the growing season were evident. It is quite possible that these initial patterns will differ over time and further long-term research is necessary to confirm these preliminary results

Various changes in soil surface properties and soil structure were found to be related to different intensities of grazing and trampling, as well as to mechanical disturbance from forestry equipment (Chap. 11). Along trails and border fences and in the tracks of tree harvesting machines, soil deformation and degradation becomes a significant fact. This is reflected in changes in organic horizon soil properties under different grazing regimes (Chap. 12). Soil cation exchange capacity was reduced by some 20 % and concentrations of available calcium and magnesium were 30–50 % lower compared to soils occurring beneath ample lichen cover. A clear relationship emerged in which decreases in soil fertility correspond to increased grazing intensity. Grazing and trampling typically affect not only aboveground parameters, like vegetation cover and structure, but also belowground processes, such as nutrient mineralization and organic matter decomposition (Chap. 13). As such, data from dwarf shrub tundra along the Finnish–Norwegian border revealed a remarkably lower rate of respiration in visibly disturbed soils, whereas the data on soil organic matter and bacterial biomass were quite variable with no significant differences between grazed and non-grazed sites.

The hygienic status of soils and surface waters were studied with an eye to the incidence of enteropathogens. Such pathogens are a potentially critical concern where tundra soils come under intensive use, such as when animals are concentrated into relatively small areas for feeding (Chap. 14). However, soil microbial communities did not seem to be influenced significantly by fecal indicators, probably because the soils themselves are not suitable reservoirs for enteric bacteria due to their low levels of nutrients, dry conditions, and low temperatures.

Together these eight chapters address many of the overarching concerns of contemporary reindeer management as they pertain to vegetation, soils, and surface waters. It is clear that herders have dynamic qualitative information that can complement the relatively static picture derived from satellite imagery and quantitative measurements of certain variables. The patterns and processes that characterize contemporary pastures in northern Fennoscandia are influenced by a diversity of factors, ranging from border fences and forestry to climate and global change (e.g., UV-B radiation). As herders and administrators attempt to adapt management to the shifting conditions, there are definite benefits to be had from closer cooperation between herders and scientists as they seek to make informed decisions to sustain reindeer herding as a viable livelihood over the long term. Herders were directly involved in the research process – developing questions, picking study sites, sampling vegetation, soils, snow cover, climate and surface waters, and employing and documenting their local knowledge. As such, the studies presented here have taken a significant step forward in creating a more participatory form of management that is flexible enough to meet future demands in the rapidly changing subarctic and boreal environments of northernmost Europe.

References

Eilertsen SM (2002) Utilization of abandoned coastal meadows in northern Norway by reindeer. Thesis, Dept Arctic Biol and Inst Med Biol, Univ Tromsø, 25 pp + 4 papers

Grellmann D (2001) Top-down and bottom-up effects in a Fennoscandian tundra plant community. Thesis, Dept Ecol Environ Sci, Umeå Univ, 24 pp + 6 papers

Kumpula J (2001) Productivity of the semi-domesticates reindeer (*Rangifer t. tarandus* L.) stock and carrying capacity of pastures in Finland during 1960-1990's. Thesis, Dept Biol, Univ Oulu, 44 pp + 6 papers

Stark S (2002) Reindeer grazing and soil nutrient cycling in boreal and tundra ecosystems. Thesis, Dept Biol, Univ Oulu, 31 pp + 5 papers

7 Defining the Quality of Reindeer Pastures: The Perspectives of Sámi Reindeer Herders

H. Kitti, N. Gunslay, and B.C. Forbes

7.1 Introduction

The present study is related to reindeer herding pastures and factors determining the quality of grazing areas. The debate on pasture conditions involves a complex set of ecological, political, cultural, and socioeconomic issues. The reindeer pastures of northern Fennoscandia have been intensively studied by biologists since the second half of the 20th century (Skuncke 1958; Kärenlampi 1972; Oksanen 1978; Väre et al. 1996; Kumpula et al. 1996; Bråthen and Oksanen 2001; Olofsson et al. 2001). Some earlier botanists have also made observations on feeding plants (Kalliola 1939; Nordhagen 1943; Dahl 1956). Biological studies (Kumpula et al. 1996, 1997) have provided much of the basis for the current management plan for the reindeer herding area. These studies are based on the concept of carrying capacity and employ satellite-based maps derived from low resolution (30 m pixel size) LANDSAT Thematic Mapper images (Chap. 8).

Reindeer herders have for a long time prior to World War II regulated reindeer herding themselves, but this is not possible to the same extent in the modern world, where state laws regulate reindeer herding. In Finland and Sweden, laws and statutes have directed reindeer herding practices from the early 20th century. Finnish law defines reindeer herding strictly as an occupational enterprise without any reference to its cultural significance to the Sámi population (Lewis 1998). In practice, the total number of reindeer in Finland is regulated by the Ministry of Agriculture and Forestry, which sets an annual maximum number of reindeer for each cooperative for the next decade. The ministry is informed by the reindeer herders' association about total number of reindeer in each district. In Norway and Sweden, current herding legislation is still premised on the goal of 'rational herd management'. Yet, as Beach (1981) has shown, this goal is a highly ambiguous one and revolves around one primary question: to what extent are the extractive industries, which

Ecological Studies, Vol. 184
B.C. Forbes, M. Bölter, L. Müller-Wille, J. Hukkinen, F. Müller, N. Gunslay, Y. Konstatinov (Eds.)
Reindeer Management in Northernmost Europe

often conflict with herding, allowed to expand at the expense of herding (Lewis 1998).

Reindeer herders have criticized both the research and the management dealing with reindeer herding (Helander 1993; Raitio 2000; Laakso 2002; Hukkinen et al. 2003; Turi 2002). Such critiques have been published earlier as well, but only recently have they received much attention among researchers, media, and administrators. Herders have been specifically critical of biological assessments because they feel this provides a one-sided picture of the state of pastures. Focusing on biological and ecological aspects does not allow for the identification of all the various determinants that may affect overall pasture condition. The current situation in management is that reindeer herders' expertise remains undervalued and unexploited, both in terms of practical planning and in the relevant legislation (Helander 1993). Participatory planning has been carried out by the Forest and Park Service in Finland, but reindeer herders feel that "they are listened to, but not heard" and that they do not have any power in decision-making (Raitio 2000). Also, the flow of information among different management organizations and between management and herders does not work in Finland (Raitio and Heikkinen 2003). Some recent research has dealt with reindeer herders' knowledge (Helander 1993; Kalstad 1996; Sipilä et al. 2000; Heikkinen 2002; Ruotsala 2002; Inga 2003), but more research should be done in the field so that reindeer herders can have their voices heard and their knowledge as practitioners can be documented and, wherever appropriate, valued as much as "scientific" knowledge.

There has been a great deal of discussion of "overgrazing" during the last few decades and reindeer have been said to cause "diminishing" of the pastures (Kumpula 1994; Torp 1999; Tuomilaakso 2001; Tuomi-Nikula 2001). Herders have been perceived as eco-criminals, transforming the mountains into "rocky areas" (Beach 2000). This perception has been relayed through the media, contributing to a large extent to the stigmatization of herders. However, there has recently been some discussion about other land uses (Burgess 1999a; Müller-Wille 1999), and their impact on the pastures have come under investigation. During last 3–5 decades the extent of other competing forms of land use, e.g., tourism, forestry, and the building of artificial lakes, has increased in many grazing areas. There is new consideration related to the impact of these developments that can lead to actual losses of pasture lands, as in the case of artificial lakes for hydropower, as well as indirect or functional losses, as in the case of tourism infrastructure (Helle and Särkelä 1993; Vistnes and Nellemann 2001). Both processes contribute to the overall diminishment of the extent of available grazing areas. As a result, the pressure from grazing and trampling on the remaining pastures increases and contributes to the observed decline. The reduction in available pastures has meant that lichen cover has decreased significantly in many areas and reindeer have begun to receive supplemental feeding during winter (January-March). In Finland this started over 10 years ago in some cooperatives and has become regular, but in some districts supplemental forage is given only in difficult years, as from harsh winters (Heikkinen 2002).

The question of the condition of pastures arises as a core issue in the current debate among Sámi herders, authorities, and scientists. There is a need to redefine possible approaches and considerations related to the assessment of pasture condition that will include integrative participation and input of herders into scientific and decision-making processes. Some researchers have emphasized the need to recognize the complexities involved in the management in social–ecological systems and to embrace rather than ignore the inherent linkages (Berkes et al. 2003). A similar approach is under discussion in the realm of conservation biology (Mascia et al. 2003).

Following this trend and the needs of herders, the present study carried out within the RENMAN project undertook to investigate further the perspective of herders related to quality of pastures. Its main aim is to provide the current debate over pastures with extra meaning that herders may contribute to the interplay of determinants of pasture conditions in a broad sense that includes the natural, socioeconomic, and cultural dimensions. We have investigated how reindeer herders themselves define the pastures. Herders' knowledge of pastures has been recognised and valued in this project by involving them in the research process.

7.2 Reindeer Herding

Reindeer herding is a traditional livelihood of Sámi people in Norway, Sweden, Finland, and Russia. In Russia other indigenous and non-indigenous peoples also herd reindeer, whereas in Finland anyone who is a citizen of the EU can own reindeer. Reindeer herding areas have been divided into reindeer herding cooperatives (56) in Finland and Sámi villages (51) in Sweden. Reindeer have free access to the pastures. The context of reindeer herding has been transformed during the last few decades. Since the early 1960s, reindeer herding has gone through profound changes in terms of the technical equipment and herding techniques used, e.g., snowmobiles, all-terrain cycles (four-wheelers), and the building of border and direction fences, which have prevented reindeer from easily leaving their area (Pelto 1973; Paine 1994). This has lead to changes in management in herding, particularly in the 1980s and 1990s in the decline of "active herding" of reindeer (Ruong 1987) in many areas.

Active herding has once again become more common in certain areas. Also economical problems are more critical now than earlier, as in a number of ways reindeer herding has been made to adapt to the modern market economy and society. The costs of reindeer herding have clearly increased, in part because of the mechanization of several aspects of reindeer herding. This has lead to more efficient meat production. Herd demography (sex and age distribution) has been changed in that most of the reindeer are now females able to produce calves, which are essential in meat production. Meanwhile, the overall number as well as percentage of adult male reindeer has diminished. But although there have been important changes related to these more market-oriented values, reindeer herding still retains many other values. It is for many herders a way of life and exemplifies values like freedom, independence and "being in nature". Reindeer herders also find it important to work in and maintain the traditional livelihood (Ruotsala 2002).

7.3 Theoretical background

Practitioners' knowledge is based on indigenous (or non-indigenous) peoples' locally developed practices in resource use (Berkes et al. 2000; Huntington 2000). Knowledge is shared between individuals and transferred from one generation to the next through the oral tradition within a sociocultural system (Brooke 1993; Jernsletten 1997; Berkes et al. 2000). Practitioners' knowledge is adaptive in nature and lives of its receivers often depend on using this information; practitioners' knowledge can thus be said to be usage oriented. It is also a cumulative body of belief as well as a dynamic process. It consists of experience in working to secure subsistence from nature (Jernsletten 1997; Berkes et al. 2000).

The types of information and concepts that define an individual's knowledge vary from person to person (and over time, e.g., generations) and there can be specialization in expertise (Brooke 1993). Language is a carrier of knowledge, and in the Sámi language there are hundreds of words for describing nature – an indication of how the Sámi regard their environment. Researchers agree that Sámi people still hold their inherited ways of thinking, which is a holistic understanding of life and nature that requires intensive study, understanding, and knowledge of the physical environment (Helander 1996; Kalstad 1996; Aikio and Müller-Wille 2002). When considering the sustainability of resources from the Sámi point of view, one is dealing with a holistic issue. Thus it is crucial to take into account that the concept of an environment consisting of several aspects, including the ecological, cultural, social, and linguistic. These factors make up the whole and must therefore be seen as a single entity (Helander 1996).

Evaluation and knowledge of the condition of pastures is an important part of the culture and knowledge of Sámi reindeer herders (Sara 1994). Herding is a joint responsibility in which discussions and decisions are related to pasture and climatic conditions, herd control, and herding strategies (Sara 1994). Among reindeer herders, knowledge is gained through practice and experience. Knowledge is also accumulated in discussions and storytelling.

Due to the holistic way of thinking within Sámi society, as many elements and factors as possible are considered before any actions are taken (Kalstad 1996). Following this approach, we consider that herd management includes certain types of knowledge that are connected to certain contexts and situations. A holistic approach is also to be strongly recommended for those who wish to understand and be involved in those matters of importance to indigenous peoples such as the Sámi. Practitioners' knowledge does not appear as a research topic but as a system of knowledge that can complement the relevant scientific knowledge and contribute to better understanding of both local and regional issues of concern. In defining pasture quality as a component of herders' knowledge, adherence to such precepts is inferred.

Discussion of the role of practitioners' knowledge in research has only recently begun in Fennoscandia, and it is still missing in reindeer management. In the Program of Sustainable Development of 1998 the Sámi parliament of Finland has recommended that the role of practitioners' knowledge should be strengthened in decision making (Sámi Parliament 1998). The same recommendation was recently made at the seminar "Practitioners' Knowledge in Reindeer Herding" held in Kautokeino, Norway, in spring 2003 (forthcoming report). However, there are some ongoing or recently completed research projects in Fennoscandia in which practitioners' knowledge is taken into account, so that the situation in the region may change in the near future (Sipilä et al. 2000; Hukkinen 2002; Aikio and Müller-Wille 2002; Inga 2003; Raitio and Heikkinen 2003; Helander and Mustonen 2004).

In other arctic regions such as northern Canada and Alaska, indigenous peoples' knowledge or practitioners' knowledge plays a visible role in the management of environmental, social, and cultural issues (Burgess 1999b; Caulfield 2000; Usher 2000). At the moment Canada is a leading country in practitioners'-knowledge-related issues and research in the Arctic (Burgess 1999b). It has become a policy there to consider and incorporate practitioners' knowledge into environmental assessment and resource management (Usher 2000). The value of such knowledge has been appreciated through various experiences with implementation of co-management. In Nunavut, which is self-governing, the Inuit have a constant role in environmental management as well as in other areas. Elders play an important role as supervisors both in management and at the community level (Burgess 1999b). In some cases, practitioners' knowledge has been found to be even more specific to local needs than scientific knowledge. For example, in a study of the population size and migration of caribou on Baffin Island, Inuit hunters had more information relevant for local needs than scientists (Ferguson et al. 1997).

7.4 Study Sites

We have been working in two field sites: the Näkkälä reindeer herding cooperative in the county of Lapland in northern Finland and Sirkas Sámi village in the county of Nordbotten in northern Sweden. In Finland there are 56 reindeer districts and in Sweden also 51 Sámi villages. Our interest in studying at two distinct sites centers on the differences that occur in management systems, in the natural environment, and in land use practices.

7.4.1 Näkkälä Reindeer Herding District

The size of Näkkälä reindeer herding district is 3,539 km^2 (Fig. 7.1) and it is divided into three smaller management areas called *siida* in Sámi language. The study area consists of two northern siida, where the reindeer herders are Sámi. The western of these is called Näkkälä siida and the eastern one is Kalkujärvi. In the southern siida, herders are mainly Finns. In the years 2000–2001 there were altogether 140 families with 163 reindeer owners in Näkkälä district The number of reindeer was 6,827 (of 8,300 allowed).

The study area lies within both the subcontinental and continental sectors, in the hemiarctic and northern boreal zones (sensu Oksanen and Virtanen 1995). The most northern part consists of treeless subarctic tundra, while further south there is a belt of mountain birch forest. In the southern part there are some pine forests, and about 30 % of the total area consists of wetland.

In the northern part of the Näkkälä district the most important land uses besides reindeer herding are tourism and local hunting and fishing. Tourism

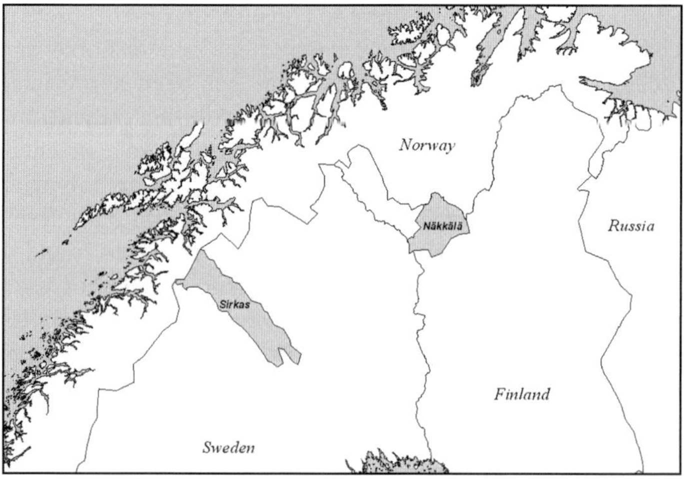

Fig. 7.1. Näkkälä and Sirkas reindeer district areas

means primarily snowmobile driving in late winter. There are designated snowmobile routes that are to be used by tourists, but during guided trips there is often a tendency to drive in other areas, too. Also local people drive snowmobiles when going fishing and ptarmigan trapping. During summer, locals use ATVs (four-wheelers) for hunting and cloudberry (*Rubus chamaemorus*) picking.

7.4.2 Sirkas Sámi Village

Sirkas Sámi village is geographically the largest Sámi village in Sweden (Fig. 7.1), covering an area of 13,370 km^2 and including maximum of 15,500 reindeer in 1998 for about 100 enterprises (Svensk Rennäring 1999).

Sirkas reindeer pastures can roughly be divided into two main areas: the summer and the winter grazing zones. Summer pastures include an area extending from the Norwegian border to about 100 km west of Jokkmokk. It is mainly a mountainous environment. The winter pastures include the forest areas surrounding Jokkmokk with a substantial extension to east. The study area lies in the continental sector.

The summer pastures are shared with the conservation administration and most of them are contained within national parks. Tourism is also one ongoing activity within the summer grazing area, including hunting and fishing in some specific places within the parks and anywhere outside its borders. Winter areas are mainly shared with forestry and hunting and fishing activities as well. Both areas include hydropower development.

7.5 Methods

Our main approach in the field consisted of interviewing and participant observation. In Näkkälä eight herders and in Sirkas nine herders were formally interviewed during the years 2001–2003, by using semi-directed interviews and open discussions. The age range of reindeer herders was between 20 and 75 years, but most of them were around 50-years-old. Interviews in Näkkälä were conducted in Sámi or Finnish depending on which language herders preferred. In Sirkas, interviews were mainly done in English. Young herders helped to translate when talking to elders. Interviews in Näkkälä were taped and transcribed. In Sirkas, most of the interviews were transcribed directly on paper due to field conditions. Notes from participant observation were also directly written down and compiled in different categories. The interviews were mainly about the pastures, their management, classification, and intensity of use. There was also some discussion of other land uses in the area.

Fieldwork was conducted in different phases. The first phase consisted of approaching and informing the herders about the project before interviewing them. The second phase dealt mainly with interviewing and discussing with herders in the field and at their homes. The third phase was to transcribe and pre-analyze the data before going back to the field and to report in the fourth phase to the herders about the preliminary results and to ask them for comments. It also provided an opportunity to conduct more interviews or to work more deeply on specific points. In each phase, we also took part in their daily work as a form of participant observation, e.g., calf marking during summer, when we also made contacts with additional herders to be interviewed. Informal time spent with herders in the field or in the village allowed the gathering of information additional to the data collected in formal interviews.

7.6 Results

7.6.1 Factors Affecting the Quality of the Pastures

According to the herders, a "good" pasture is a place where reindeer get enough forage and where reindeer can graze in peace. Forage conditions are dependent both on the quality and the quantity of vegetation, but also on weather conditions such as snow, wind, and ice cover. Peaceful grazing conditions are essential for reindeer; they graze in peace and feed properly when disturbance is minimal. In addition to reindeer needs, herders pointed out that their own needs should also be met. Herders value an area where it is relatively easy to herd and otherwise manage reindeer and which is easily accessible and is not too far away from their home.

Herders talked about a number of factors (Table 7.1) that affect the availability of forage, peace, and herders needs. We organized them into two main categories. First, there are environmental factors such as biological, geographical, and climatic factors. Second, there are anthropogenic factors such as socioeconomic and management-related aspects.

Results from Näkkälä and Sirkas were quite similar, but there are differences mainly because of dissimilar natural conditions, such as prevailing topography, and some different land-use practices. In this text, we point out the differences by mentioning which factors apply to Näkkälä and which to Sirkas. If the factors are same, the respective areas are not mentioned.

7.6.1.1 Biological

Vegetation is the factor that is usually emphasized in the natural sciences. It is an important factor affecting reindeer pastures, but according to the herders

Table 7.1. Factors affecting the quality of the pastures according to herders in Näkkälä and Sirkas

Environmental	Biological	Vegetation Predators Insect harassment
	Geographical	Natural borders Altitude Aspect Exposition
	Climatic	Temperature Snow depth Icing of ground Wind
Anthropogenic	Socioeconomic	Hunting and fishing Motor vehicles Walkers/skiers Forestry Dams Conservation
	Management	Number of reindeer Sex and age distribution Location Political decisions

it is more often a question of access to the forage than the amount of forage as such. As explained by one herder: "Natural obstacles like cliffs or wide rivers could make the access to forage difficult, but fences, dams and roads could also be obstacles" (Herder 2, Sirkas).

Reindeer feed on different forage in different seasons and graze in different habitats. In summer the diet consists of fresh herbs, willow (*Salix* spp.), and birch (*Betula* spp.) leaves, sedges, and grasses. In late summer and autumn reindeer continue partly on the same diet but they also consume mushrooms when available. The quicker the pace of phenological change during the growing season, the better it is for the reindeer, and especially if it happens at different times in different places, so that they can find fresh herbs and graminoids in sufficient quantity as they move from one area to another throughout the summer: "There are lot of rivers around, small rivers. The surroundings of rivers are often really good places because plants are growing up there very quickly and at different times during the summer, depending on the place" (Herder 4, Sirkas).

In summertime access to food is not usually difficult, because in both Näkkälä and Sirkas pastures are considered to be good and the snow does not prevent moving from place to place. Winter is the more critical time for reindeer. A herder from Sirkas said: "...[T]he yearly pasture could be compared to the shape of a bottle, the main body of the bottle representing the summer pastures and the bottleneck representing the winter pastures" (Herder 5 Sirkas). But herders agree that the biggest reason for that is other than vegetation. Herders point out that the access to the forage is more critical than simply the amount of lichens. According to reindeer herders, there has been too much emphasis placed on the importance of lichen cover, as already pointed out by Laakso (2002) and Raitio (2000). Referring to winter diet, one herder attacked: "Researchers do not have down-to-earth experience of reindeer" (Herder 3, Näkkälä). Reindeer herders reported that reindeer eat virtually everything on the ground, e.g., dwarf shrubs and standing dead grass. One reindeer herder said in reference to stomach contents: "I have often found dwarf shrubs [e.g., lingonberry, *Vaccinium vitis-idaea*] when slaughtering reindeer" (Herder 3, Näkkälä).

According to the herders the most common and "harmful" predators in the study areas are wolverine and eagle. There are also some bears, wolves, and lynx. Predators affect reindeer in two ways: they kill animals but they can also stress them so that they cannot graze in peace. With predators in the vicinity, reindeer move around a lot within a given feeding area and are generally restless to the extent that their feeding and accumulation of fat reserves for winter are disturbed. The predators also affect herders' income by reducing the total number of reindeer and thus decreasing meat production. Further, they affect the demographic distribution within herds in that there are fewer calves, the most vulnerable to predation. Also the quality of meat can be affected if animals are too stressed over a prolonged period.

Insect harassment is a potentially serious problem for reindeer during summer, again because of disturbed feeding. On the other hand, mosquitoes help reindeer herders by keeping the reindeer in flocks (in which diminished individual insect attack occurs), which makes it easier to collect and manage the animals for calf marking.

7.6.1.2 Geographical Factors

Natural borders help the herders keep their reindeer from mixing with neighboring herds, which avoids conflicts between herders among different herding units and districts. Natural borders also decrease the necessity of active herding. Lakes, large rivers, cliffs, and high mountains are all elements that may be used as borders. In other places, it is often necessary to construct fences to delimit the pastures of each unit and district, which is the case in both Sirkas and Näkkälä.

Some geographical features function as obstacles for reindeer movement, such as steep slopes and deep valleys in Sirkas. Moving there is difficult and tiring for reindeer, because the food is mainly concentrated in the valleys. These kinds of slopes also include fewer feeding locations. Indeed the diversity of geographical features such as gently inclined slopes with various aspects and altitudes contribute greatly to the quality of the pastures. In such environments, the pace of plant phenology during the growing season is staggered at different times and places and so secures a continuous stock of high quality forage throughout the summer.

Snow melts and evaporates faster on south-facing slopes, which means that forage is generally more easily accessible there earlier in the growing season. This also has special significance in late winter when there is lots of snow in most places, and reaching forage is difficult. In the beginning of summer plants in exposed habitats are the first to grow: "Melting season time is very important for early summer. If snow melts soon enough, then the gap between winter pastures and summer pastures is not so big. Reindeer can then find rapidly some fresh herbs after winter" (Herder 2, Sirkas)

7.6.1.3 Climatic Factors

Weather (short-term) and climate (long-term) conditions are important for the accessibility of the forage, the growth season and the movements of reindeer.

Hard snow or ice, on the ground or on top of snow cover, may prevent reindeer from reaching enough forage. Therefore, an understanding of snow is essential in reindeer herding (Sara 1994). The main issue for reindeer herders during winter is to estimate whether there is *ealat*, which means that reindeer cannot dig through the snow and get to the food on the ground (Sara 1994). In that case, reindeer should be moved to areas where the snow cover is thinner and softer.

Icing events usually occur in the autumn, but can also occur during warm periods in winter, if there is rain followed by cold and snow. "The ground gets so hard that not even the knife goes through it" (Reindeer herder 4, Näkkälä). Some years the snow surface is frozen hard throughout much of the winter and the area where reindeer can graze is subsequently very limited. This caused high mortality and low birth rates of reindeer in Näkkälä in the 1990s. The same can occur if there is lots of snow or the snow gets hard in early spring.

In late spring, snowmelt determines the advent of the growing season. In summer, ideal conditions include mild temperatures and windiness. Good summer pastures should include areas where the snow is melting early enough so that the grasses, sedges, and herbs will have already emerged by the time the herd moves onto the summer pastures. If there is too much snow and if the melting season continues until late spring, it may negatively affect the quality of the summer pastures in terms of available food stock.

Temperature is another important factor, especially during the summer. If the weather is too hot reindeer tend to stay on the windy slopes and the tops of the mountains the whole day and come down to the valleys to feed only in the evening and at night. In Sirkas, where there are high mountains, reindeer often lay on the remaining snow patches. In Näkkälä reindeer also go to the peninsulas of lake Pöyrisjärvi. The peninsulas have also been natural calf marking areas because they form a natural fence.

Reindeer are disturbed by insect harassment in summer. Reindeer prefer to stay in windy areas, such as mountain and fell tops and on peninsulas in lakes to avoid them. The animals also tend to walk against the wind because of predators and insects during summer. This can cause difficulties for herding practices, because reindeer might not go where herders wish to direct them.

If it is exceptionally warm during the spring, the snow in the mountains melts too fast, with the result that the water level and flow of the rivers becomes so high so that reindeer cannot cross and have to wait until the flood waters recede. This can prevent reindeer from moving from place to place.

7.6.1.4 Socioeconomic Factors

Prominent among other land users in both study areas are tourists, hunters, and fishermen. In Sirkas, forestry and dams are important additional factors. In both areas there are also conservation or protected areas.

As stated, peaceful grazing conditions are highly important in that reindeer only feed properly when disturbance is minimal. In spring, pregnant dams can suffer from harassment (e.g., predators, tourists). In summer and fall, reindeer accumulate reserves for the winter, and at this time the calves are still small and need the highest quality grazing conditions to facilitate their growth. Too much disturbance when they are feeding tends to weaken them; thus free access to feeding spots is very important. Sources of stress that disturb feeding habits of reindeer can affect both animal health and meat production.

In Näkkälä and Sirkas, tourism mainly consists of summer hiking and winter skiing and snowmobile safaris. There is no massive tourism, but reindeer herders state that even a few tourists are enough to disturb reindeer. One reindeer herder pointed out in Näkkälä: "This area is a very good place, but reindeer cannot graze there because there are often skiers there" (Herder 1, Näkkälä). The tourists or safari associations either ignore or do not have information about the basic rules when driving or hiking in reindeer herding areas and often scare reindeer from their grazing positions in certain areas. Some still consult with reindeer herders to get information about which areas should be avoided. Even one reindeer moving away is enough to induce the whole herd to move away as well as pointed out by one herder from Sirkas: "When hikers see a herd, they want to get as close as possible from it so that they can take pictures of calves. But it does not require more than one hiker to induce the herd to move away. If one reindeer is scared and runs away, the other ones follow" (Herder 6, Sirkas). Another problem caused by snowmobile driving are the tracks left by the machines. Once the snow has sintered and hardened, reindeer tend to follow them, which can make herding practices more difficult.

Reindeer herders use ATVs (four wheelers) as a working tool, but other locals and tourists also use them. Driving ATVs causes physical disturbance on the land by destroying vegetation and soil layers (Olli 1995), so that if one drives 20 km with an ATV, 1 ha of pasture is degraded as a direct result. There have been several routes established for four wheelers, which people usually follow. In the wetlands one often has to make new tracks, because old ones get too soft, so that the route continually widens over time. Although the wet sites appear to be severely damaged they tend to recover reasonably well and quickly if left alone for several years. But moderately dry and severely dry ground is also easily damaged and takes much longer to recover (Forbes et al. 2001). Some herders are worried about the influence of the routes on the quality of the pastures, but not much has been done about it yet.

One problem during autumn is ptarmigan and elk hunting with dogs. The dogs run free and are not under control, and so they may chase reindeer and break the flock, which frightens and stresses them. This can also disturb the reindeer herders' work if they have gathered the flock together. Fishing safaris and fishermen walking where the herd is grazing can provoke the same disturbances.

Forestry reduces the winter pastures and changes the vegetation layer. Reindeer do not graze in the areas where the trees have been cut down, except during the first year when the trees are laying down and arboreal lichens are easily reachable (Sirkas). After that the grass *Deschampsia flexuosa* usually starts to grow and the area loses its value as winter pasture for many years. Another problem is that the tree trunks and branches left on the ground kill the lichens under them.

In Sirkas, the area near the Luleå River has been used for spring and fall pasture. In the 1960s the construction of dams inundated some of these areas. The dam construction forced the herders to use the summer and winter pastures in spring and autumn, leading to overuse of the available forage in these areas. One of the main effects of dam construction is that the rivers during winter either remain open or are covered with ice cover too weak to support the animals during migration. This affects the route of the reindeer and most often herders have to carry them by truck, which increases the costs of herd management and also the stress on the animals.

Most of the summer pastures in Sirkas lie within national parks. Portions of the Näkkälä pastures are protected by nature conservation laws (Wilderness Act 62/1991). External disturbances in Sirkas summer pastures in the national park are low, insofar as the parks regulations allow only the Sámi to use the natural resources, drive motor vehicles, and to fish and hunt within the border of the parks. Herders state that conservation is positive because it prevents land users other than herders from exploiting the areas.

7.6.2 Management and Economic Questions

According to the concept of carrying capacity, the abundance and density of reindeer is argued to be one of the main factors affecting the quality of the pastures. Reindeer herders both in Sirkas and Näkkälä agree that there were too many largely unsupervised reindeer in 1980s, which contributed to some "overgrazing" problems, mainly loss of lichen cover from the combination of winter grazing and summer trampling (see Chaps. 1, 8, 9, 11–13). The "quality" of the pastures is reported to have begun to recover, which in this case means that lichen cover has increased in some areas beginning in the 1990s (Kumpula et al. 2004). Reasons may include the smaller number of reindeer and the greater efforts via active herding to keep the animals away from lichen grounds in summer.

The size of the pastures is one of the factors affecting the quality. The larger the pastures are, the more potential feeding spots there are at different times of the year. In both study areas, winter pastures are more limited than summer pastures. A herder from Näkkälä pointed out, "This is a very good area, but it just should be wider" (Herder 1, Näkkälä), meaning that it is an area that does not easily develop an ice cover in years when the ground freezes. The demographic distribution of herds has also changed because of the orientation toward meat production, so that there are more females and calves in the herd now than earlier. Earlier there were more male animals, which were often strong enough to dig through the hard and deep snow cover. This afforded weaker animals at least some possibility of obtaining forage.

Due to economic costs reindeer herders say that the pastures should be located relatively close to the villages in winter, because they need to herd the reindeer daily, which increases costs via fuel consumption. On the other hand, the pastures should not be accessible to other people too easily, so that reindeer can graze in peace. Development of a road network and bus transportation to and within national parks in Sirkas may well lead to an increase of tourism activities and hunting and fishing trips in the area.

Fishing and hunting resources are also essential components of what defines a good pasture. Herding requires a certain level of flexibility in order to manage and control the course of herding under changing conditions and to ensure sufficient income. In Sirkas, herders spend not only summer but also some of their time in autumn within the summer grazing areas where they fish and hunt, respectively. These activities provide herders with additional income or fish and meat for their own consumption. It is traditionally an important part of the economy of the household as well as of the social relations among families.

7.6.3 Interactions Among the Different Factors

The factors identified in this chapter do interact with each other. For instance, there are biological factors (such as predators) and management-related factors (such as environmental policies) that can together affect the overall quality of pastures and the conditions for grazing peace. The number of predators allowed is determined by national environmental policies. This has a number of effects: reindeer move away from areas where there are high numbers of predators. An area can be a good forage spot but it becomes useless if it cannot be used, because it contains too many predators. As mentioned, predators affect also the meat production by killing mostly calves. In addition to the loss of animals, predators affect the daily course of management. Keeping the herd within pastures inhabited by predators requires the herders remain closer to the herd, increasing the functional cost.

Both local and national economic development policies determine tourism activities in the reindeer herding area, which in turn affect grazing peace and forage conditions. According to the herders, reindeer are easily frightened by humans, motor vehicles, and dogs, and they move away from the disturbed areas (Tyler 1991; Aastrup 2000). Tourism infrastructure has measurable influence on animal movement and pasture condition (Helle and Särkelä 1993; Vistnes and Nellemann 2001). Another effect of tourism can be trampling. The effects of footpaths and four-wheelers can potentially extend well beyond the paths and trails they make, because when these are used continuously the surface organic layers are lost and the mineral soils beneath eventually begin to erode away (Olli 1995; Forbes et al. 2004).

What we can see by these examples is a chain of interactions through which the overall quality of pastures is determined. In these particular cases, we can see how both environmental and anthropogenic factors can influence pasture quality, and we also see how the eventual impact of environmental factors are potentially mitigated or exacerbated by anthropogenic input, such as through environmental and development policies.

7.7 Scientific Knowledge and Practitioners' Knowledge

The questions of attributing equivalent value to practitioners' knowledge in comparison to scientific knowledge, and how to proceed, remain critical. In this study, herders' knowledge has provided information about several factors and their contributions to pasture quality. Some of this information corresponds to that provided by the scientists (Table 7.2). Other aspects represent a different dimension and expose the need to reconsider the exclusively scientific approach as it relates to studies of pasture quality. It is not a question of

Table 7.2. Matching scientific knowledge with practitioners' knowledge

Study subject	References
Feeding habits and forage plants of reindeer	Nieminen and Heiskari (1989); Warenberg et al. (1997)
Behavior of reindeer/caribou	Espmark (1970); Skogland (1986)
Effect of insect harassment on reindeer/caribou	Helle and Tarvainen (1984); Downes (1986); Toupin (1996)
Effect of snow cover on pastures/reindeer/caribou	Helle (1984); Telfer and Kelsall (1984); Adamczewski et al. (1988); Collins and Smith (1991); Kumpula (2001b); Kumpula and Colpaert (2003)
Effect of tourism on pastures/reindeer	Helle and Särkelä (1993); Helle (1995); Nellemann et al. (2000); Weladji and Forbes (2002)
Effect of forestry on pastures	Terry (2000); Kumpula (2003); Kumpula et al. (2003)
Effect of ATVs on pastures	Olli (1995)
Effect of infrastructure on reindeer/caribou	Horejsi (1981); Curatolo and Murphy (1986); Vistnes and Nellemann (2001); Nellemann et al. (2003)

contrasting the two types of knowledge. Scientific and practitioners knowledge sometimes overlap significantly. For example, biologists use calf weight as one indirect indicator of pasture conditions (Kumpula 2001a). In a similar way, caribou hunters (Kofinas et al. 2003) and herders use body fats (back fat, stomach fat, and marrow) as indicators for the condition of individual animals, as well as the herd and the range as a whole. Both types of knowledge provide indirect information about environmental factors and animal condition and, in turn, the condition of the pastures. Reindeer herders also have their own classification scheme and terminology for different intensities of use (Table 7.3).

As presented here, it seems that different kinds of knowledge do match in some respects. However, the main difference between scientists and herders centers on methods of approaching and then defining quality (herders), rather than on merely identifying different factors (scientists). Most of the pasture studies related to the latter are based on biology, emphasizing satellite mapping of land cover classes, and corresponding calculations of carrying capacity based on meat production. As such, they do not account for the anthropogenic and other natural dimensions of quality, either in terms of time or context. Herders' own classification of pasture types differs considerably from those

Table 7.3. Reindeer herders' classification (in Sámi language) for differently grazed areas

Guorba eanan	The soil is usually so thin that lichens do not grow.
Varas eanan	The area where reindeer have not been that year or after growing period.
Fieski	The size of the area. The area has not been too much used.
Cilvi	Grazed summer area that does not do for grazing the same year.
Doldi, Smurvi	Heavily grazed summer area that does not do for grazing at all.
Duolmmus eana	Trampled summer area that does not do for grazing at all.

derived from satellite imagery. As shown in the results, the herders´ classification takes into account several factors encompassing nature, economics and culture, whereas in the prevailing biological approach these factors are not acknowledged. They mainly deal with vegetation, soil, meat production, and animal health. These studies provide a snapshot of a certain point in time but do not really encompass the complexity of the situation concerning overall grazing conditions and the dynamic ways in which they develop.

At present, the decision-making process underlying reindeer management policies is driven almost exclusively by scientific studies and conclusions (Chaps. 1, 3, 16, 17). During the present study, one herder questioned the knowledge and the decision-making process in the following way: "We are the herders, we are the ones who know herding, but we are not asked what is good or bad for reindeer and for us" (Herder 5, Sirkas). From the herders' perspective, this indicates the lack of communication and information that exists among herders, scientists, and decision makers, as well as the need to bridge the gap by developing new tools and instruments that would contribute to greater consideration of herders knowledge and provide a means for herders to participate with influence in the process. Such a bridge would also highlight the need to provide for greater resilience for contemporary herding. As explained by herders, grazing conditions are changing all the time. Some changes are not predictable, such as the variability of weather conditions and climate over time, whereas the anthropogenic factors are at least potentially largely predictable. In any case, herders have no choice but to respond to the changes. Reduced resilience means a reduced capacity for active adaptation and engenders passive adaptation. Herders are integral components of the reindeer management system. When reindeer are stressed, herders need to respond to the stress. For that, they need a certain flexibility in managing for the short, medium, and long term.

There are environmental effects, which include biological, geographical, and climatic ones, and anthropogenic effects, including socioeconomic, and management-related ones. Together these influences act as quality determinants.

7.8 Discussion and Conclusions

"What is good here is not good there"; "What is good today might not be good tomorrow". By using these quotations, we summarize the dual aspect of pasture quality, as pointed out by herders. Quality is contextual, and it is dynamic.

Herders define quality according to basic grazing circumstances such as peace and forage conditions. However, they emphasize that these conditions may be influenced by several interacting factors. There are environmental effects, which include biological, geographical, and climatic ones, and anthropogenic effects, including socioeconomic and management-related ones. Together these impacts act as quality determinants. The determining process might be either direct or indirect and result from a set of interactions between the factors that are not only often coexistent but also interrelated. Reindeer might not use an area with good vegetation conditions and good pasture due to predators, whose number depends to a large extent on environmental policies (Beach 1997, 2000; Torp 1999). The quality of pastures is therefore potentially dependent on several chains of interrelated factors. Studying them requires thus the identification of all potential factors. Their impacts cannot be assessed independently from the relations between all the factors. In this perspective, focusing selectively on either environmental or anthropogenic influences provides a limited picture of the state of the pasture. Quality as defined by the conditions that meet the need of both the reindeer and the herders, and being the result of a complex system of interactions between environmental and anthropogenic factors, implies that investigations related to pasture quality must combine the study of environmental and anthropogenic aspects. Assessing pasture quality requires a holistic approach allowing identification and understanding of all potential quality determinants in any given context.

Arising from interactions between several factors, quality is also a dynamic process since quality determinants are changing not only in space but also in time. A regulation can remain unchanged for years, whereas the environmental conditions may change considerably over much shorter periods. Ideally, regulations should account for such changing conditions and help to maintain or build resilience within the system, rather than restrict it (Folke et al. 2002; Chapin et al. 2004; Forbes et al. 2004). Rapidly changing winter circumstances, such as amount of snow and icing of the ground, comprise one example. The extent of spring or winter pastures in Sirkas has become smaller in some areas mainly because of hydroelectric dams. Pasture land is lost due to inundation by artificial lakes or reservoirs or because of forest clear cuts, obliging the herders to move their reindeer earlier to summer pastures or to keep them longer in fall pastures. The temporal dimension needs to be considered when approaching the question of intensity of use. Herders have been accused of allowing "overgrazing" of the pastures (Beach 2000). However, studying the history of an individual place may contribute to a better understanding of the factors that have been involved in the process and to an identification of the roles of each of these factors in determining the quality of pastures at that particular location.

This study has shown that the quality of pastures is dynamically and contextually determined. It has also demonstrated that quality is defined via a complex set of interactions between environmental and anthropogenic factors. However, for the relevant government authorities, quality has been often, and is still to a large extent, determined by a carrying-capacity model, which compares the number, density and slaughter weights of reindeer against prevailing land-cover classes visible in low resolution satellite imagery. By focusing on these biological indicators, such an approach overlooks several other influences that are used on an ongoing basis by herders for determining the overall "quality" of pastures – or, perhaps more appropriately, grazing conditions.

If decision-making in reindeer herding is based solely on the knowledge of biologists, economists, engineers, and administrators, reindeer herding in the future will resemble a specialized type of agriculture and private business aimed solely at profit, rather than as both an important livelihood and a source of income (Kalstad 1996). This is in fact very much the case now. Björklund (1994) noted that the Sámi pastoral management system had for the previous 10 years or so been the object of a rather deliberate integrative effort by the Norwegian national authorities. Local people need to be involved in the decision-making process concerning how local resources are to be exploited under current conditions, which possibly also allows projections into the future (Helander 1993; Kalstad 1996; Turi 2002).

Sámi reindeer herding is a culturally based and way of living. Sámi people have kept, managed, and herded reindeer in their way for decades and even centuries (see Chap. 2). Herders have faced intensive modernization and have had to adjust to it. Part of the practitioners' knowledge has remained relatively unchanged, while other components have evolved and developed anew during this adaptation process. To be able to talk about Sámi reindeer herding, a significant amount of the responsibility for the decision-making should be returned to districts and the herders themselves (Kalstad 1996, see also Chaps. 1, 2 and 17). In recent decades nation-states have progressively appropriated decision-making capacities from the reindeer districts, so that at present the state makes virtually all decisions concerning reindeer management.. The main trend of the political agenda today is to avoid "overgrazing" by reducing the number of the animals as well as the overall number of herders, and therefore herding families. There is very little, if any, consideration given to soliciting the participation and input of herders, or to integrating them into the process.

In addition to the factors that have been dealt with in our study, there is one more essential facet to consider related to the loss of practitioners' knowledge. Practitioners' knowledge has diminished and changed as cultural assimilation into the dominant societies of Sweden and Finland has escalated (see Chap. 2). In the process of adapting to the modern technology some knowledge of the weather, of snow and animals has obviously been lost (Helander 1993). At the same time, new knowledge has developed about, e.g., supplemental feeding, which is a relatively new technique in herding. Helander (1993) points out threats to maintenance of knowledge connected to assimilation, such as: transition of herding to meat production; loss of language-based knowledge; limitation in control over education; overall reduction in land area available for reindeer management; and knowledge exchange among researchers and herders. In this standpoint, social and cultural changes may also be perceived as factors that can affect the quality of pastures, primarily by removing from the community essential knowledge for herding. As it is derived from active practices, knowledge is learned and taught in a dynamic sociocultural environment. Changes in this situation might affect the continuation and the transmission of knowledge. To a significant extent, maintaining sustainable pasture quality is related to social and cultural cohesion.

References

Aastrup P (2000) Responses of West Greenland caribou to the approach of humans on foot. Polar Research 19:83–90

Adamczewski J, Gates C, Soutar B, Hudson R (1988) Limiting effects of snow on seasonal use and diets of caribou (*Rangifer tarandus groenlandicus*) on Coats Island, Northwest Territories, Canada. Can J Zool 66:1986–1996

Aikio MS, Müller-Wille L (2002) Living at the timberline: the Sámi and the mountain birch forests in northernmost Europe. In: Kankaanpää S, Müller-Wille L, Susiluoto P, Sutinen M-L (eds) Northern timberline forests: environmental and socio-economic issues and concerns. Finnish Forest Research Institute, Jyväskylä, pp 40–56

Beach H (1981) Reindeer-herd management in transition: the case of Tuorpon Saameby in northern Sweden. Almqvist and Wiksell Int, Uppsala, 542 pp

Beach H (1997) Negotiating nature in Swedish Lapland: ecology and economics of Saami reindeer management. In: Smith A, McCarter J (eds) Contested arctic: indigenous peoples, industrial states, and the circumpolar environment. University of Washington Press, Seattle, pp 122–149

Beach H (2000) Reindeer pastoralism politics in Sweden: protecting the environment and designing the herder. In: Hornborg A, Pálsson G (eds) Negotiating nature: culture, power and environmental argument. Lund University Press, Lund, pp 179–212

Berkes F (1999) Sacred ecology: traditional ecological knowledge and resource management. Taylor and Francis, Philadelphia, 209 pp

Berkes F, Colding J, Folke C (2000) Rediscovery of traditional ecological knowledge as adaptive management. Ecol Appl 10:1251–1262

Berkes F, Colding J, Folke C (eds) (2003) Navigating social-ecological systems. Cambridge Univ Press, Cambridge, 393 pp

Björklund I (1994) Sámi livelihood: indigenous resource management in a changing world. Majority–Minority Relations. In: Helander E (ed) The case of the Sámi in Scandinavia. Report. Diedut 1:70–74

Brooke LF (1993) The participation of indigenous peoples and the application of their environmental and ecological knowledge in the Arctic Environmental Protection Strategy, vol 1. Inuit Circumpolar Conference, Ottawa, 119 pp

Bråthen KA, Oksanen J (2001) Reindeer reduce biomass of preferred plant species. J Veg Sci 12:473–480

Burgess P. (1999a) Human environmental interactions in Upper Lapland, Finland. Arctic Centre Rep 27. Arctic Centre, University of Lapland, Rovaniemi, 80 pp

Burgess P (1999b) Traditional knowledge. Arctic Council Indigenous Peoples' Secretariat, Copenhagen, 118 pp

Caulfield RA (2000) Political economy of renewable resources in the Arctic. In: Nuttall M, Callaghan TV (eds) The Arctic: environment, people, policy. Harwood Academic Publishers, Amsterdam, pp 485–513

Chapin, FS III, Angelstam P, Apps M, Berkes F, Folke C, Forbes BC, Juday G, Peterson O (2004). Vulnerability and resilience of high-latitude ecosystems to environmental and social change. Ambio 33:344–349

Collins W, Smith T (1991) Effects of wind-hardened snow on foraging by reindeer. Arctic 44:217–222

Curatolo JA, Murphy SM (1986) The effects of pipelines, roads and traffic on the movement of caribou, *Rangifer tarandus*. Can Field Nat 100:218–224

Dahl E (1956) Mountain vegetation in south Norway and its relation to the environment. Det Norske Videnskaps Akademi Oslo I Mat Naturv Klasse 3, 374 pp

Downes CM (1986) The influence of insects on distribution, microhabitat choice, and behaviour of the Burwash caribou herd. Can J Zool 64:622–629

Espmark Y (1970) Abnormal migratory behaviour in Swedish reindeer. Arctic 23:199–200

Ferguson MAD, Messier F (1997) Collection and analysis of traditional ecological knowledge about a population of arctic tundra caribou. Arctic 50:17–28

Folke C, Carpenter S, Elmqvist T et al (2002) Resilience and sustainable development: building adaptive capacity in a world of transformations. Ministry of the Environment, Stockholm, 74 pp

Forbes BC, Ebersole JJ, Strandberg B (2001) Anthropogenic disturbance and patch dynamics in circumpolar arctic ecosystems. Conserv Biol 15:954–969

Forbes BC, Fresco N, Shvidenko A, Danell K, Chapin FS III (2004) Geographic variations in anthropogenic drivers that influence the vulnerability and resilience of high latitude social-ecological systems. Ambio 33:377–382

Heggberget TM, Gaare E, Ball JP (2002) Reindeer (*Rangifer tarandus*) and climate change: importance of winter forage. Rangifer 22:13–32

Heikkinen H (2002) The adaptation of reindeer husbandry in post-industrial functional environment in the western reindeer husbandry area in Finland 1980–2000 (in Finnish). SKS, 369 pp

Helander E (1993) The role of Sámi traditions in sustainable development. In: Käkönen J (ed) Politics and sustainable growth in the Arctic. Aldershot, Darmouth pp 67–79

Helander E (1996) Sustainability in the Sámi area: the X-file factor. In: Helander E (ed) Awakened voice. Diedut 4. Nordic Sámi Institute, Kautokeino, pp 1–6

Helander E, Mustonen T (2004) Snowscapes, dreamscapes: snowchange book on community voices of change. Fram Oy, Vaasa, 562 pp

Helle T (1995) Reindeer husbandry and hunting. In: Hytönen M (ed) Multiple use forestry in the Nordic countries. The Finnish Forest Research Institute, Helsinki, pp 157–190

Helle T (1984) Foraging behaviour of semi-domesticated reindeer (*Rangifer tarandus tarandus*) in relation to snow in Finnish Lapland. Reports from the Kevo Subarctic Research Station 19:35–47

Helle T, Tarvainen L (1984) Effects of insect harassment on weight gain and survival in reindeer calves. Rangifer 4:24–27

Helle T, Särkelä M (1993) The effects of outdoor recreation on range use by semi-domesticated reindeer. Scand J For Res 8:123–133

Horejsi BL (1981) Behavioral response of barren ground caribou to a moving vehicle. Arctic 34:180–185

Hukkinen J, Jääskö O, Laakso A, Müller-Wille L, Raitio, K eds (2002) (Reindeer herders speak: problems, potential solutions and research needs of reindeer herding in Finnish Lapland from the viewpoint of reindeer herders) in Finnish. Technology, Society, Environment 1/2002. Espoo: Helsinki University of Technology Laboratory of Environmental Protection.

Hukkinen J, Jääskö O, Laakso A, Müller-Wille L, Nevalainen S, Raitio K (2003) The value of the culture of reindeer husbandry in Finland: challenges for administration, guidance and control: the final report of 1st work package dealing with reindeer herding of EU's RENMAN-project (in Finnish). Helsinki University of Technology, Espoo, Laboratory of Environment Protection Publication. Technology, Society, Environment 1/2003, 44 pp

Huntington HP (2000) Using traditional ecological knowledge in science: methods and applications. Ecol Appl 10:1270–1274

Inga B (2003) Traditional knowledge among Sámi reindeer herders of northern Sweden. In: Soppela P, Ruth W, Åhman B, Riseth JÅ (eds.) Reindeer as a keystone: species in the north – biological, cultural and socio-economic aspects. Arctic Centre Reports 38. Rovaniemi, pp 55–56

Jernsletten N (1997) Sámi traditional terminology: professional terms concerning salmon, reindeer and snow. In: Gaski H (ed) Sámi culture in a new era: the Norwegian Sámi experience. Karasjok, Norway, Davvi Girji, pp 86–108

Kalliola R (1939) Plant ecological research in the alpine belt of Finnish Lapland (in German). Ann Bot Soc Fennicae Vanamo 13(2), 328 pp

Kalstad JK (1996) The modern challenge facing knowledge in Sámi subsistence. In: Helander E (ed) Awakened voice: the return of Sámi knowledge, vol 4. Nordic Sámi Institute, Diedut, pp 21–30

Kärenlampi, L (1972) Factor analysis studies on the vegetation of the surroundings of the Kevo Subarctic Station. Rep Kevo Subarc Res Sta 9:66–72

Kofinas G, Lyver P, Russell D, White R, Nelson A, Flanders N (2003) Toward a protocol for community monitoring of caribou body condition. Rangifer Special Issue 14:29–38

Kumpula J (2001a) Productivity of the semi-domesticated reindeer (*Rangifer t. tarandus* L.) stock and carrying capacity of pastures in Finland during 1960–1990s. PhD Thesis, University of Oulu, 44 pp

Kumpula J (2001b) Winter grazing of reindeer in woodland lichen pasture: effect of lichen availability on the condition of reindeer. Small Ruminant Res 39(2):121–130

Kumpula J (2003) Effects of forest handling on reindeer pastures (in Finnish w/English summary). Fish and Game Report 286. Finnish Game and Fisheries Research Institute, Kaamanen, 60 pp

Kumpula J (1994) Inventory of natural fall and winter pasture resources with the help of LANDSAT-5 satellite imagery and their effect on reindeer herding in different reindeer herding zones (in Finnish). Liscenciate thesis, University of Oulu, Department of Zoology

Kumpula J, Colpaert A (2003) Effects of weather and snow conditions on reproduction and survival of semi-domesticated reindeer (*Rangifer t .tarandus*). Polar Res 22:225–233

Kumpula J, Colpaert A, Kumpula T, Nieminen M (1996) Pasture inventory of the northern part of the reindeer husbandry region (in Finnish). Fish and Game Research Institute, Reindeer Research Station, Kaamanen, and University of Oulu, Oulu

Kumpula J, Colpaert A, Nieminen M (1999) Summer pasture resources of the Finnish reindeer management area (in Finnish w/English summary). Finnish Game and Fisheries Research Institute, Kaamanen, 54 pp

Kumpula J, Colpaert A, Nieminen M (2003) Effects of forestry and snow conditions on the use of pastures by reindeer in the Ivalo herding district (in Finnish w/English summary). Fish and game report 271. Finnish Game and Fisheries Research Institute, Kaamanen, 39 pp

Kumpula J, Colpaert A, Kumpula T, Nieminen M (1997) Winter pasture resources of the Finnish reindeer management area (in Finnish w/English summary). Finnish Game and Fisheries Research Institute, Kaamanen, 42 pp

Kumpula J, Colpaert A, Anttonen M, Nieminen M (2004) The repeated reindeer pasture inventory in the northernmost part (13 districts) of the reindeer management area during 1999–2003 (in Finnish w/English summary). Finnish Game and Fisheries Research Institute, Helsinki, 39 pp

Laakso A (2002) The anomalies of contemporary reindeer herding management and supporting reindeer research: the need for a new paradigm. Pro gradu Thesis, University of Lapland, Rovaniemi, 127 pp

Lewis D (1998) Indigenous rights claims in welfare capitalist society: recognition and implementation: the case of the Sámi people in Norway, Sweden and Finland. Arctic Centre Reports 24. University of Lapland, Rovaniemi, 247 pp

Mascia M, Brosius JP, Dobson T, Forbes BC, Nabhan G, Tomforde M (2003) Conservation and the social sciences. Conserv Biol 17:649–650

Mathiesen SD, Haga ØE, Kaino T, Tyler NJC (2000) Diet composition, rumen papillation and maintenance of carcass mass in female Norwegian reindeer (*Rangifer tarandus tarandus*) in winter. J Zool 251:129–138

Müller-Wille L (1999) Human environmental interactions: issues and concerns in Upper Lapland, Finland. Arctic Centre Reports 26. Arctic Centre, University of Lapland, Rovaniemi, 82 pp

Nellemann C, Vistnes, I, Jordhøy P, Strand O (2003) Progressive impact of piecemeal infrastructure development on wild reindeer. Biol Conserv 113:307–317

Nellemann, C, Jordhøy P, Støen O-G, Strand O (2000) Cumulative impacts of tourist resorts on wild reindeer (*Rangifer tarandus tarandus*) during winter. Arctic 53:9–17

Nieminen M, Heiskari U (1989) Diets of freely grazing and captive reindeer during summer and winter. Rangifer 9:17–34

Nordhagen R (1943) Sikils valleys and Norway's mountain pastures (in Norwegian). Bergens Museums Skriften 22, 607 pp

Oksanen L (1978) Lichen grounds of Finnmarksvidda, northern Norway, in relation to summer and winter grazing by reindeer. Rep Kevo Subarctic Res Stat 14:64–71

Oksanen L, Virtanen R (eds) (1995) Geographical ecology of northernmost Fennoscandia. Acta Bot Fenn 153:1–110

Olli A (1995) Off-road traffic and possibilities to control it in the Pöyrisjärvi wilderness. In: Sippola A-L, Alaraudanjoki P, Forbes BC, Hallikainen V (eds) Northern wilderness areas: ecology, sustainability, values. Arctic Centre Publ 7. University of Lapland, Rovaniemi, pp 354–364

Olofsson J, Kitti H, Rautiainen P, Stark S, Oksanen L (2001) Effects of summer grazing by reindeer on composition of vegetation, productivity and nitrogen cycling. Ecography 24:13–24

Paine R (1994) The herds of the tundra: a portrait of Saami reindeer pastoralism. Smithsonian Institution, Washington, DC, 242 pp

Pelto P (1973) The snowmobile revolution: technology and social change in the Arctic. Benjamin/Cummings, Menlo Park, California, 225 pp (revised Waveland Press, Prospects Heights, IL 1987, 234 pp)
Raitio K (2000) Conflicts and power in land use management in the Sámi area in Finland (in Finnish). Pro gradu-tutkielma. University of Helsinki, 99 pp
Raitio K, Heikkinen H (2003) More initiative, less complaining. The views of administration of development of the institutions that are engaged in reindeer husbandry (in Finnish). Helsinki University of Technology Laboratory of Environment Protection Publication. Technology, Society, Environment 4/2003, 48 pp
Ruong I (1987) Sámi institute in Sámi peoples' future thoughts (in Sámi). Sámi instituhta, Kautokeino, 201 pp
Ruotsala H (2002) Changing reindeer herding districts. Living, work and environment in Kittilä´s Kyrö reindeer herding district and Kola peninsula's reindeer husbandry collective in years 1930–1995 (in Finnish). Ethnological archive 49. Antiquity Association of Finland, 471 pp
Russell DE, Martell AM, Nixon WAC (1993) Range ecology of the Porcupine caribou herd in Canada. Rangifer special issue no 8:1–168
Sámi Parliament (1998) The program of sustainable development of Sámi people. Prepared in co-ordination with Environmental Ministry (in Finnish). Accepted in the meeting of Sámi Parliament 28.03.1998. Sámi Parliament, Finland, 5 pp
Sara MN (1994) The pastures (in Sámi). In: Eira NI (ed) Being with the reindeer. DAT, Kautokeino, 151 pp
Sipilä P, Magga H, Aikio P (2000) Arboreal lichen inventory in Lappi paliskunta (in Finnish). Poromies 6:5–8
Skjenneberg S, Slagsvold L (1979) Reindeer husbandry and its ecological principles. Universitetsforlaget, Oslo
Skogland T (1986) Movements of tagged and radio-instrumented wild reindeer in relation to habitat alteration in the Snohetta region, Norway. Rangifer 1: 267–272
Skuncke F (1958) Reindeer pastures and their classification (in Swedish). Lappväsendet, Renforskningen, Uppsala, 204 pp
Svensk Rennäring (1999). Statistic Sweden, Halmstad, 149p
Telfer ES, Kelsall JP (1984) Adaptation of some large North American mammals for survival in snow. Ecology 65:1828–1834
Terry EL (2000) Winter habitat ecology of mountain caribou in relation to forest management. J Appl Ecol 37:589–602
Torp E (1999) Reindeer herding and the call for sustainability in the Swedish mountain region. Acta Borealia 1:83–95
Toupin B (1996) Effect of insect harassment on the behaviour of the Riviére George caribou. Arctic 4:375–382
Tuomilaakso M (2001) The views of reindeer herders and reindeer researchers oppose each other again. Depressed in Skalluvaara's round-up area (in Finnish). Lapin Kansa 25.10.2001
Tuomi-Nikula H (2001) Overgrazing threatens reindeer management (in Finnish). Lapin Kansa 30.12.2001, p 4
Turi JM (2002) The world reindeer herder livelihood: current situation, threats and possibilities. In: Kankaanpää S, Müller-Wille L, Susiluoto P, Sutinen ML (eds) Northern timberline forests: environmental and socio-economic issues and concerns. Finnish Forest Research Institute, Jyväskylä, pp 70–75
Tyler NJC (1991) Short-term behavioural responses of Svalbard reindeer *Rangifer tarandus platyrhynchus* to direct provocation by a snowmobile. Biol Conserv 56:179–194
Usher PJ (2000) Traditional ecological knowledge in environmental assessment and management. Arctic 53:183–193

Vistnes I, Nellemann C (2001) Avoidance of cabins, roads, and power lines by reindeer during calving. J Wildlife Manage 65:915–925

Vistnes I, Nellemann C, Jordhøy P, Strand O (2001) Wild reindeer: impacts of progressive infrastructure development on distribution and range use. Polar Biol 24:531–537

Väre H, Ohtonen R, Mikkola K (1996) The effect and extent of heavy grazing by reindeer in oligotrophic pine heaths in northeastern Fennoscandia. Ecography 19:245–253

Warenberg K, Danell Ö, Gaare E, Nieminen M (1997) Reindeer pasture plants (in Finnish). Ladbruksforlaget, Helsinki, 112 pp

Weladji RB, Forbes, BC (2002) Disturbance effects of human activities on *Rangifer tarandus* habitat: implications for life history and population dynamics. Polar Geogr 26:171–186

8 Very High Resolution Remote Sensing Data in Reindeer Pasture Inventory in Northern Fennoscandia

T. Kumpula

8.1 Introduction

Reindeer (*Rangifer tarandus*) management is one of the oldest forms of land use and livelihood in northern Fennoscandia. During the past four decades reindeer husbandry has gone through significant changes. The most remarkable change in the study area came in the mid 1950s when a reindeer fence three meters high was built along the Finnish–Norwegian border in Jauristunturit. Originally, during the 19th century, the closing of national borders was due to the border legislation agreement between Russia (of which Finland was then a part), Sweden, and Denmark (of which Norway was then a part). Finally in 1852 Sámi herders were forbidden to cross the national border with their reindeer and practice traditional herding migration from coastal summer pastures to inland winter pastures. They were now forced to choose to be either "Finnish" or "Norwegian" Sámi herders (Näkkäläjärvi 2003; Pennanen 2003).

Motorization, fencing and settlement patterns and practices in the late 1960s changed the traditional herding culture. The development led to a system whereby during part of the year reindeer were allowed to graze freely inside the reindeer fences that now surround the northern herding districts of Finland. In Norway, the traditional migration from coastal summer pastures to inland winter pastures has continued. The reindeer population increased drastically, especially in the 1980s, and as a result of the grazing and trampling of the large herds the winter pastures deteriorated significantly. This, coupled with a series of hard winters simultaneously contributed to the starvation of reindeer. These problems with reindeer and their pastures came to be controversial topics during the 1990s (Heikkinen 2002; Pennanen 2003; Chaps. 1, 3, 9). A need was appreciated for spatial knowledge about the quantity and quality of the pastures. Remote Sensing and Geographical Information System

Ecological Studies, Vol. 184
B.C. Forbes, M. Bölter, L. Müller-Wille, J. Hukkinen, F. Müller, N. Gunslay, Y. Konstatinov (Eds.)
Reindeer Management in Northernmost Europe

(GIS) methods, together with botanical pasture evaluations, have been found to provide means of producing reliable inventories of the extensive reindeer herding area in northern Fennoscandia.

The reindeer pastures of Finland and Norway were studied and classified by using remote sensing and ground measurements in the 1990s. The Finnish and Norwegian reindeer herding areas (123,000 and 90,000 km^2 respectively) were inventoried with Landsat TM-5 images (Colpaert et al. 1995, 2003; Johansen et al. 1995; Kumpula et al. 1997). Pasture inventories of caribou, musk oxen and elk have been done also in arctic Alaska and Canada since the 1970s. Landsat and SPOT images were the data source in these studies (George et al. 1977; Adams 1978; Thomson and Klassen 1980; Ferguson 1991; Pearce 1991; Arsenault et al. 1997; Hansen et al. 2001; Théau et al. 2005). Also in Russian Artic, in Nenets autonomous okrug, Rees et al. (2003) studied changes of reindeer pastures using Landsat TM data. Landsat TM and SPOT images seem to be suitable for large-scale inventories, but the coarse resolution (30 m multispectral resolution) hinders detailed study of vegetation cover and types of land use.

The latest very high resolution (VHR) satellite systems were launched at the turn of the millennium. IKONOS-2 and Quickbird-2 represent the VHR satellite systems that produce multispectral images with less than five meter pixel size. Both systems also have panchromatic sensors that have a resolution of one meter or less. The spatial cover of VHR satellite images is 80–100 km^2. The resolution of the new VHR images is very close to that of aerial photographs, where it is usually less than one meter. The spatial cover of aerial photographs varies from approximately 5 × 5 to 20 × 20 km.

8.1.1 Aims of the Research

The first goal of this research was to make a detailed classification of reindeer pastures in the Jauristunturit area. Scientists and reindeer herders chose the area, which includes both winter and summer pastures, because of mutual interest (Chap. 1).

The second aim was to investigate the suitability of the new IKONOS-2 images for reindeer pasture classification in general. The study focused on the vegetation types detectable, the accuracy of the classifications, and the benefits of VHR remote sensing.

The third aim was to detect if it is possible to find out and classify the intensity of reindeer trampling and grazing in Jauristunturit by using IKONOS-2 images. Special focus was on the lichen forage sources of the winter pastures. The aim was to test if it is possible to estimate lichen biomass from IKONOS-2 image on the basis of coupled image classification and field measurements.

8.2 Research Area

The Jauristunturit research area is located partly in the Finnish Näkkälä reindeer herding district and partly in the Norwegian West Finnmark reindeer herding area (68°45' N, 24°00' E; Fig. 8.1). The area of the Finnish Näkkälä herding district is 3,500 km^2. In Näkkälä there were 8,700 reindeer in 2002 (2.5 per km^2 Paliskuntain yhdistys 2003). The Norwegian West Finnmark reindeer herding cooperative has an area of 24,000 km^2 and 60,000 reindeer (2.5 per km^2). The minimum size of the IKONOS-2 image formed a research area of 100 km^2.

The research area is divided by a fence three meters high, and it mainly follows the national border of Finland and Norway. This fence follows the water divide of the Gulf of Bothnia and the Arctic Ocean. The Norwegian part of the area is used only as a winter pasture for a few months per year, whereas the Finnish part is used as an early summer pasture and a reserve winter pasture. Reindeer management is the main form of land use on both sides of the fence.

The Jauristunturit area is situated in the hemiarctic zone (Oksanen and Virtanen 1995). Mean annual temperature is 1.5 °C and annual precipitation is less than 450 mm. The area has the most continental climate in Fennoscandia (Helminen 1987). The topography of the area is relatively smooth. Elevation varies from 350 to 588 masl.

The Jauristunturit area is treeless tundra dominated by dwarf shrubs (*Betula nana*, *Empetrum nigrum*, *Vaccinium vitis-idea*, *Vaccinium myrtillus*) and lichen heaths (*Cladina* sp.). In more moist habitats, lichens are replaced by mosses (*Polytrichum* sp., *Dicranum* sp., *Pleurozium schreberi*) and grasses (*Deschampsia flexuosa*, *Festuca ovina*). Scattered individuals of mountain birch (*Betula pubescens* subsp. *tortuosa*) can be found. Juniper (*Juniperus communis*) is the most common conifer. Some isolated Scots pine (*Pinus sylvestris*) can be found even above 500 meters. Peatlands in the area belong to the Northern palsa aapa mire zone (Eurola and Vorren 1980).

Reindeer lichens (*Cladina* sp.) and (*Cetraria* sp.) are important fodder for the reindeer in winter when no green vegetation is available. Reindeer can dig lichens out from under almost one meter of snow (Helle 1980). In treeless tundra – as in Jauristunturit – wind accumulates snow in the depressions and gets packed hard. In winter pastures with more than meter-thick snow cover, it is impossible for reindeer to graze.

8.3 Material

8.3.1 Field Data

Field work was carried out during the summers of 2001, 2002, and 2003. Each test site inventoried for the classification represented a homogenous land cover type. The size of the test sites was set at a minimum of 20 m × 20 m. This size represents about 30 IKONOS-2 multispectral pixels. However, some test sites were smaller. For example, some exposed sand and gravel patches on glaciofluvial terrain were selected as individual test sites even though they were less than 10 m × 10 m. The test sites were selected randomly although all the main vegetation types and land use patterns in the research area were included in the field data.

On each test site, the vegetation and pasture type as well as the intensity of the grazing were estimated. In dry biotopes five to ten vegetation quadrants (50 cm × 50 cm) were inventoried and in wet biotopes five vegetation quadrants. On these sites, the cover of plant species was defined in percentages to the lowest possible taxon. Bare soil cover, aphytic humus layer cover, and the average height of the main shrubs were measured. Reindeer lichens were inventoried with special attention paid to estimating the percentages of cover and average height of each species. The intensity of trampling/grazing was estimated as slight, moderate, or heavy according to the cover percentages and height of reindeer lichens (*Cladina* sp.) and cover of bare exposed mineral ground. The coordinates of the test sites were measured with GPS.

8.3.2 Remote Sensing and GIS Data

The remote sensing platform used in this study was IKONOS-2. Totally cloud-free and good-quality multispectral imagery was acquired on 28 June 2001 (Fig 8.1). In 2001, June was very warm and by the end of June the growing season was already at its peak.

The IKONOS-2 satellite was launched 29 September 1999. IKONOS-2 has sun-synchronous orbit at an altitude of 681 kilometers. The IKONOS-2 sensor can acquire four-meter resolution multi-band (blue, green, red, and infra-red) images (Table 8.1). Data are collected in 11-byte format with 2,048 gray levels (Dial et al. 2003).

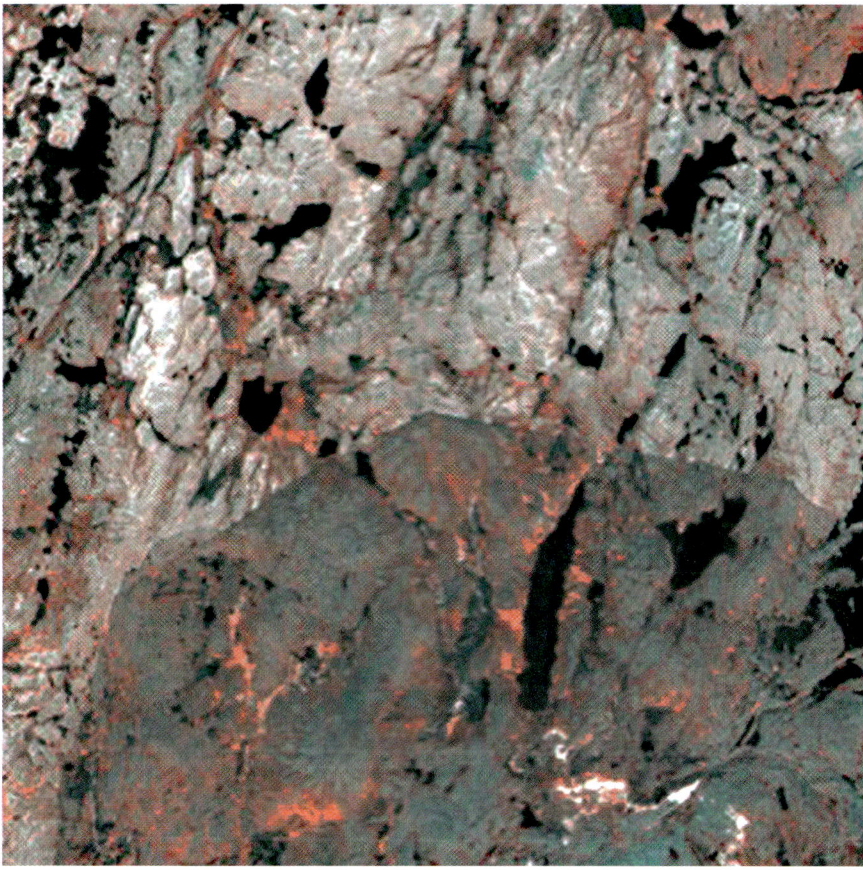

Fig. 8.1. IKONOS-2 false color image from Jauristunturit, acquired 28 June 2001. The reindeer fence on Finnish–Norwegian border is visible because of higher reindeer lichen coverage in Norway (*upper part* of the image). Clear white areas *down right* must not be confused with lichen areas; they represent aeolian blowouts and sand patches on Pöyrisharju esker formation. Image coverage is 10 km^2

Table 8.1. The spectral bands of IKONOS-2 and resolution at nadir

Spectral band	Wavelength μm	Resolution (at nadir)
1 (Blue)	0.455 – 0.516	4 m
2 (Green)	0.506 – 0.595	4 m
3 (Red)	0.632 – 0.698	4 m
4 (NIR[a])	0.757 – 0.853	4 m
Panchromatic	0.45 – 0.90	1 m

[a] Near infrared

8.4 Methods

8.4.1 Lichens as Indicators of Grazing and Trampling Intensities

Reindeer lichens are mostly of white and yellow-white color. To the human eye the separation of lichens from other vegetation is relatively easy. Differences in lichen cover and height can be detected accurately in the field by an experienced field worker. In the field of remote sensing of reindeer pastures, lichens offer an interesting field of study. In dry boreal forest and arctic tundra habitats lichens can cover up to 90 % of the terrain. As lichens are an important source of winter fodder for reindeer, their occurrence in certain biotopes can be an indicator of the condition of pastures. Reindeer lichens (*Cladina stellaris*, *Cladina mitis*, *Cladina rangiferina*, *Cetraria nivalis*, *Cladonia uncialis*) are also sensitive to both zoogenic and anthropogenic trampling. The spectral reflectances of lichens and green vegetation are different. Sollheim et al. (2000) studied reflectances of *Cladina stellaris* and *Cetraria nivalis* and proved that spectral differences allow separation of lichen-rich and lichen-poor biotopes. This promotes the utility of lichens in reindeer pasture studies and also in environmental pollution studies.

The vegetation types of the Jauristunturit were classified into reindeer pasture classes on the basis of the literature and field work observations (see Haapasaari 1988; Oksanen and Virtanen 1995).

On dry lichen heaths in the Jauristunturit the differences in the grazing and trampling pressure between Norway and Finland are clearly visible both in the field and remote sensing images. The reindeer fence on the border divides the two grazing systems where the Norwegian pastures are used during the winter period February–April. (Fig. 8.2) The Finnish pastures have been used as early summer pastures and formerly also as winter pastures.

The five reindeer lichens are important winter fodder for reindeer. In the Jauristunturit the different grazing and trampling intensities were studied by measuring the cover and height of the lichen species. A high percentage of bare soil or aphytic humus layer also indicated intensive grazing and trampling.

8.4.2 Digital Image Processing and Accuracy Assessment

The ERMapper 6.3 image processing software (ERMapper, West Leederville, Australia) was used for the image processing. The image was rectified into the Finnish coordinate system using digital maps produced by Finnish National Survey, and ground control points were collected from the field. The rectifications' RMS error was less than 1.5.

Fig. 8.2. Reindeer grazing trail on *Cladina stellaris* type lichen heath on the Norwegian side of Jauristunturit. This lichen has a coverage of 85 percent and height of 55 mm. Grazing has occurred through approximately 50–80 cm of deep snow. Width of the picture on the ground is about 1.5 m in the lower part, width of the reindeer grazed track is about 30 cm

Supervised classification methods were used in creating the classification. Field-test sites were displayed on top of RGB false color image composition, and a suitable training area for the classification was marked for each class. ERMapper's minimum distance and maximum likelihood methods were used in the classification. Classifications were first visually evaluated and then imported to the ArcGIS 8.3 (ESRI, Redlands, CA, USA) program. There the accuracy of the classification was tested by using field test sites, and the field classification of the field test sites was compared with image classification.

8.5 Results

8.5.1 Classification of Reindeer Pasture Types

8.5.1.1 Winter Pastures

In general, the lichen/dwarf shrub pastures of the Jauristunturit can be divided into five main classes, which correspond to different grazing and trampling intensities. The reindeer lichen cover and height are the main parameters used to distinguish between classes of grazing and trampling intensities. These classes represent the major dry biotopes of the Jauristunturit that are typically lichen dominated.

1. *Cladina stellaris* lichen heath. In *Cladina stellaris* lichen heath, the grazing and trampling intensities are the lowest. This type of heath is found only on the Norwegian side, and *Cladina stellaris* and *Cetraria nivalis* are the main species and their cover was over 75 %. The average height of *Cladina stellaris* is ≥ 5 cm and for *Cetraria nivalis* it was found to be ≥ 4 cm. This biotope also includes shrubs, especially *Betula nana* (cover 10–40 %, height 25–50 cm), *Empetrum nigrum* (1–15 %, 1–8 cm), *Vaccinium vitis-idea* (1–5 %), and *Vaccinium myrtillus* (0–5 %).
2. *Cetraria nivalis* lichen heath. Intermediate grazing and trampling intensities are found mainly on the Norwegian side. In this pasture type, the dominant lichen species are *Cetraria nivalis* and *Cladina stellaris*. *Cetraria nivalis* cover and height averaged 30–40 %, 2.5–4 cm, respectively. *Cladina stellaris* cover was 15–25 % and height was about 3.5 cm. Shrubs of this type are *Betula nana* (10–30 %, 15–35 cm), *Empetrum nigrum* (1–15 %, 1–5 cm), *Vaccinium vitis-idea* (1–5 %), and *Vaccinium myrtillus* (0–1 %).
3. *Cetraria nivalis/Empetrum nigrum* heath. Intermediate grazing and trampling intensities are found also on the Finnish side of the fence. This type is mainly located in an area where the reindeer fence was moved 250 meters northward about eight years ago. This former Norwegian pasture is

now part of the Finnish grazing system. The total area of this "new" Finnish pasture land is about 220,000 m^2 (22 ha). Here, *Cetraria nivalis* and *Cladina stellaris* are the main lichen species. The total cover of reindeer lichens can be as high as 45 %, *Cetraria nivalis* cover being about 20–30 %, height 2–3 cm, with *Cladina stellaris* cover 10–20 % and height 2–3 cm. The most common shrubs in this vegetation type are (cover 10–40 %, height 15–35 cm), (1–15 %, 1–5 cm), *Vaccinium vitis-idea* (1–5 %), and *Vaccinium myrtillus* (0–1 %).

4. *Empetrum nigrum*-dominated lichen heath. *Empetrum nigrum* dominated lichen heaths were seen to be associated with high grazing and trampling intensity. This type is mainly found on the Finnish side. *Empetrum nigrum* type heaths are located in very dry places usually on the tops and acclivities of hills and ridges. Reindeer lichen covers 10–20 %, and as high as 30 % in some places. The main reindeer lichen species are *Cetraria nivalis* and *Cladina mitis*. More lichens of the *Cladonia* genus were found here than in the previous heaths. This indicates higher grazing and trampling pressure, which *Cladonia* species seem to tolerate better. The cover of bare soil was also found to be higher than in the preceding three (1–3 %). *Empetrum nigrum* is the most dominant shrub species here; other shrubs are *Betula nana*, *Phyllodoce caerulea* and *Vaccinium vitis-idea*.
5. *Cetraria nivalis*/*Cladonia* sp. heath. The main reindeer lichen species are *Cetraria nivalis* (cover 15 %, height 1–1.5 cm) and *Cladina mitis* (5 %, 1–2 cm). *Cladonia* species cover is the highest in this type of heath, 5–10 %. This type is typical on the Finnish side on the tops of hills and ridges. In the areas that are more intensively grazed and trampled, the crust lichens and ground lichens become more dominant. *Sphaerophorus globularis* lichen is one of the species that can only be found in these more intensively grazed areas. Bare ground cover was noted at about 1–5 %. Dominant shrubs are *Empetrum nigrum* (cover 10–35 %) and *Betula nana* (15–30 % and 10–25 cm)
6. *Betula nana*/*Vaccinium myrtillus* heath. This type is more moist than the lichen heath types. Reindeer lichen cover was still found at about 10 %. The main reindeer lichen species are *Cladina stellaris*, *Cladina mitis* and *Cladina rangiferina*. *Betula nana* is the most dominant shrub (cover 25–60 %, height 30–60 cm), other dominant shrubs being *Vaccinium myrtillus*, *Phyllodoce caerulea*, *Vaccinium vitis-idea* and *Empetrum nigrum*. Moss cover can be over 60 %. Grasses such as *Deschampsia flexousa* and *Festuca ovina* as well as sedges (*Carex* spp.) become more dominant than in other heath types. Forbs were also found in this heath, for example, *Solidago virgaurea* and *Pedicularis lapponica*.

8.5.1.2 Summer Pastures

Reindeer summer fodder consists mainly of green vegetation such as grasses, sedges, forbs, and shrubs. In summer pastures the percentage of lichens is naturally low, not due to grazing or trampling. Summer pastures are located in valleys, moist slopes, riparian zones, and peatlands.

7. *Betula nana* dominated heath. Compared with the previous types, this vegetation type contains more moisture. Grasses and forbs are clearly more plentiful, and only few reindeer lichens are to be found. *Betula nana* is the most dominant shrub (cover 30–80 %, height 30–70 cm); other shrubs are *Vaccinium myrtillus*, *Phyllodoce caerulea*, *Vaccinium vitis-idea*, and *Empetrum nigrum*. Moss cover can be over 80 %. *Deschampsia flexuosa* and *Festuca ovina* cover 5–20 %. Forb cover is 5–20 % (*Solidago virgaurea*, *Pedicularis lapponica*, *Bartsia alpina*, *Cirsium helenioides*, *Equisetum* spp., *Epilobium* spp., *Galium boreale*). This type of heath represents suitable summer pasture for reindeer.
8. Wet peatland. This type is open mire with *Carex* spp., *Sphagnum* spp., *Eriophorum angustifolium* and other species adapted to wet biotopes.
9. Dwarf shrub peatland with hummocks. This type is hummocky (formed by frost activity) with dwarf shrub, mainly *Betula nana*. Usually this peatland type is found in the marginal zone between open and wet peatlands. Hummocks can be from 30 cm to over one meter high. Shrub cover varies from 15 to 60 %. In some places areas between hummocks are wet, or they can be bare mineral ground.
10. Willow-dominated peatland. Peatland dominated by willow (*Salix* sp.), and mainly found on the Norwegian side of the fence, can also be quite hummocky. The main shrub species are *Salix glauca*, *S. phylicifolia*, and *Betula nana*. In the densest places, Salix cover can be 70 % with a height of 60–80 cm. On the Finnish side this type has been grazed quite extensively and Salix has been reduced (see Chap. 9). Cover of *B. nana* is greater where willow cover has been reduced. Some forbs and grasses can be found in this type (*Solidago virgaurea*, *Pedicularis lapponica*, *Bartsia alpina*, *Cirsium helenioides*, *Equisetum* spp., *Epilobium* spp., *Galium* spp.).

8.5.1.3 Other Classes

11. Sand and gravel areas. The southeastern part of the image covers a small part of the Pöyrisjärvi esker formation. In Näkkälä, it is typical that in the vicinity of eskers there are dune fields originally developed after the glacial retreat. Eskers with glaciofluvial material and dunes with aeolian material are very vulnerable to erosion. In old dune fields there are some relatively large patches of 5–15 ha which are pure sand. Smaller sand

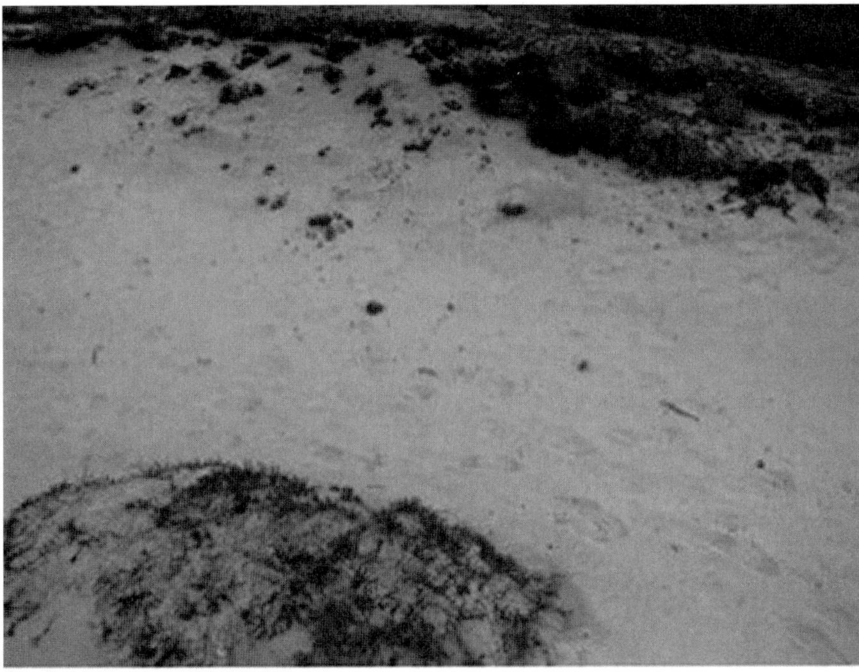

Fig. 8.3. Aeolian sand blowout patch in the southeastern part of the research area on the northern end of Pöyrisharju esker

patches are also typical in dune areas (Fig. 8.3). On eskers, where the material is coarser sand, gravel patches are smaller. Sand and gravel patches are mainly found on the Finnish side because on the Norwegian side there are not so many glaciofluvial and aeolian deposits.

12. Water. Lakes and rivers and some parts of very wet peatlands comprise this type.

The results of the IKONOS-2 image classifications indicate that in the Jauristunturit area on the Finnish–Norwegian border it is possible to distinguish five classes of reindeer lichens, which correspond to grazing and/or trampling intensities. Especially on the Norwegian side, the difference between the ungrazed and heavily grazed lichen-dominated vegetation type is clear (Fig. 8.4). In the depressions on the Norwegian side where snow accumulates to over 1 m lichen cover is hardly grazed. Where the snow cover is thinner, the grazing pressure is higher. On the Finnish side, snow accumulation did not make such a big difference to lichen cover and height because of the effects of summer trampling.

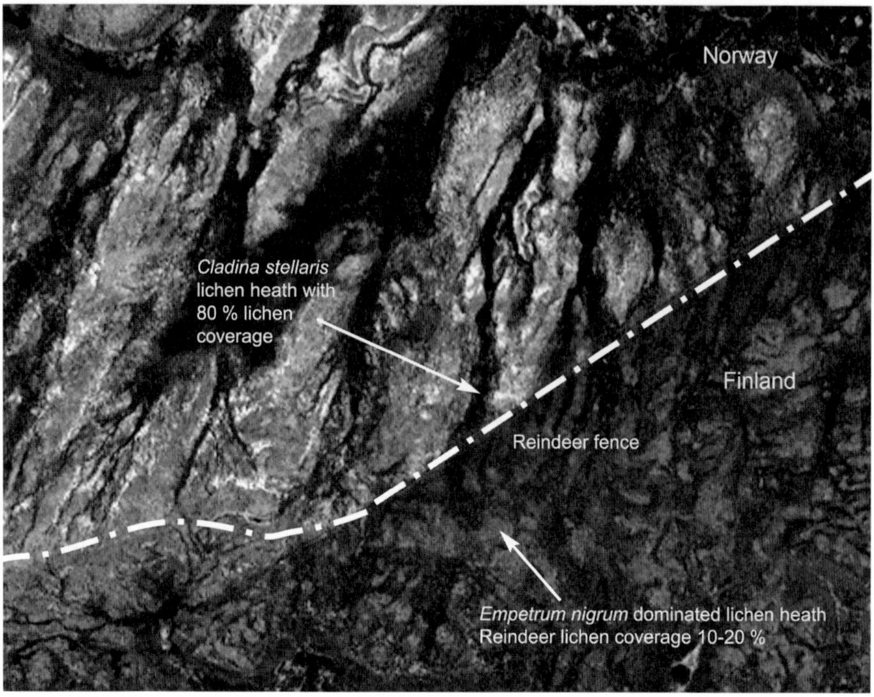

Fig. 8.4. IKONOS-2 zoom showing lichen coverage differences on Norwegian and Finnish side of the research area. Image (zoom area) covers 1.5 × 1.1 km

8.5.2 Accuracy Assessment

The accuracy assessment was done using 219 test sites collected from the field. Each test site classification was compared with the image classification manually on the computer screen. To avoid possible coordinate errors due to the rectification process and GPS measurements, each test site was checked by using a 3 × 3 grid. When more than half of the nine pixels were the same as the field classification, the image classification of that test site was considered accurate. Water class was tested by randomly checking five water points from the topographic map.

Field data points were not equally distributed between the classes which meant that some of the classes had just a few sites and some more than 20. Accuracy estimation is more valid when the number of samples is high. During the field work, field sites were chosen semi-randomly, but the aim was to cover all possible sites with a similar number of test sites. In practice, the infrequent occurrence of some reindeer pasture types limited the number of test sites. The reindeer pasture types that have only a few test sites are spatially insignificant classes and represent less than three % of the total image area.

The final accuracy achieved was 88 % (Table 8.2). If the accuracy %age is 80 or more the endeavor can be evaluated as satisfactory. The sand and gravel class was classified with 100 % accuracy. Many of the field test sites were quite small with a size of just a few pixels. This is because sand and gravel reflectance is very strong and differs notably from the surrounding vegetation.

All lichen-dominated winter pasture types were classified with an accuracy of 83 % or more. The lowest accuracy of lichen types was in the *Cetraria*

Table 8.2. Accuracy of the classification

	Class	1	2	3	4	5	6	7	8	9	10	11	12	Total	User's Accuracy (%)
1	Willow (*Salix* sp.) dom. peatland	10	1				2							13	77
2	*Betula nana* dominated heath	1	24		3		1							29	83
3	*Cetraria nivalis/Empetrum nigrum* heath			9		1				2				12	83
4	*Betula nana/ Vaccinum myrt.* heath	1	2		17									20	85
5	*Empetrum nigrum* dom. lichen heath			1		22					1			24	91
6	Dwarf shrub peatland with hummocks	2	2				16							20	80
7	Wet peatland							14	1					15	93
8	Water								5					5	100
9	*Cetraria nivalis/ Cladina* sp. heath									12	1			13	92
10	*Cetraria nivalis* lichen heath									1	11	1		13	84
11	*Cladina stellaris* lichen heath									1	1	20		22	90
12	Sand and gravel areas												17	17	100
	Total	14	29	10	20	23	19	14	6	15	14	21	17	203	
	Mean producer's accuracy	71	83	90	85	93	84	100	83	80	78	95	100		88

nivalis/Empetrum nigrum heath type, 83 %. This type was mainly found in the area where the fence was moved in 1994–1995 and was a spatially small-scale class with only 18 field test sites. In this type, lichen cover and height values were between the values in *Empetrum nigrum* and *Cetraria nivalis* lichen heaths, where three field test sites were misclassified. The rest of the reindeer lichen classes were classified higher and they also had a greater number of field test sites.

Dwarf shrub peatland with hummocks (76 % accuracy) and Salix dominated peatland (77 %) classes were not very satisfactorily classified. This was because they were partly mixed together and also misclassified as *Betula nana*-dominated types. It was expected that dwarf shrub peatland with hummocks has vegetation similar to that of a *Betula nana*-dominated heath. *Betula nana/Vaccinium myrtillus* heath (85 %) and *Betula nana*-dominated heath (83 %) were also mixed together.

8.5.3 Reindeer Lichen Biomass on Winter Pasture Types

Calculation was based on field measurements of reindeer lichen cover and height. The biomass of reindeer lichens was estimated by using a formula developed by Colpaert et al. (2003).

$lbm = (0.61434 \times lc \times lh) + 0.000038075 \times lh2 \times lc2$
lbm = lichen biomass (kg ha^{-1} dry mass)
lc = lichen cover in %
lh = lichen height

Reindeer lichen biomass in winter pasture types varied from 225 to 3,436 kg ha^{-1} (Table 8.3). *Cladina stellaris* lichen heath was the most lichen-rich type. This type was only found on the Norwegian side. *C. stellaris* lichen heath has the highest cover and height values where the snow cover seems to be thickest. But even in the “best” places there are scattered signs of grazing. This is probably due to the fact that snow cover varies from year to year. In some years when snowfall is less than average, parts of usually deep snow areas are available for reindeer to dig and graze.

The *Cetraria nivalis/Cladonia* sp. type lichen heath had the lowest value of lichen biomass. This type is mainly found on the Finnish side, but some patches were also present on the Norwegian side. This occurs typically on the highest points of hills and ridges. These are the spots that reindeer graze first every winter because of the thin snow cover.

Table 8.3. Reindeer lichen biomass values of pasture types

	Class	Lichen height mm	Lichen coverage %	Lichen biomass kg/ha
1	Peatland (*Salix*)	28	1	17.2
2	*Betula nana* dominated heath (fresh)	28	5	86.6
3	*Cetraria nivalis* (FIN intermediate)	28	45	833.3
4	*Betula nana/Vacc. myrt.* heath	23	10	143.1
5	*Empetrum* dom. lichen heath	20	30	381.7
6	Dwarf shrub peatland with hummocks	0	0	0.0
7	Wet peatland	0	0	0.0
8	Water	0	0	0.0
9	Intensive grazed *Cetraria nivalis* heath (FIN)	18	20	225.8
10	*Cetraria nivalis* lichen heath (NOR)	35	60	1,456.1
11	*Cladina stellaris* lichen heath (NOR)	55	80	3,436.1
12	Sand	0	0	0.0

8.6 Discussion

8.6.1 IKONOS-2 Image Classification and Accuracy

The aim of the research was to study the utility of IKONOS-2 data in reindeer pasture inventory. Firstly, the IKONOS-2 multispectral image with four-meter resolution from the Jauristunturit area was of good quality. Four-meter resolution image is comparable to an aerial photograph, and during field work it was used as a map and in planning each day's work. In comparing the IKONOS-2 multispectral with a typical aerial photograph with a 63-centimeter resolution, the IKONOS-2 image was visually interpreted to be less affected by shadows. The terrain in the Jauristunturit is relatively smooth, except for some ridges and eskers.

Already the visual interpretation of the image gave a good idea of the research area's vegetation types and field conditions. It was obvious that lichen distribution could be studied. For the classification image, 280 field test sites were collected. The reindeer pasture types were developed to present the most important pasture types. Twelve classes were divided into winter and summer pastures, and additional classes such as water and sand were included. The classification was an overall success, with 88 % accuracy. Peatland classes were the most misclassified, however, the study focused more on winter pasture types with reindeer lichens. Reindeer lichen classes were all classified satisfactorily.

8.6.2 Grazing and Trampling Intensities in the Jauristunturit

It was possible to distinguish five different lichen pasture types. In the Jauristunturit treeless tundra, the reindeer fence between Norway and Finland allows the investigation of grazing and trampling intensity. In this matter lichen cover and height were used as indicators of grazing and trampling. As a result of the field work and IKONOS-2 image classification, five grazing intensities were identified. Lichen biomass estimations also gave results that indicated grazing and trampling intensities. Generally, the pastures on the Finnish side were more grazed and trampled and the lichen biomass values were significantly lower than in Norway.

What are the reasons for such a disparity? The two main ones are: different management of pastures and traditions for their use, and different snow conditions in regard to protection of the lichen. Norwegian herders use the area only in late winter for a few months and there is no trampling of lichen pastures in the summer. The area is so remote that there is no significant activity other than reindeer herding. The Finnish side of the Jauristunturit have been used for a long time as an early summer pasture (Chap. 9). In the period of free grazing in the late 1970s to mid-1990s, some of Näkkälä's reindeer also grazed in the Jauristunturit in the winter. In summer, reindeer graze on peatlands, moist biotopes and, when moving from place to place, they cross lichen-rich winter types and trample them. Especially in summer and when it is dry, reindeer lichen is easily damaged by trampling. Winter grazing and summer trampling in the Finnish Jauristunturit have degraded lichen pastures more than those pasture types that persist in Norway. The loss of visible lichen cover in hollows may be attributed to actual consumption (grazing), versus trampling on exposed ground, or a combination of both.

Summer grazing has also affected willows (*Salix glauca* and *Salix phylicifolia*) on the Finnish side of the fence (Chap. 9). This is clearly visible in some points along the reindeer fence and it is also visible in the IKONOS-2 image. Naturally, this is due to the fact that on Norwegian side there are no reindeer to graze willows in summer. There was no difference in cover of *Betula nana* and other shrubs between the Finnish and Norwegian Jauristunturit.

Especially in the Norwegian area, the distribution of snow affects the lichen species cover and height. Snow cover varies from year to year but it seems that some areas have thicker snow cover every year and in certain areas, such as ridges, snow cover is relatively thin every year. Snow is already accumulating in valleys and small depressions in early winter. When Norwegian reindeer come to the area in February they do not dig and graze in places where the snow depth is over one meter, and in general they minimize their digging efforts (Helle 1980). The result of that is that these depressions are generally not grazed year after year. If in the depressions snow depth is over 1.5–2 m, the snow melts very late in summer, and those places are not favoured by reindeer lichens and are represented by snowbed vegetation (*Salix herbacea*, *Cassiope hypnoides*, etc.).

8.6.3 Very High Resolution Remote Sensing Data in Reindeer Pasture Inventory

Very High Resolution images are a useful but expensive data source for pasture inventories, but can be recommended for the investigation of small areas. VHR allows more detailed study of pastures than Landsat or SPOT. For example, VHR data combined with field measurements could be used in monitoring pasture conditions. Together with new satellite VHR images, old aerial photographs could be used in monitoring the development of pasture conditions. Allard (2003) used an IKONOS-2 image printout from 2000 and aerial photograph from 1975 in a study of vegetation degradation due to the grazing and trampling on Swedish mountainous heaths and achieved good results in the detection of patches of erosion. Stow et al. (2004) used IKONOS-2 images as a new data source in change detection studies on arctic tundra. The capability of VHR as a method of detecting grazing and trampling intensities should be further investigated. IKONOS-2 images have great potential, especially in studying lichens with respect to grazing, trampling, and pollution. IKONOS-2 also has great potential for use in arctic areas where no aerial photographs are available. With an image resolution of 2–4 m it is almost possible to detect the exact pixel of the field site with GPS.

Small spatial cover is the main limiting factor in the application of VHR images. They cannot be used, for example, in studying the entire Finnish or Norwegian reindeer pasture areas. VHR images could be used as reference material when classifying larger scale images (Landsat TM, SPOT), and in those studies VHR images could best be used for more detailed work, as for example in the study of very critical pasture areas.

8.7 Conclusions

The main thrust of this research was testing the Very High Resolution IKONOS-2 image in classification of reindeer pastures. The results indicate success in this effort in the Jauristunturit research area. With IKONOS-2 image it was possible distinguish 12 classes in treeless tundra. In the investigation of reindeer grazing and trampling intensities it was possible to detect five different grazing/trampling intensities. IKONOS-2 has great potential in lichen pastures studies in circumpolar arctic areas. Small spatial cover and high costs are the disadvantages of IKONOS-2 and other VHR data usage.

References

Adams GD (1978) Remote sensing for wildlife habitat analysis in Canada, an overview. Pecora II Symp, Sioux Falls, SD, USA

Allard A (2003) Detection of Vegetation Degradation on Swedish Mountainous Heaths at an Early Stage by Image Interpretation. Ambio 38:510–519

Arseneault D, Villeneuve N, Boismenu C, Leblanc Y, Deshaye J (1997) Estimating lichen biomass and caribou grazing on the wintering grounds of northern Quebec: an application of fire history and Landsat data. J Appl Ecol 34:65–78

Colpaert A, Kumpula J, Nieminen M (1995) Remote sensing as a tool for reindeer range land management. Polar Record 31(177):235–244

Colpaert A, Kumpula J, Nieminen M (2003) Reindeer pasture biomass assessment using satellite remote sensing. Arctic 56(2):147–158

Dial G, Bowen H, Gerlach F, Grodecki J, Oleszczuk R (2003) IKONOS satellite, imagery, and products. Remote Sensing Environ 88:23–36

Eurola S, Vorren K-D (1980) Mire zones and sections in Northern Fennoscandia. Aquilo Ser Bot 18:10–15

Ferguson RS (1991) Detection and classification of muskox habitat on Banks Island, Northwest Territories. Using Landsat Thematic Mapper Data. Arctic 44 [Supp 1]:66–74

George TH, Stringer WJ, Preston JE, Fibich WR, Scorup R (1977) Reindeer range inventory in western Alaska from computer-aided digital classification of Landsat data. Proceedings 28th Alaska science conference, Anchorage, Alaska, pp 33–42

Haapasaari M (1988) The oligotrophic heath vegetation of northern Fennoscandia and its zonation. Acta Bot Fenn 135:1–219

Hansen MJ, Franklin SE, Woudsma CG, Peterson M (2001) Caribou habitat Mapping and fragmentation analysis using Landsat MSS, TM and GIS data in the North Columbia Mountains British Columbia, Canada. Remote Sensing Environ 77:50–65

Heikkinen H (2002) Sopeutumisen mallit – Poronhoidon adaptaatio jälkiteolliseentoimintaympäristöön Suomen läntisellä poronhoitoalueella 1980–2000 (only in Finnish). Doctoral thesis. University of Oulu, SKS, 369 pp

Helle T (1980) Habitat and food selection of the wild forest reindeer (*Rangifer tarandus fennicus lönn*) and semi-domestic reindeer (*Rangifer tarandus tarandus L.*) in Finland. Acta Universitatis Ouluensis, Series A, Scientiae rerum naturalium 107, Oulu, Finland

Helminen V (1987) Lämpöolot. In: Alalammi P (ed) Atlas of Finland (ed.), Folia 131, Climate, National Board of Survey and Geographical Society of Finland, Helsinki, pp 4–10.

Johansen B, Karlsen SR (2002) Finnmarksvidda - changes in lichen cover 1987-2000. In: Haugerud RE (ed) Rangifer report, vol 6, pp 65–66

Johansen B, Johansen M-E, Karlsen SR (1995) Vegetasjons- og beitekartlegging I Finnmark og Nord-Troms. Norut informasjonsteknologi AS, 60 pp

Kumpula J, Colpaert A, Kumpula T, Nieminen M (1997) Suomen poronhoitoalueen talvilaidunvarat. Kala- ja Riistaraportteja 93, Riista- ja kalatutkimus

Näkkäläjärvi K (2003) The siida, or Sámi Village, as the basis of community life. In: Jukka P, Näkkäläjärvi K (eds) Siiddastallan. From Lapp Communities to Modern Sámi Life. Siida Sámi Museum Inari. Pohjoinen, Gummerrus, pp 114–121

Oksanen L, Virtanen R (1995) Topographical, altitudinal and regional patterns in continental and suboceanic heath vegetation of northern Fennoscandia. Acta Bot Fenn 153:1–80

Paliskuntain yhdistys (2003) Porotilastot (reindeer statistics) 2002. Poromies. Paliskuntain yhdistys, Rovaniemi, Finland

Pearce CM (1991): Mapping muskox habitat in the Canadian High Arctic with SPOT Satellite Data. Arctic 44:49–57

Pennanen J (2003) Large-scale reindeer herding brings changes to the siidas. In: Jukka P, Näkkäläjärvi K (eds) Siiddastallan. From Lapp Communities to Modern Sámi Life. Siida Sámi Museum Inari. Pohjoinen. Gummerrus, pp 138–139

Rees WG, Williams M, Vitebsky P (2003) Mapping land cover change in a reindeer herding area of the Russian Arctic using Landsat TM and ETM+ imagery and indigenous knowledge. Remote Sens Environ 85:441–452

Sollheim I, Engelsen O, Hosgood B, Andreoli G (2000) Measurement and modelling of the spectral and directional reflection properties of lichen and moss canopies. Remote Sens Environ 72:78–94

Stow DA, Hope A, McGuire D, Verbyla D, Gamon J, Huemmrich F, Houston S, Racine C, Sturm M, Tape K, Hinzman L, Yoshikawa K, Tweedie C, Noyle B, Silapaswan C, Douglas D, Griffith B, Jia G, Epstein H, Walker D, Daeschner S, Petersen A, Zhou LM, Myneni R, (2004) Remote sensing of vegetation and land-cover change in Arctic Tundra Ecosystems. Remote Sens Environ 89:281–308

Théau J, Peddle RD, Duguay CR (2005) Mapping lichen in a caribou habitat of northern Quebec, Canada, using a enhancement-classification method and spectral mixture analysis. Remote Sens Environ 94:232–243

Thompson DC, Klassen GH (1980) Caribou habitat mapping in the southern district of Keewatin NWW: an application of digital Landsat data. J Appl Ecol 17:125–138

9 Vegetation: Structure, Cover, and Biomass of Subarctic Tundra Wetlands Used as Summer Pastures

H. Kitti and B.C. Forbes

9.1 Introduction

Grazing is an important factor influencing the structure, cover, and production of vegetation at both local and landscape levels and over short- and long-term time scales. Recent overviews and experimental studies treating herbivore interactions with vascular and non-vascular arctic and alpine plants and soils are many (Oksanen and Virtanen 1995; Manseau et al. 1996; Suominen and Olofsson 2000; Grellmann 2001; Olofsson et al. 2001, 2004; van der Wal et al. 2001; Eilertsen 2002; Bråthen 2003; den Herder et al. 2003, 2004).

Reindeer grazing profoundly influences the composition of vegetation (Oksanen and Virtanen 1995; Manseau et al. 1996; Austrheim and Eriksson 2001). It can also alter plant species' mineral composition (Väre et al. 1996; Olofsson et al. 2001), change nutrient mineralization rates and primary production (Stark et al. 2000; Olofsson et al. 2001, 2004), and increase ground temperature (Olofsson et al. 2001; Chap. 11). Reindeer consume several dozen species of plants, with a profound seasonal dietary shift from mainly vascular taxa during spring, summer, and autumn to a heavy winter dependence, under natural conditions, on arboreal and ground lichens (Warenberg et al. 1997). Some plant species are generally avoided by reindeer, while others are selected as forage (van der Wal et al. 2001; Cooper and Wookey 2003; den Herder et al. 2003). Factors involved in forage selection by *Rangifer* spp. include the amount of available biomass, secondary compounds, palatability/digestibility and especially plant nutrition (Julkunen-Tiitto 1989; Nieminen and Heiskari 1989; Tahvanainen et al. 1991; Staaland and Sæbø 1993; Mårell et al. 2002). In migratory wild herds, patterns of movement and habitat use in spring reveal that animals precisely exploit the "green wave" represented by rapidly advancing plant phenology (Lent and Klein 1988). However, the semi-domestic herds of modern day Fennoscandia are largely restricted by fences within and between countries and individual districts. Nevertheless, in summer the animals feed highly selectively (Nieminen and Heiskari 1989).

Ecological Studies, Vol. 184
B.C. Forbes, M. Bölter, L. Müller-Wille, J. Hukkinen, F. Müller, N. Gunslay, Y. Konstatinov (Eds.)
Reindeer Management in Northernmost Europe

Responses of forage species to grazing will vary according to the plant's ability to regenerate after damage, the effects of plant competition, the abiotic conditions, and the type and intensity of herbivory (Crawley 1983; Bryant et al. 1991). Herbivores, including reindeer, can thus modify competitive interactions of plants leading to an almost complete change in the species composition of a community (Manseau et al. 1996). There has been lively debate over whether deciduous shrubs are better at recovering from defoliation than evergreen shrubs, but the evidence is mixed (Tolvanen et al. 1992; Olofsson 2001).

Research on reindeer pastures has mostly concentrated on biological/ecological aspects, primarily on lichen-dominated winter pastures (Kalliola 1939; Nordhagen 1943; Dahl 1956; Oksanen 1978; Kumpula et al. 1996; Väre et al. 1996), but lately summer pastures have also been studied (Kumpula et al. 1996; Bråthen and Oksanen 2001; Olofsson et al. 2001; Eilertsen 2002). There is evidence that summer grazing is seldom permanently destructive to moist subarctic tundra (Klein 1968) and mountain birch forest understory vegetation (Thannheiser et al. 2005). In one recent experimental manipulation in northern Fennoscandia, fertilizers were applied to reindeer pastures to gauge top-down versus bottom-up controls on the plant communities. The productivity of the targeted plant communities was clearly nutrient limited but increases in biomass, with and without fertilization, were kept in check by herbivory (Grellmann 2001).

The most obvious overall effect of reindeer on sub-arctic Fennoscandian vegetation has been the decrease in lichen cover in both winter and summer pastures due to the combined effects grazing (winter) and trampling (summer; Oksanen 1978; Suominen and Olofsson 2000; Chaps. 1, 8, 10). However, it has also been shown that animals can reduce the biomass of their preferred forage species in summer pastures (Bråthen and Oksanen 2001), with implications for both vegetation and animal productivity (Post and Klein 1999). Related to this is the pattern by which the losses among preferred plants are offset by gains among less palatable species, so potentially affecting pasture quality. On drier, nutrient-poor soils, lichens tend to be replaced by ruderal mosses, whereas on wetter, nutrient-rich substrates heavy grazing leads to a switch in dominance from mosses to graminoids (Zimov et al. 1995; Olofsson 2001). The aim of our study was to investigate the effects of long-term (40–50 years) reindeer grazing on the vegetation structure, cover, and biomass of summer tundra pastures in northern Finnish Lapland.

9.2 Study Site and Methods

The study was carried out in Jávrresduottar (Jauristunturit; Fig. 9.1). It is located in a continental area along the border between Finland and Norway,

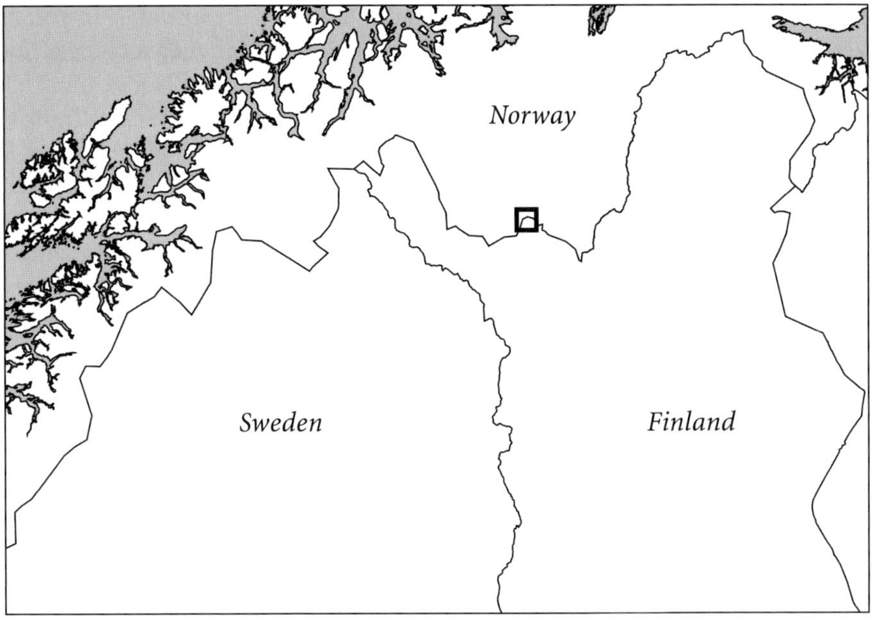

Fig. 9.1. Map showing the location of the study area at Jávrresduottar (Jauristunturit) along the Finnish-Norwegian border

Enontekiö commune and Finnmark commune, between 450 and 510 m asl The study area was bordered by latitudes 68°47.00'N and 68°50.25'N and longitudes 23°45.25'E and 23°55.25'E. It is characterized by exposed fjells and heaths of the *Betula-Cladina* type, but these are interspersed with willow-and-graminoid-dominated vegetation, with soils that remain saturated virtually throughout the growing season (Chaps. 8, 10).

This investigation, therefore, differs from recent work on summer pastures in the vicinity of Fennoscandian border fences (Olofsson et al. 2001). The latter study did not encompass *Salix-Eriophorum-Carex*-dominated vegetation, which has a patchy distribution generally restricted to low-lying and riparian habitats, but rather concentrated primarily on the prevailing *Empetrum-Dicranum* and *Betula nana-Cladina* vegetation types. The imposition of the border fence means that these small areas of wetland constitute important summer habitat for the Näkkälä herding district since the animals can no longer move to the Norwegian coast in spring/summer.

The fence between Finland and Norway was built in the end of 1950s to prevent reindeer from crossing the border. It separates winter pastures of Norwegian reindeer and summer-late winter pastures of Finnish reindeer. The general pattern is for the animals to arrive at the Finnish side when the first mosquitoes appear, typically in mid June because the fjells in the area help provide wind exposure and thus some relief for the animals. In recent years animals have also returned to this area in late July after earmarking (for ownership) takes place just to the south. It means that on the Norwegian side there is no grazing during summertime because animals must leave the area no later than 1 April. The result is that the Norwegian side can be used as a control site to compare with the Finnish site, which is fairly heavily grazed, mainly during the summer. There are about nine or ten animals per square kilometer during the few weeks they are using this particular site. In winter habitats in this region the density of reindeer is lower, about five per square kilometer. In the past, animals were occasionally present during winter on the Finnish side, but only in years when the snow was too deep in other areas. This has not been the case for many years now. At any rate, even when present in winter the animals do not feed on *Salix* spp. but graze mainly on the more exposed lichen grounds.

The information for the management of the area was provided by reindeer herders through interviews and participant observation. The precise study sites in Finland were selected in conjunction with reindeer herders during preliminary interviews and field excursions in spring/summer 2001 and late winter 2002. The sites sampled in Norway were comparable to those in Finland in terms of slope, aspect, bedrock, and soils (Chaps. 10, 13, 15) and vegetation types (Chap. 8). The habitats were quite rich in nutrients, the most prominent vascular plants being *Eriophorum angustifolium*, *Salix lapponum*, *Carex rotundata*, *C. rostrata*, and *C. aquatilis*. Dominant mosses included *Straminergon stramineum*, *Warnstorfia exannulata*, *W. sarmentosa* and *Paludella squarrosa*. There were five different-sized study areas of same biotope on both sides of the border (Fig. 9.2), situated in an area about 30 km^2.

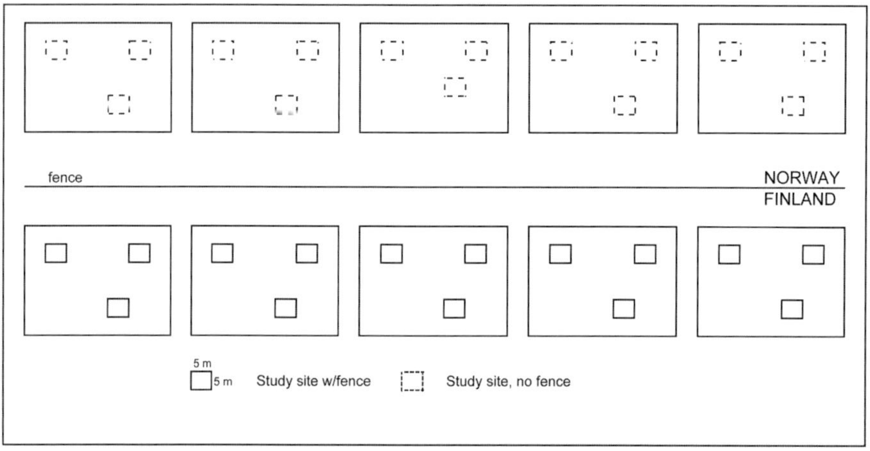

Fig. 9.2. Experimental design with plots arrayed in wet riparian habitats on either side of the international border

In each study area there were three randomly sampled plots of 25 m^2 within which were randomly selected 25 subplots of 0.25 m^2 (50 x 50 cm).

9.2.1 Plant composition

In each plot four subplots were randomly sampled. The data were collected following the protocol of the International Tundra Experiment or ITEX (Walker 1996) by using a point frame of 0.25 m^2, which was placed according to stratified random sampling protocols. There were 100 ocular points at 5 cm intervals within 50 x 50 cm squares in each quadrat. A pin (diameter 3 mm) was dropped from each point and all taxa with which it came into contact were recorded on sheets that mapped each hit according to vertical and horizontal location within the quadrat. For narrow-leaved ericoid dwarf shrubs, only contacts with separate branches were counted. For other vascular plants each contact was counted. Bryophytes and lichens were considered to grow in one layer, so the maximum was one count per species at each hole. Species found in a plot but not hit by any pin were counted as 0.1. The composition, cover, and frequency were measured for vascular plants. Species' frequencies and cover estimates were provided for bryophytes and lichens. Heights of *Salix lapponum, Eriophorum angustifolium*, and *Carex* species were measured in each plot, five measurements per species. All areas were photographed. Any species that were not identified in the field were dried and afterwards identified at the Universities of Oulu and Helsinki.

9.2.2 Biomass

Material was collected at the peak of the growing season, between 12–19 July 2002. From each study area all the plants were clipped from three randomly sampled subplots in each randomly sampled plot. Altogether, 15 subplots were clipped in Norway and 15 in Finland. Species were sorted and bagged separately, except for *Carex* spp., since most of these were sterile. Plants were put into paper bags and air dried in a heated hut. Material was then dried in the laboratory at +60 °C for 72 hours. Then it was weighed with an electronic scale.

9.2.3 Data analysis

The homogeneity of the measured factors (vegetation composition, biomass, and height) between the Finland and Norway and among the study areas within the areas was tested using analysis of variance (ANOVA), in which area was nested within regions. Before analyzing each dataset, the homogeneity of variance was tested with the Levene test and the normal distribution with the Kolmogorov-Smirnov test.

9.3 Results

The vegetation was generally similar in wetland habitats on both sides of the fence separating Finland and Norway in terms of species composition, cover, and structure. The main exception was *Salix lapponum,* which differed in all respects. The cover of *S. lapponum* was significantly lower (P=0.02) in the summer-grazed site in Finland compared to the winter-grazed site on the Norwegian side of the fence (Fig. 9.3). Total vegetation cover also differed, but not quite significantly (P=0.06). Biomass of *S. lapponum* (grazed site 6.0 ± 8.4 g/m^2; ungrazed site 17.6 ± 26.6 g m^{-2}) was also significantly lower in the summer-grazed site (P=0.05; Table 9.1), but total biomass and total number of

Table 9.1. Effects of grazing in an area and study area (nested within area) on the plant cover, biomass, and height

		Plant Cover			Biomass			Height		
		df	F	P	df	F	P	df	F	P
Vascular plants	area	1	0	0.95	–	–		–	–	
	Study area (area)	8	2.39	0.05	–	–		–	–	
Grasses	area	1	0.78	0.4	–	–		–	–	
	Study area (area)	8	0.55	0.81	–	–		–	–	
Dwarf shrubs	area	1	0.55	0.48	–	–		–	–	
	Study area (area)	8	2.96	0.02	–	–		–	–	
Cyperaceae	area	1	0.15	0.71	–	–		–	–	
	Study area (area)	8	5.31	<0.01	–	–		–	–	
Mosses	area	1	2.41	1.6	–	–		–	–	
	Study area (area)	8	4.27	<0.01						
Salix lapponum	area	1	8.08	0.02	–	–		1	4.41	0.07
	Study area (area)	8	1.84	0.13	–	–		8	1.5	0.24
Salix sp.	area	–	–		1	5.15	0.05	–	–	
	Study area (area)	–	–		8	1.48	0.24	–	–	
Erioph. ang.	area	1	0.12	0.74	1	0.01	0.93	–	–	
	Study area (area)	8	5.04	<0.01	8	4.5	<0.01	–	–	
Betula nana	area	1	0.05	0.83	–	–		–	–	
	Study area (area)	8	5.04	<0.01	–	–		–	–	
Carex sp.	area	1	0.26	0.62	1	0.39	0.55	–	–	
	Study area (area)	8	5.11	<0.01	8	5.2	<0.01	–	–	

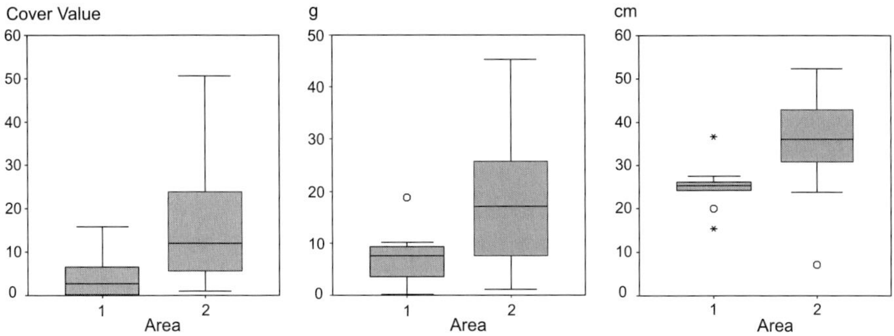

Fig. 9.3. Plant cover, biomass, and height of *Salix lapponum*. *Area 1* = Finland, no summer grazing; *Area 2* = Norway, summer grazing. The *horizontal line* in the *middle of the box* marks the median of the sample. The *edges of each box* mark the 25th and 75th percentiles. The *whiskers* show the range of values that fall within 1.5 box-lengths of the edges. Cases that have values more than 3 box-lengths below the lower edge or above the upper edge are marked by an asterisk (*), while cases that have values between 1.5 and 3 box-lengths outside the edges are marked by an open circle (*O*)

Fig. 9.4. Reindeer fence along the border separating a riparian habitat in Finland (*left*) from Norway (*right*). Under contrasting management regimes, animals are present on the Finnish side mainly in summer but on the Norwegian side only in winter. Thickets comprised of upright shrubs of *Salix lapponum* are clearly visible on the slopes in the right middle background of the photo. *S. lapponum* is present in comparable habitat on the Finnish side but not readily visible. The individual plants are generally much smaller and display evidence of heavy browsing compared to those on the Norwegian side of the fence. *Photo: B.C. Forbes*

species did not differ. Height of *Salix lapponum* (grazed site 25.1 ± 5.4 cm; ungrazed site 35.7 ± 7.8 cm) did not differ significantly (P=0.07), but a clear pattern revealed that plants were consistently shorter in riparian areas on the Finnish side (Table 9.1, Fig. 9.4). And no significant increase in graminoid cover was detected there.

9.4 Discussion

Herbivory can and usually does modify the productivity of plants (Ouellet et al. 1994). In the present study, the cover, biomass, and height of *Salix lapponum* were substantially lower on the Finnish side of the border fence subject to summer grazing. Many other studies also indicate that grazing reduces the amount and reproductive success of *Salix* sp., which as a woody plant is sensitive to grazing (Ouellet et al. 1994; Manseau et al. 1996; Crête et al. 2001; Moen and Danell 2003; den Herder et al. 2004). On the other hand *Salix* sp. is often able to compensate for the consumed parts quite quickly (Bryant et al. 1991). Compensation depends on, among other things, the intensity and timing of grazing and the supply of nutrients, especially of nitrogen (Houle and Simard 1996). Other influential factors are climatic conditions and the length of growing season. In the study area, one reason for slow compensation growth is probably the short growing period. Other plant groups or species were not significantly affected by grazing, so that no plant group or species compensated for the loss of biomass of *Salix lapponum*. The loss of preferred forage species over the long term is discussed by Ouellet et al. (1994) and Bråthen and Oksanen (2001).

Numerous studies have been made on different aspects of herbivore impacts in northern Fennoscandia (Moen 1990; Staaland et al. 1993; Väre et al. 1996; Virtanen et al. 1997; Virtanen 1998; Kumpula 2001; Olofsson 2001; Bråthen and Oksanen 2001; Eilertsen 2002; den Herder and Niemelä 2003). We could not find specific mention of the responses of *Salix lapponum* to grazing, but it is known to be an important forage species in Fennoscandia (Nieminen and Heiskari 1989; Staaland and Sæbø 1993). Other research has shown that, in general, willows (e.g., *Salix herbacea*, *S. glauca*, *S. phylicifolia*, *S. planifolia*) do decrease under prolonged grazing pressure by *Rangifer* spp. (Thing 1984; Manseau et al. 1996; Crête et al. 2001; den Herder and Niemelä 2003; den Herder et al. 2004), whereas the productivity and reproductive capacity of clonal rhizomatous graminoids (e.g., *Eriophorum angustilfolium*, *Carex aquatilis*, *C. membranacea*) can increase significantly (Tolvanen et al. 2001; Walde 2001).

In the riparian areas near the border fence in this study, the soil surface remained intact despite relatively heavy long-term grazing pressure. However, Moen and Danell (2003) reported significant damage to mire habitats along fences in the Swedish mountains, including erosion on the steeper slopes. On the other hand, we did observe an overall reduction in albedo, or surface reflectance, on the Finnish side compared to Norway (Chap. 8). This is generally due to the loss of fruticose lichen mats in mesic and dry habitats, which is somewhat simplistically attributed to "overgrazing" (Chap. 1). But albedo reduction may also be partly attributable to the lower cover and biomass of *Salix lapponum* in wetland habitats, which represent a much smaller surface area than dry heaths in this area, yet still comprise important summer habitat.

The situation in wetlands contrasts sharply with that in the adjoining heath habitats, in which lichen cover, biomass, and composition differ greatly on either side of the border fence and in experimental exclosure plots (Oksanen 1978; Olofsson 2001; Chaps. 8, 10). Our results show that subarctic wetlands subject to long-term grazing in summer are rather resistant to overall changes in vegetation composition, cover and structure, even after some 50 years of moderately heavy use. These findings support those from recently published long-term research in the nearby mountain birch forest. Detailed plant sociological studies there have concentrated primarily on the dry birch forest type because floristic change was not evident in the understory of wetter birch forests after nearly 40 years of moderately heavy grazing and trampling (Thannheiser et al. 2005; Wehberg et al. 2005).

The measured differences in the cover, biomass, and height of *Salix lapponum* on the Finnish side of the border fence appear to be directly dependent on grazing pressure. There is evidence that willows are relatively resilient among northern shrubs and can tolerate herbivory by compensating for tissue loss (Wolff 1978; Bryant et al. 1991). This is believed to be particularly the case in wet, nutrient-rich habitats (Houle and Simard 1996), such as the riparian zones sampled for this study (Chap. 10). We therefore expect that a release of grazing pressure would lead to a fairly rapid regrowth of *Salix*, at least in wetter nutrient-rich sites, to levels of cover, height, and biomass comparable to those on the Norwegian side of the border fence. As long as the summer density of animals in the area remains similar to that at present, we expect the cover, productivity, and height of *Salix* to remain at comparatively low levels.

References

Austrheim G, Eriksson O (2001) Plant species diversity and grazing in the Scandinavian mountains: patterns and processes at different scales. Ecography 24:683–695

Bråthen KA (2003) Arctic and alpine plants and their interactions with herbivores – from individuals to the community. PhD thesis, Department of Biology, University of Tromsø

Bråthen KA, Oksanen J (2001) Reindeer reduce biomass of preferred plant species. J Veg Sci 12:473–480

Bryant JP, Danell K, Provenza F, Reichardt PB, Clausen TA, Werner RA (1991) Effects of mammal browsing on the chemistry of deciduous plants. In: Tallamy DW, Raupp MJ (eds) Phytochemical induction by herbivores. Wiley, New York, pp 135–154

Cooper EJ, Wookey PA (2003) Floral herbivory of *Dryas octopela* by Svalbard reindeer. Arct Antarct Alp Res 33:369–376

Crawley MJ (1983) Herbivory: the dynamics of animal–plant interactions. Univ California Press, Berkeley

Crête M, Ouellet J-P, Lesage L (2001) Comparative effects on plants of caribou/reindeer, moose and white-tailed deer herbivory. Arctic 54:407–417

Dahl E (1956) Rondane: mountain vegetation in south Norway and its relation to the environment. Nygaard , Oslo

Den Herder M, Niemelä P (2003) Effects of reindeer on the re-establishment of *Betula pubescens* subsp. *czerepanovii* and *Salix phylicifolia* in a subarctic meadow. Rangifer 23:3–12

Den Herder M, Kytöviita M-M, Niemelä P (2003) Growth of reindeer lichens and effects of reindeer grazing on ground cover vegetation in a Scots pine forest and a subarctic heathland in Finnish Lapland. Ecography 26:3–12

Den Herder M, Virtanen R, Roininen H (2004) Effects of reindeer browsing on tundra willow and its associated insect herbivores. J Appl Ecol 41:870–879

Eilertsen SM (2002) Utilization of abandoned coastal meadows in northern Norway by reindeer. PhD thesis, Department of Arctic Biology and Institute of Medical Biology, University of Tromsø

Eilertsen SM, Schjelderup I, Mathiesen SD (2002) Early season grazing effects on birch, grass, herbs and plant litter in coastal meadows used by reindeer: a short term case study. In: Eilersen SM, Utilization of abandoned coastal meadows in northern Norway by reindeer. Thesis, Dept Arctic Biology and Inst medical Biology, Univ Tromsø, Norway

Grellmann D (2001) Top-down and bottom-up effects in a Fennoscandian tundra plant community. Thesis, Department of Ecology and Environmental Science, Umeå Univ, Umeå

Hämet-Ahti L (ed; 1998) Excursion flora (in Finnish). Botanical Museum, Helsinki

Houle G, Simard, G (1996) Additive effects of genotype, nutrient availability and type of tissue damage on the compensatory response of *Salix planifolia* ssp. *planifolia* to simulated herbivory. Oecologia 107:373–378

Julkunen-Tiitto R (1989) Distribution of certain phenolics in *Salix* species (Salicaceae). University of Joensuu. Publ Sci 15:1–29

Kalliola R (1939) Plant ecological research in the Finnish alpine belt region (in German). Ann Bot Soc Fenn Vanamo 13(2), 328 pp

Klein DR (1968) The introduction, increase, and crash of reindeer on St. Matthew Island. J Wildl Manage 32:350–367

Kumpula J (2001) Productivity of the semi-domesticated reindeer (*Rangifer t. tarandus* L.) stock and carrying capacity of pastures in Finland during 1960–1990´s. Thesis, Univ Oulu, Oulu

Kumpula J (2003) Effects of forest handling on reindeer pastures (in Finnish with English abstract). Fish and Game Report 286. Fish and Game Research Institute, Kaamanen, 60 pp

Kumpula J, Colpaert A (2003) Effects of weather and snow conditions on reproduction and survival of semi-domesticated reindeer (*Rangifer t. tarandus*). Polar Res 22:225–233

Kumpula J, Colpaert A, Kumpula T, Nieminen M (1996) Pasture inventory of the northern part of the reindeer husbandry region (in Finnish). Fish and Game Research

Institute, Reindeer Research Station, Kaamanen and University of Oulu, Department of Geography

Kumpula J, Colpaert A, Nieminen M (2003) Effects of forestry and snow conditions on the use of pastures by reindeer in the Ivalo herding districts (in Finnish with English abstract). Fish and game report 271. Fish and Game Research Institute, Kaamanen, 39 pp

Lent PC, Klein DR (1988) Tundra vegetation as a rangeland resource. In: Tueller PT (ed) Vegetation science applications for rangeland analysis and management. Kluwer Academic, Dordrecht, pp 307–337

Manseau M, Huot J, Crête M (1996) Effects of summer grazing by caribou on composition and productivity of vegetation: community and landscape level. J Ecol 84:503–513

Mårell A, Ball JP, Hofgaard A (2002) Foraging and movement paths of female reindeer: insights from fractal analysis, correlated random walks, and Lévy flights. Can J Zool 80:854–865

Moen J (1990) Summer grazing by voles and lemmings upon subarctic snow-bed and tall herb meadow vegetation: an enclosure experiment. Holarct Ecol 13:316–324

Moen J, Danell Ö (2003) Reindeer in the Swedish mountains: an assessment of grazing impacts. Ambio 32:397–402

Nieminen M, Heiskari U (1989) Diets of freely grazing and captive reindeer during summer and winter. Rangifer 9:17–34

Nordhagen R (1943) Sikilsdalen og Norges fjellbeiter. Bergens Museums Skriften 22, 607 pp

Oksanen L (1978) Lichen grounds of Finnmarksvidda, northern Norway, in relation to summer and winter grazing by reindeer. Rep Kevo Subarctic Res Stat 14:64–71

Oksanen L, Virtanen R (eds; 1995) Geographical ecology of northernmost Fennoscandia. Acta Botan Fenn 153, 110 pp

Olofsson J (2001) Long term effects of herbivory on tundra ecosystems. PhD thesis, Department of Ecology and Environmental Science, Umeå University, Umeå

Olofsson J, Kitti, H Rautiainen P, Stark S, Oksanen L (2001) Effects of summer grazing by reindeer on composition of vegetation, productivity and nitrogen cycling. Ecography 24:13–24

Olofsson J, Stark S, Oksanen L (2004) Reindeer influence on ecosystem processes in the tundra. Oikos 105:386–396

Ouellet JP, Boutin S, Heard DC (1994) Responses to simulated grazing and browsing of vegetation available to caribou in the Arctic. Can J Zool 72:1426–1435

Post E, Klein D (1999) Caribou calf production and seasonal range quality during a population decline. J Wildl Manage 63:335–345

Staaland H, Sæbø S (1993) Forage diversity and nutrient supply of reindeer. Rangifer 13:169–177

Staaland H, Scheie JO, Grøndahl FA, Persen E, Leifseth AB, Holand Ø (1993) The introduction of reindeer to Brøggerhalvøya, Svalbard: grazing preference and effect on vegetation. Rangifer 13:15–19

Stark S, Wardle, DA, Ohtonen R, Helle T, Yeates GW (2000) The effect of reindeer grazing on decomposition, mineralization and soil biota in a dry oligotrophic Scots pine forest. Oikos 90:301–310

Suominen O, Olofsson J (2000) Impacts of semi-domesticated reindeer on structure of tundra and forest communities in Fennoscandia: a review. Ann Zool Fenn 37:233–249

Tahvanainen J, Niemelä P, Henttonen H (1991) Chemical aspects of herbivory in boreal forests: feeding by small rodents, hares and cervids. In: Palo RT, Robbins CT (eds) Plant defences against mammalian herbivory. CRC Press, Boca Raton, pp 115–131

Thannheiser D, Tømmervik H, Wehberg J (2005) The vegetation changes and recent impact on the mountain birch forest during the last 40 years. In: Wielgolaski F-E (ed) Plant ecology, herbivory, and human impact in Nordic mountain birch forests. Ecological studies 180. Springer, Berlin Heidelberg New York, pp 235–253

Thing H (1984) Feeding ecology of the West Greenland caribou (*Rangifer tarandus groenlandicus*) in the Sisimiut-Kangerlussuaq region. Danish Rev Game Biol 12(3):1–53

Tolvanen A, Laine K, Pakonen T, Saari E, Havas P (1992) Compensatory responses of a deciduous dwarf shrub, the bilberry (*Vaccinium myrtillus* L.) to simulated herbivory – some comparisons with the evergreen lingonberry (*Vaccinium vitis-idaea* L.). Acta Oecol 13:607–615

Tolvanen A, Schroderus J, Henry GHR (2001) Demography of three dominant sedges under contrasting grazing regimes in the High Arctic. J Veg Sci 12:659–670

Van der Wal R, Brooker R, Cooper E, Langvatn R (2001) Differential effects of reindeer on high arctic lichens. J Veg Sci 12:705–710

Väre H, Ohtonen R, Mikkola K (1996) The effects and extent of heavy grazing by reindeer in oligotrophic pine heaths in Northeastern Fennoscandia. Ecography 19:245–253

Virtanen R (1998) Impact of grazing and neighbour removal on a heath plant community transplanted onto a snowbed site, NW Finnish Lapland. Oikos 81:359–367

Virtanen R, Henttonen H, Laine K (1997) Lemming grazing and structure of a snowbed plant community – a long-term experiment at Kilpisjärvi, Finnish Lapland. Oikos 79:155–166

Walde R (2001) Effects of simulated grazing and nutrient availability on asexual reproduction in the clonal plant *Eriophorum angustifolium* ssp. *angustifolium* Honckeny. Master's thesis, University of Tromsø

Walker MD (1996) Community baseline measurements for ITEX studies. In: Molau U, Mølgaard P (eds) International Tundra experiment. Danish Polar Centre, Copenhagen, pp 39–41

Warenburg K, Danell Ö, Gaare E, Nieminen M (1997) Reindeer pasture vegetation (in Finnish). Paliskuntain yhdistys, Rovaniemi

Wehberg J, Thannheiser D, Meier K-D (2005) Vegetation of the mountain birch forest in northern Fennoscandia. In: Wielgolaski F-E, Karlsson PS, Neuvonen S, Thannheiser D (eds) Plant ecology, herbivory, and human impact in Nordic mountain birch forests. Ecological studies 180, Springer, Berlin Heidelberg New York, pp 35–52

Wolff JO (1978) Burning and browsing effects on willow growth in interior Alaska. J Wildl Manage 42:135–140

Zimov SA, Chuprynin VI, Oreshko AP, Chapin FS III, Reynolds JR, Chapin MC (1995) Steppe-tundra transition: a herbivore-driven biome shift at the end of the Pleistocene. Am Naturalist 146:765–794

10 The Chemical Response of Reindeer Summer Pasture Plants in a Subarctic Peatland to Ultraviolet (UV) Radiation

P. Soppela, M. Turunen, B. C. Forbes, P. Aikio, H. Magga, M-L. Sutinen, K. Lakkala, and C. Uhlig

10.1 Introduction

Reindeer management is an important livelihood in northern Fennoscandia. There are about 0.7 million reindeer in Finland, Sweden and Norway and approximately 300,000 calves are born every year. The survival of reindeer is highly dependent on renewable natural resources, or ecological preconditions provided by natural pastures (Helle et al. 1990; Reimers 1997; Kumpula et al. 1998). Summer pastures play a central role in the growth of reindeer. Reindeer calves are born in spring and their growth is most rapid during the first few months of life when they graze on summer pastures. Reindeer are mainly slaughtered during autumn. Most of the slaughtered animals are calves (>70 %), and the productivity of reindeer management and income of reindeer herders is highly dependent on the growth success of the calves during the summer. Body mass and fat stores that reindeer are able to accumulate on the summer pastures significantly affect the condition of reindeer and their survival over winter (Helle et al. 1987; Soppela 2000; Soppela and Nieminen 2002).

The diet of reindeer is markedly different between summer and winter. During summer, reindeer feed on green vegetation such as grasses, sedges, shrubs, herbs, and leaves of deciduous trees (Warenberg et al. 1997). This diet has a high content of energy, protein, and minerals (Nieminen and Heiskari 1989; Staaland and Sæbo 1993) and it enables rapid growth of reindeer and accumulation of muscle and fat. During autumn and early winter, reindeer gradually change to a diet consisting mainly of lichens and wintergreen parts of shrubs, sedges, and grasses (Warenberg et al. 1997). The main winter feeds in many areas are ground lichens (*Cladina* spp.; Kumpula 2001). Winter diet has a low content of nitrogen and minerals (Nieminen and Heiskari 1989; Staaland and Sæbo 1993; Danell et al. 1994; Storeheier et al. 2002). Lichens

Ecological Studies, Vol. 184
B.C. Forbes, M. Bölter, L. Müller-Wille, J. Hukkinen, F. Müller, N. Gunslay, Y. Konstatinov (Eds.)
Reindeer Management in Northernmost Europe

contain easily digestible carbohydrates and supply enough energy to maintain the reindeer over winter, but do not prevent undernutrition in which a negative balance of nitrogen occurs (Ryg and Jacobsen 1982; Soppela et al. 2000). Availability of winter diet depends very much on snow and weather conditions and varies throughout the winter. The diets of reindeer in both winter and summer have been studied in various respects, including chemical composition (Nieminen and Heiskari 1989; Staaland and Sæbo 1993), digestion (Utsi 1998; Mathiesen 1999; Nilsson 2003), and nutrional value (Danell et al. 1994; Kumpula 2001; Storeheier et al. 2002). The results of these studies emphasize seasonal, regional, and plant-species-specific variation in the diet of reindeer.

The most important summer habitats of reindeer in Finland are peatlands, open mountains, and riverbanks. Reindeer select their forage plants carefully (Hofmann 1985), preferring plants with high nitrogen content and avoiding those containing a high content of phenolic compounds (Danell et al. 1994). During spring and early summer, when calving and early lactation occurs, reindeer depend greatly on peatland plant species such as early sprouts of cotton grass (*Eriophorum vaginatum*) or bogbean (*Menyanthes trifoliata*) due to their high protein content. Later on, other vascular plants such as *Eriophorum angustifolium*, *Equisetum fluviatile*, *Rubus chamaemorus*, *Carex* sp., and leaves of birch (*Betula* spp.) and willow (*Salix* spp.), and other deciduous trees and shrubs such as *Vaccinium myrtillus* also become important (Warenberg et al. 1997).

As part of global change, stratospheric ozone depletion and increasing levels of ultraviolet (UV) radiation had been discovered over the Arctic by the mid-1990s. It has been predicted that subarctic regions will be subjected to a 14% maximum increase in the annual UV dose in the years 2010–2020 relative to 1979–1992. Increases in solar UV-B radiation (280–320 nm) have been estimated to continue until the 2050s in the boreal and subarctic regions with an abundant peatland cover (Taalas et al. 2000, 2002). In northern ecosystems plants and animals live under harsh climatic conditions at or near their adaptation limits, and may be sensitive to additional stress from increased UV-B radiation (Barnes et al. 1987; Gwynn-Jones 1999; Turunen et al. 1999a; Turunen and Latola 2005). Changes in plant productivity or in their production of chemical defense compounds due to increased UV-B radiation may affect the quality of plants, and thus have indirect impact on the preconditions of northern sources of livelihood such as reindeer management.

Longer-term field studies investigating the plant response to UV radiation have thus far involved either UV enhancement experiments that can predict future UV effects (Johanson et al. 1995; Weih et al. 1998; Gwynn-Jones 1999; Laakso et al. 2000; Björn 2002) or UV exclusion experiments that study the effect of present-level UV radiation on plants (Turunen et al. 1999a,b, 2002). The effect of UV radiation on plants depends strongly on the efficiency of UV-B protection and acclimation, and the repair mechanisms of the damage caused (Jansen et al. 1998; Jordan 2002; Turunen and Latola 2005). Plants can avoid UV-B radiation by e.g., increasing leaf reflectivity, leaf thickness, UV-B absorbing compounds, such as phenolics, flavonoids, tannins, and lignin or by phenological timing related to reproductive effort and biomass production (Jansen et al. 1998; Turunen et al. 1999a,b; Laakso et al. 2000; Björn 2002; Close and McArthur 2002; Jordan 2002; Tegelberg 2002; Turunen and Latola 2005). A review of 62 field research papers indicated that accumulation of UV-B-absorbing compounds was the most consistent response to enhanced UV-B radiation on plants, and less response was found in plant height and leaf mass per area (Searles et al. 2001). At present, UV-induced effects to the physiology, structure, and nutritive quality of the subarctic and arctic plants and to plant growth and ecological interactions are intensively studied. These studies include investigations on the effects of increased concentrations of UV-absorbing compounds in plant–herbivore interactions (Lahtinen et al. 2004), e.g., the effects on nutritive values of reindeer forage plants, and plant–microorganism interactions.

The objective of this research was to investigate whether increasing UV radiation affects the quality of summer pasture plants of reindeer. The effects of present-day levels of UV radiation on the chemical composition and digestibility were studied on a few of the most important summer pasture plants of reindeer. We especially studied peatland species, *Menyanthes trifoliata* and *Eriophorum russoleum* since they were the two most dominant vascular plant species in our study site and are important grazing plants for reindeer. We focused on the following questions:

1. Are there differences in the chemical composition of summer pasture plants of reindeer among different plants species?
2. Does the chemical composition of summer pasture plants, such as *Eriophorum* species or *M. trifoliata*, change due to UV radiation? Do the plants, for example, produce higher contents of defense compounds, such as soluble phenolics?
3. Does the digestibility of those pasture plants that are important in the diet of reindeer change due to the effects of UV radiation on their chemical composition?

The studies were conducted in a natural peatland ecosystem with a UV-B filtration experiment in a reindeer pasture of the Lappi Reindeer Herding District in Vuotso, Eastern Lapland, Finland during 2002 and 2003. Part of the results of this contribution have been previously published in the final report of RENMAN project (Forbes 2004) and in conference abstracts (Soppela et al. 2002a,b, 2003).

10.2 Materials and Methods

10.2.1 Study Site

The infrastructure of the study site was set up on a peatland near Vuotso (67° N, 27° E) in summer 2001 in cooperation with reindeer herders from the Lappi Reindeer Herding District. With their knowledge, herders made an important contribution to the selection of the study site and of the plant species studied, as well as to development of the infrastructure for the experiment. The study site (100 × 100 m in size) was fenced from reindeer in 2001.

Vegetation composition and cover were determined from the plots once during summer according to the protocol employed by the International Tundra Experiment ("ITEX", Walker 1996). The peatland was treeless and oligotrophic, and featured a *Sphagnum*-rich ombrogenous raised bogs (Table 10.1), characterised by sedges (*Eriophorum* sp., *Carex* spp., *Scheuchzeria palustris*), herbs (*M. trifoliata, R. chamaemorus*), and shrubs (*Vaccinium oxycoccus, Andromeda polifolia, B. nana*). The peatland is used as a summer pasture for reindeer of the Lappi Reindeer Herding District. The two most dominant vascular plant species in the field site were *M. trifoliata* and *E. russoleum*. They were selected to study the possible change of chemical composition of plants in the UV filtration experiment. Plant species to be studied included also *B. nana, E. angustifolium, R. chamaemorus*, and *Carex* spp. from the unmanipulated locations inside the study site.

10.2.2 UV Filtration Experiment

The UV filtration experiment was conducted in 2002 and 2003 with three treatments. Each treatment was replicated ten times (altogether 30 plots, sizes 120 × 120 cm; Fig. 10.1). The plots in the natural peatland ecosystem were fenced with wooden frames and covered with different varieties of plastic filter. Treatments were: (1) UV-B exclusion under a clear polyester plastic, 10 plots; (2) control for the effect of plastics under a clear cellulose acetate, 10 plots; and (3) ambient natural solar UV irradiance with no plastic filters, 10 plots. The purpose of the experiment was to study the effect of natural, present-day-level UV radiation (treatments 2 and 3) on plants as compared to filtered UV-B radiation (treatment 1). The filters were placed at about 20 cm above the surface of a *Sphagnum* layer. The filters were placed in early summer before vegetation started to grow and they were raised a few times during summer according to plant growth. The experimental period lasted from 6 June to 5 September in 2002, and from 9 June to 3 September in 2003. The transmittance of the plastic filters used in the experiment is presented in Fig. 10.2.

Table 10.1. Soil characteristics of the *Sphagnum*-rich ombrogenous raised bog from experimental site in Vuotso

Horizon	Depth (cm)	Humification/ texture[a]	Soil color[b]	pH	C (TOC)[c]	N	C/N	NH_4	NO_3	P_{AL}[d]	Ca_{AL}[d]	Mg_{AL}[d]	K_{AL}[d]
					%	%		mg/kg	mg/kg	mg/kg	mg/kg	mg/kg	mg/kg
H_1	0–30	H3	10YR 6/8	4.2	41.5	0.64	65	82.9	<0.2	45	1,680	568	133
H_2	30–90	H6	7.5YR 3/2	4.5	50.2	1.19	42	18.5	<0.2	10	888	258	35
IIC	>90	Silty sand	5Y5/1	5.7	0.9	0.04	26	2.5	<0.2	10	336	128	21

[a] Von Post (1937)
[b] Munsell Color (1990)
[c] Total organic carbon
[d] Ammonium-lactate extracted nutrient (i.e., plant-available)

Fig. 10.1. View of the UV filtration experiment in Vuotso, Finland

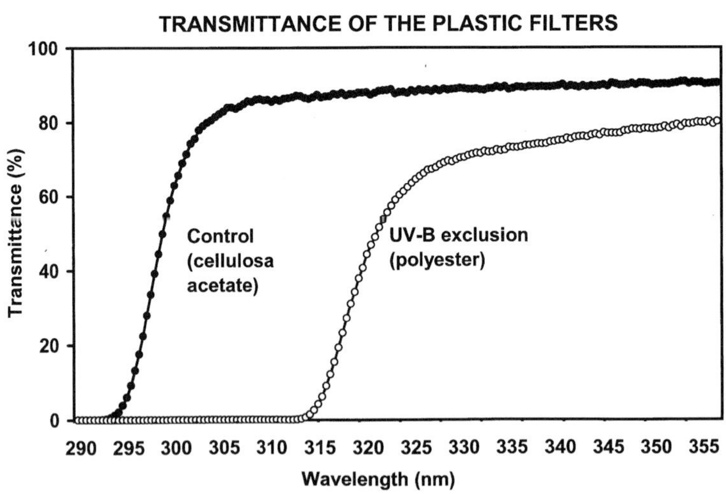

Fig. 10.2. Transmittance (%) of the plastic filters: cellulose acetate (control) and polyester (UV-B filtration) used in the UV filtration experiment (FMI 2001–2003b)

To control the warming effect caused by the plastic filters, temperature measurements were conducted at the plots with 12 Tiny Talk II Miniature Temperature Data loggers (range –40 °C to 75 °C; OTLM Software; Orion Components Ltd, Chichester, UK), placed at 10 cm above the surface of *Sphagnum* layer. The data loggers were placed in the plots each summer at the same time as the filters and they registered the temperature at 60-min intervals throughout the growing season.

In 2002 there were no significant differences in temperature measured from the ambient, control, or UV-B exclusion plots, but in 2003 the mean temperature of the ambient plots was 1.1 °C lower than that measured from the control plots (under cellulose acetate plastic film). However, no significant differences were found between the temperature measured in control (cellulose acetate) and UV-B exclusion plots (polyester) in either year. No additional water was needed since the water level on the peatland was high in all plots during the experiment period in both years. The filters were slightly raised in the center by the upthrust of a central vertical pole so that rainfall drained to the sides of the plot and not into it.

Weather and UV irradiance data were provided by the Finnish Meteorological Institute, Sodankylä, Finland (FMI 2001–2003a,b). Within the fenced area of the study site a simple soil profile description was performed during August 2003, and replicate soil samples from three different locations were sampled into polyethylene bags according to soil horizons (Table 10.1). Soil samples for which results are reported here (Table 10.1) have been analyzed for acidity (pH_{H20}), total organic carbon (TOC) (Norwegian-European Standard 13137), Kjeldahl-N (Bremner 1960), NO_3 and NH_4 (Norwegian Standard 4746), and plant-available Ca, Mg, K, and P (Egnér and Riehm 1960) at the Norwegian Centre for Soil and Environmental Research (Jordforsk).

10.2.3 Sampling and Analyses

Menyanthes trifoliata and *E. russoleum* were chosen to study in the UV filtration experiment as they were two most dominant vascular plant species in the study site, and are important grazing plants for reindeer. Plants were sampled from the experimental plots at least twice during the growing season: 6–13 August and 5 September in 2002; and 19 May, 31 July–1 August and 3 September in 2003. Mainly green leaf tissue and stem of plants were sampled. Samples for the analyses of phenolics were immediately stored with liquid nitrogen, and those for the analyses of nutrients and for in vitro digestibility were dried at 60 °C for 3 days and fine ground for analyses. Samples for both liquid nitrogen and air drying were taken from *M. trifoliata* from the experimental plots. *E. russoleum* was sampled only for phenolics (liquid N). In addition to the UV filtration experiment, plant samples were collected from 10 randomly chosen unmanipulated locations outside the experimental plots.

Plant samples used for the analysis of phenolics were stored at –40 °C. Phenols were extracted with methanol and total content of soluble phenolics was determined by the Folin-Dennis method at the Finnish Forest Research Institute in Kolari, Finland (Martin and Martin 1982). Chemical analyses of nutrients and digestibility of the plants were conducted at the Chemical Analysis Laboratory of the Holt Research Centre, Tromsø, Norway. Major nutrients (Kjehldal-nitrogen, water soluble carbohydrates), fibers (cellulose, lignin), macro minerals (P, K, Mg, Na, Ca, S), and some trace elements (Fe, Cu, Mn, Zn) were measured by standard methods (AOAC 1980). In vitro digestibility was analysed by using a cattle-rumen-fluid method modified to reindeer forages according to Tilley and Terry (1963).

10.3 Results

10.3.1 Weather and Irradiance Conditions

Growing season 2002 was colder than 2003; the mean temperature that year in June was very low, 0.2 °C (FMI 2001–2003a). The precipitation in June/July 2002 was over twice as high as in 2003, 222 mm vs. 100 mm. The Brewer measurements of UV irradiance from 2001–2003 show that biologically effective UV-B radiation (UV-B_{BE}) during summer months was slightly higher in 2002 than in 2003 (Fig. 10.3). The maximum monthly UV-B_{BE} value was 0.1184 Wm^{-2} in June 2002 and 0.1164 Wm^{-2} in July 2003 (FMI 2001–2003b).

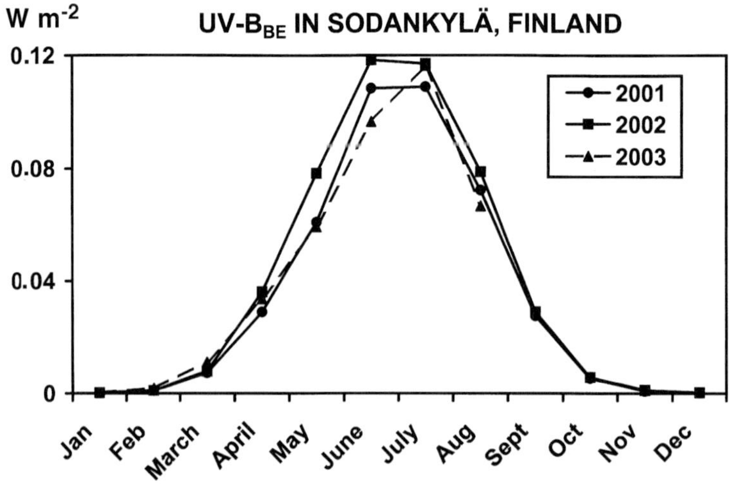

Fig. 10.3. Biologically effective monthly UV-B radiation values (UV-B_{BE}) during 2001–2003 in Sodankylä (FMI 2001–2003b)

10.3.2 Content of Soluble Phenolics in Different Plant Species

Total content of soluble phenolics varied a lot among the plant species (Fig. 10.4). The highest content of phenolics was found in leaves of *B. nana* and *R. chamaemorus*. The content of phenolics in these species was 10–20 times higher than in *M. trifoliata*, and 2–8 times higher than in *E. russoleum*. The content of soluble phenolics in stem and leaves of *E. russoleum* was lowest in May, increased in July or early August and declined in September (data not shown).

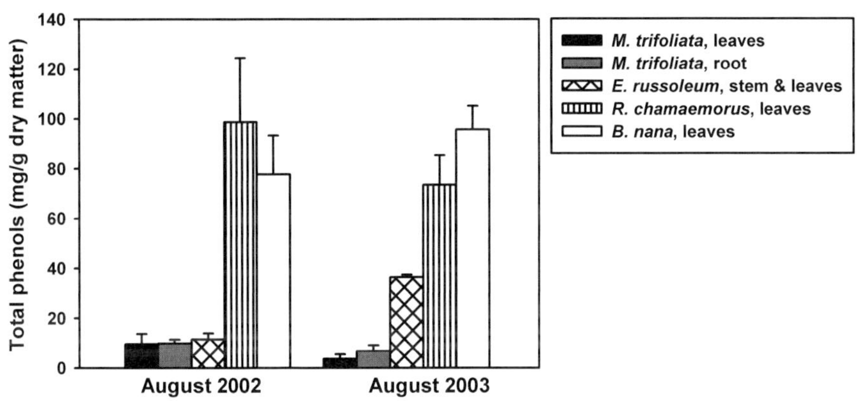

Fig. 10.4. Total content of soluble phenolics (mean ± standard deviation, SD, n=10 for all species except for *Eriophorum russoleum* n=5) in different plant species collected from unmanipulated peatland areas in Vuotso, Finland

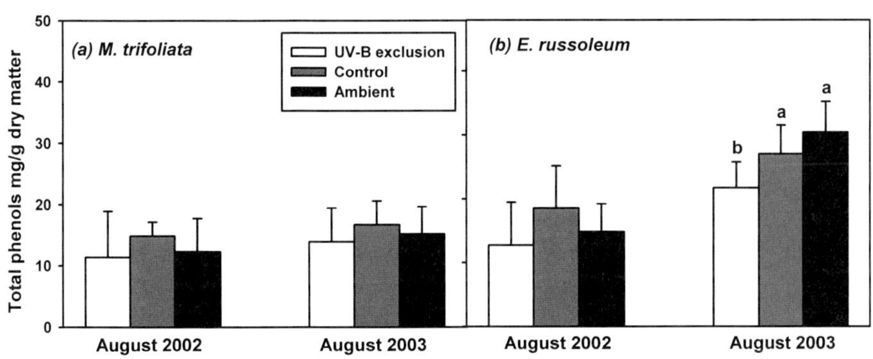

Fig. 10.5. Total content of soluble phenolics (mean ± SD, n=10) (*a*) in leaves of *Menyanthes trifoliata* and (*b*) in stem and leaves of *Eriophorum russoleum* in different treatments of the UV-filtration experiment 2003 in Vuotso. Significant differences among treatments in August 2003 are shown by *different superscript letter* (one-way ANOVA $P<0.01$; followed by LSD test)

In the UV filtration experiment in summers 2002 and 2003, total content of soluble phenolics was the lowest in both *M. trifoliata* and *E. russoleum* under UV-B exclusion treatment (Fig. 10.5a, b). In summer 2003, total content of soluble phenolics of *E. russoleum* responded more sensitively in the UV-B exclusion treatment than *M. trifoliata*. Total content of soluble phenolics in *E. russoleum* was significantly higher in ambient plots (under natural UV radiation) and in the control plots studying the effects of a plastic filter ($P<0.01$, one-way ANOVA) than in plants growing under UV-B exclusion (Fig. 10.5b).

10.3.3 Content of Nitrogen and Fibers in Different Plant Species

The content of nitrogen varied significantly among the different plants species (Fig. 10.6). Nitrogen content was highest in leaves of *M. trifoliata*, followed by *R. chamaemorus, B. nana*, *E. angustifolium*, and *E. russoleum.* Leaves and root of *M. trifoliata* also had high contents of water-soluble carbohydrates as compared to other species (data not shown).

There were significant differences also in the content of fibers among the different plant species (Fig. 10. 7a, b). The content of cellulose was highest in *E. russoleum* and that of lignin highest in *B. nana.* In the UV-filtration experiment, there were no significant differences in nitrogen content and fiber fractions between the UV-treatments in *M. trifoliata* and *E. russoleum* (data not shown).

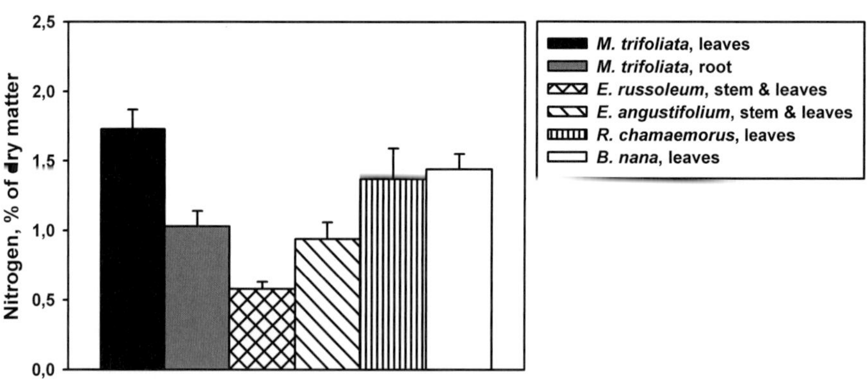

Fig. 10.6. Content of nitrogen (mean ± SD, n=10) in different plant species collected from unmanipulated peatland areas in Vuotso, Finland

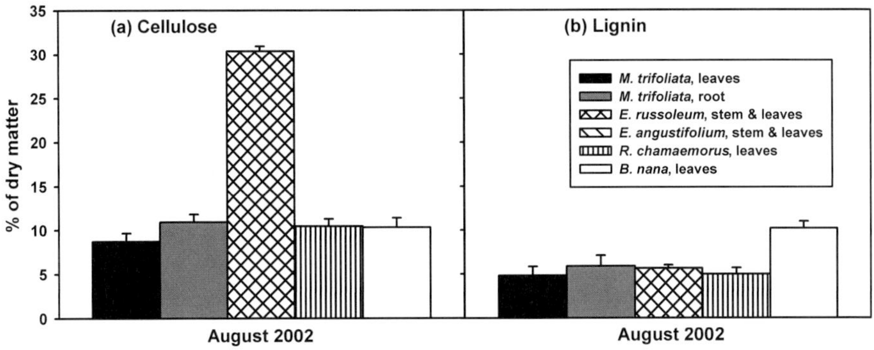

Fig. 10.7. Content of cellulose and lignin (mean ± SD, n=10) in different plant species collected from unmanipulated peatland areas in Vuotso, Finland

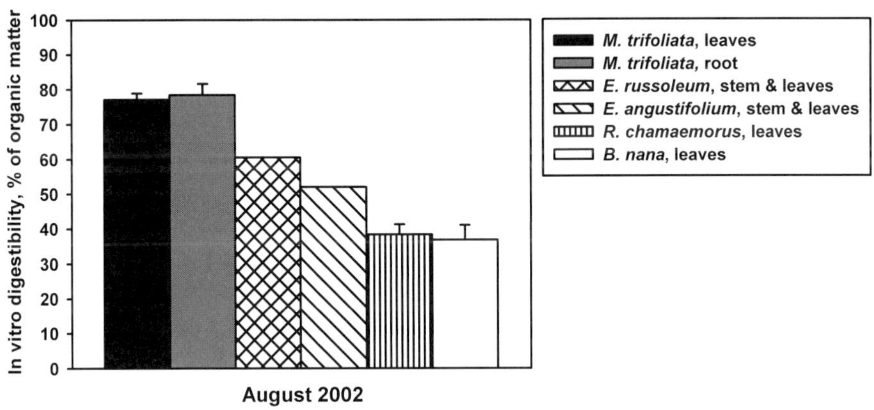

Fig. 10.8. In vitro digestibility of different plant species (mean ± SD, n=10 for all species except *Eriophorum russoleum* and *E. angustifolium*, where pooled samples of 10 single samples were used) collected from unmanipulated peatland areas in Vuotso, Finland

10.3.4 Digestibility of Different Plant Species

In vitro digestibility of the leaves and roots of *M. trifoliata* was very high, and significantly higher (2 times) than that of leaves of *R. chamaemorus* and *B. nana* (Fig. 10.8). In the UV-filtration experiment, there were no statistical differences in digestibility of plant parts between UV treatments of *M. trifoliata* and *E. russoleum* (data not shown).

10.4 Discussion

The results from the UV-B filtration experiment in Vuotso during the summer of 2003 show that UV-B radiation induced production of UV-absorbing phenolics in *E. russoleum*. The plants that grew under natural, present-day level UV radiation, produced significantly higher contents of soluble phenolics than the plants grown under UV-B exclusion. The response indicates that UV-B radiation is essential for the synthesis of soluble phenolics and the content of phenolics is increased with increasing UV radiation.

Increasing solar UV-B radiation represents a potential risk for the growth, physiology, and quality of subarctic and arctic plants. There have been relatively few studies about the effects of enhanced or natural UV radiation on the chemical composition of the summer pasture plants of reindeer (Johanson et al. 1995; Turunen et al. 1999b, 2002; Julkunen-Tiitto and Zon 2001). Many studies have reported increased concentration of phenolics due to elevated UV radiation (Gwynn-Jones 1999; Turunen et al. 1999a,b; Searles et al. 2001; Tegelberg 2002), and there may be also influences on leaf nitrogen and fibre content (Gwynn-Jones 1999). The UV-induced increase of total phenolics of *E. russoleum* in our study agree with similar responses documented in *B. nana* (Julkunen-Tiitto and Zon 2001) and *V. myrtillus* (Johanson et al. 1995; Turunen et al. 1999b). However, there were no responses in nitrogen, lignin, cellulose content, and in vitro digestibility in the UV experiment in our study.

It is worth noting that different phenolic compounds may have different responses to UV radiation. The methanol extraction, which was used in this study to determine the content of total soluble phenolics, is a rather crude method, and does not allow separation of individual phenolic compounds, which may differ in their natural seasonal variation and response to UV exclusion.

There was no UV-induced response in total phenolics of *M. trifoliata* in our study. This suggests that the chemical responses to UV radiation are not the same in all reindeer forage plant species, and/or can vary in different years and during different times of the growing period. Varying responses of plants to UV radiation during different years may be due to varying irradiance, temperature, and moisture conditions during the growing seasons of different years. Altogether, the effects of UV radiation were not great in these plant species, and further studies are necessary to confirm the results. The experiment was of relatively short duration for this kind of investigation, only two growing seasons. It is important to get continuous, updated information from the UV experiments conducted during the course of several summers, because UV responses in plants may be cumulative.

In addition to chemical composition of the plants, UV radiation may affect the plant growth and gas balance in peatlands. Niemi et al. (2002a) found that leaf cross section and the percentage of CH_4-conducting aerenchymatous tissue in *E. vaginatum* were significantly reduced by elevated UV-B radiation. Their findings indicate that increasing UV-B may have more substantial effects on gas exchange in peatlands than previously thought. As for peatland ecosystems, it has been reported that UV-B radiation can alter the concentration of UV-B-absorbing compounds, and affect the membranes and the growth of peat mosses (*Sphagnum* spp.), the structure of the vascular plant species(e.g., *E. vaginatum*), peatland microfungal community, and the fluxes of CO_2 and CH_4 between the peat and the atmosphere (Niemi et al. 2002a,b; Rinnan et al. 2003; Robson et al. 2003, 2004).

Large species-specific differences in different nutrients and fibers of plants found in our study in the unmanipulated sites agree with previous studies (Nieminen and Heiskari 1989; Staaland and Sæbo 1993). In our study, nitrogen content of leaves of *M. trifoliata* and *B. nana* was much higher than that of *Eriophorum* sp. in late summer. However, in early spring the sprouts of *Eriophorum* sp. have high nitrogen content and are the preferred forage for reindeer (Warenberg et al. 1997). Nitrogen content of green forage plants of reindeer is commonly decreased in late summer and winter while fiber content is increased (Staaland and Sæbo 1993; Warenberg et al. 1997).

There was high variation in in vitro digestibility among different plant species in our study. The reindeer is a mixed feeder and the availability of easily digestible feed is important (Hofmann 1985; Mathiesen 1999; Nilsson 2003). High nitrogen content and in vitro digestibility of *M. trifoliata* as compared to *Eriophorum* sp. and *B. nana* agrees with previous studies (Nieminen and Heiskari 1989). In general, herbs and grasses are more digestible for reindeer than leaves of trees and shrubs, which often contain poorly digestible secondary compounds (Warenberg et al. 1997). Reindeer utilize fibrous grasses poorly (Utsi 1998; Nilsson 2003). The method used for measurement of in vitro digestibility in this study was modified from the method used in common practice for cattle and sheep. The suitability of the method for reindeer forages was not tested by in vivo comparisons. A similar method (VOS, rumen organic matter digestibility) has been recently tested for reindeer (Åhman et al. 2001; Wallsten 2003), and it suits well for digestibility determination of vascular plants used by reindeer as long as the digestibility is relatively high. However, it may not be sensitive enough to detect differences among treatments.

The effects of UV radiation on the chemical composition of reindeer pasture plants and on foraging of reindeer are still ambiguous and require further research. Reindeer select their forage very carefully, choosing the most fresh, nutritious and digestible parts of plants during the growth season. The excess of phenolic defense compounds in plants may deteriorate reindeer forage and affect feeding behavior of reindeer. Previous studies have shown that reindeer prefer a low phenolic content in its plant selection for feeding (Danell et al. 1994). Availability of high protein forage, containing high quality fats and minerals during short northern summer, is crucial for lactation of hinds, growth of calves, and recovery of adult reindeer from the rigors of winter.

10.5 Conclusions

The overall goal of the present research was to get information at the basic level as to whether increasing UV radiation poses a threat to the quality of summer pastures and hence, to the foraging and nutrition of reindeer. The information could be used in predicting the use of pastures. Our conclusions are:

(1) There are differences in the phenolic content of different plant species and at different times in the growing season.

(2) UV radiation induced production of UV-B-absorbing phenolic compounds in *E. russoleum* during the second summer of research. The content of soluble phenolics in *E. russoleum* was significantly higher when natural UV radiation was allowed to reach these plants than when it was excluded.

(3) The effect of UV radiation on the concentration of UV-absorbing phenolics of *M. trifoliata* was not significant.

(4) Whether or not UV radiation was allowed to reach the studied species in the UV filtration experiment had no significant influence on nitrogen or fiber content or on digestibility.

The intensity of UV radiation depends, e.g., on the process of ozone depletion, which is affected by the global warming due to increasing contents of CO_2 and other greenhouse gases. The trend of increasing UV radiation is predicted to last for at least the next 50 years before the positive effects of international agreements can be seen (Taalas et al. 2000, 2002). The present UV filtration experiment in Lapland is to our knowledge the first one conducted in a natural subarctic peatland ecosystem. Similar experiments in peatland ecosystems have been conducted so far only in Tierra del Fuego under the Antarctic ozone hole where natural UV irradiance is much higher than here (Robson et al. 2003, 2004). Further, long-term research is necessary to confirm preliminary results, and further studies of UV-B radiation effects on plant growth, vegetation cover, and productivity in reindeer summer pastures are needed.

Our experiences working in this subproject, together with members of other disciplines and local people were positive and encouraging for further studies. In particular, the local reindeer herders' partnership in and contribution to our subproject was prominent.

Acknowledgements. We are thankful for the excellent co-operation to reindeer herders of the Lappi Reindeer Herding District, in Vuotso. Special thanks to Inga-Briitta Magga, Väinö Aikio, Iisakki Magga, Kari Magga, Nilla Hirvasvuopio and Osmo Pokuri for their help in the establishment of the experiments and infrastructure. We thank Ulla Heiskari, Marek Krywult and Beata Krzewicka for plant cover measurements and Merja Viitala for helping in the field work. Thanks for technical solutions for experimental arrangements go to Raimo Pankkonen, for skillful plant analyses to Anneli Kurkkio and Lisa Torske, and for administration to Tuija Holm. We are specially thankful for the analysis of soluble phenolics to the Finnish Forest Research Institute, Kolari Research Station. We thank Dr. R. Rinnan, Dr. B. Åhman, Dr. S. Peth, Dr. L. Heikkilä, and Prof. Dr. M. Bölter for their useful comments during the preparation of the manuscript.

References

AOAC (1980) Official methods of analysis of the Association of Analytical Chemists, 13th edn (Horwitz W ed). AOAC, Washington, DC

Åhman B, Wallsten J, Danell Ö, Udén P (2001) Digestibility of lichens and silage in reindeer rumen fluid. Rangifer Rep 5:105

Barnes PW, Flint SD, Caldwell MM (1987) Photosynthesis damage and protective pigments in plants from a latitudinal arctic/alpine gradient exposed to supplemental ultraviolet-B radiation in the field. Arct Alp Res 19:21–27

Björn LO (2002) Effects of ultraviolet-B radiation on terrestrial organisms and ecosystems with special reference to the Arctic. In: Hessen DO (ed) UV radiation and arctic ecosystems. Springer, Berlin Heidelberg New York, pp 93–121

Bremner JM (1960) Determination of nitrogen by the Kjeldahl method. J Agric Sci 55:11–33

Close D, McArthur C (2002) Rethinking the role of many plant phenolics – protection from photodamage not herbivores? Oikos 99:166–172

Danell K, Utsi PM, Palo T, Eriksson O (1994) Food plant selection by reindeer during winter in relation to plant quality. Ecography 17:153–158

Egnér HH, Riehm WR (1960) Untersuchungen über die chemische Boden-Analyse als Grundlage für die Beurteilung des Nährstoffzustandes der Böden. Royal Agricult High School Ann 26:199–215

FMI (2001–2003a) Temperature and precipitation measurements in Vuotso. Finnish Meterological Institute, Arctic Research Centre, Sodankylä, Finland

FMI (2001–2003b) Brewer measurements. Finnish Meterological Institute, Arctic Research Centre, Sodankylä, Finland

Forbes BC et al (2004) RENMAN final report. The challenges of modernity for reindeer management: integration and sustainable development in Europe's subarctic and boreal regions. Arctic Centre Rep 41, 100 pp

Gwynn-Jones D (1999) Enhanced UV-radiation and herbivory. Ecol Bull 47:77–83

Helle T, Pulliainen E, Aspi J (1987) Size-related changes in winter condition of male calves in reindeer. Rangifer 7:2–11

Helle T, Kilpelä S-S, Aikio P (1990) Lichen ranges, animal densities and production in Finnish reindeer management. Rangifer, Spec Iss 3:115–121

Hofmann RR (1985) Digestive physiology of the deer – their morphophysiological specialization and adaptation. In: Fennessy PF, Drew KR (eds) Biology of deer production. Royal Soc New Zealand Bull 22:393–407

Jansen MAK, Gaba V, Greenberg BM (1998) Higher plants and UV-B radiation: balancing damage, repair and acclimation. Trends Plant Sci 3:131–135

Johanson U, Gehrke C, Björn LO, Callaghan TV, Sonesson M (1995) The effects of enhanced UV-B radiation on a subarctic heath ecosystems. Ambio 24:106–111

Jordan BR (2002) Molecular response of plant cells to UV-B stress. Funct Plant Biol 29:909–916

Julkunen-Tiitto R, Zon J (2001) The trade-off between growth and secondary metabolism of Betula nana (Betulaceae) seedlings? The Arctic on Thinner Ice. Symposium on the state of the art of our knowledge of rapid changes in natural and social systems and how they are likely to affect livelihoods in the Nordic Arctic, 10–11 May 2001, Oulu, Finland. Abstracts, p 57

Kumpula J (2001) Winter grazing of reindeer in woodland lichen pasture. Effect of lichen availability on the condition of reindeer. Small Ruminant Res 39:121–130

Kumpula J, Colpaert A, Nieminen M (1998) Reproduction and productivity of semidomesticated reindeer in Northern Finland. Can J Zool 76:269–277

Laakso K, Sullivan JH, Huttunen S (2000) The effects of UV-B radiation on epidermal anatomy in loblolly pine (*Pinus taeda* L.) and Scots pine (*Pinus sylvestris* L.). Plant Cell Environ 23:461–472

Lahtinen M, Salminen JP, Kapari L, Lempa K, Ossipov V, Sinkkonen J, Valkama E, Haukioja E, Pihlaja K (2004) Defensive effect of surface flavonoid aglycones of Betula pubescens leaves against first instar Epirrita autumnata larvae. J Chem Ecol 30:2257–2268

Martin JS, Martin MM (1982) Tannin analyses in ecological studies: lack of correlation between phenolics, proanthocyanids and protein-precipitating constituents in mature foliage of six oak species. Oecologia (Berlin) 54:205–211

Mathiesen SD (1999) Comparative aspects of digestion in reindeer. Dr Phil thesis, Univ Tromsø, Norway

Munsell Color (1990) MACBeth Divison of Kollmorgen Instruments Corporation, 2441 North Calvert Street, Baltimore, Maryland 21218

Niemi R, Martikainen PJ, Silvola J, Sonninen E, Wulff A, Holopainen T (2002a) Responses of two *Sphagnum* moss species and *Eriophorum vaginatum* to enhanced UV-B in a summer of low UV intensity. New Phytologist 156:509–515

Niemi R, Martikainen PJ, Silvola J, Wulff A, Turtola S, Holopainen T (2002b) Elevated UV-B radiation alters fluxes of methane and carbon dioxide in peatland microcosms. Glob Change Biol 8(4):361–371

Nieminen M, Heiskari U (1989) Diets of freely grazing and captive reindeer during summer and winter. Rangifer 9:17–34

Nilsson A (2003) Adaptation of semi-domesticated reindeer to emergency feeding. Acta Univ Agr Sueciae, Agraria 399:1–40 + appendeces I–V

Reimers E (1997) *Rangifer* population ecology: a Scandinavian perspective. Rangifer 17:105–118

Rinnan R, Impiö M, Silvola J, Holopainen T, Martikainen PJ (2003) Carbon dioxide and methane fluxes in boreal peatland microcosms with different vegetation cover-effects of ozone or ultraviolet-B exposure. Oecologia 137:475–483

Robson TM, Pancotto VA, Flint SD, Ballare CL, Sala OE, Scopel AL, Caldwell MM (2003) Six years of solar UV-B manipulations affect growth of Sphagnum and vascular plants in a Tierra del Fuego peatland. New Phytol 160:379–389

Robson TM, Pancotto VA, Ballare CL, Sala OE, Scopel AL, Caldwell MM (2004) Reduction of solar UV-B mediates changes in the *Sphagnum capitulum* microenvironment and the peatland microfungal community. Oecologia 140:480–490

Rozema J, van den Staaij J, Björn LO, Caldwell M (1997) UV-B as an environmental factor in plant life: stress and regulation. Tree 12:22–28

Ryg M, Jacobsen E (1982) Seasonal changes in growth rate, feed intake, growth hormone, and thyroid hormones in young male reindeer (*Rangifer tarandus tarandus*). Can J Zool 60:15–23

Searles PS, Flint SD, Caldwell MM (2001) A meta-analysis of plant field studies simulating stratospheric ozone depletion. Oecologia 127:1–10

Soppela P (2000) Fats as indicators of physiological constraints in newborn and young reindeer (*Rangifer tarandus tarandus* L.). Dr Phil thesis. Acta Universitatis Ouluensis A 349:1–64 + 61 p (http://www.herkules.oulu.fi/isbn9514256891)

Soppela P, Heiskari U, Nieminen M, Salminen I, Sankari S, Kindahl H (2000) The effects of a prolonged undernutrition on serum lipids and fatty acid composition in reindeer calves during winter and spring. Acta Physiol Scand 168:337–350

Soppela P, Nieminen M (2002) The effect of moderate undernutrition on the fatty acid composition of white adipose tissue in young reindeer (*Rangifer tarandus tarandus* L.). Compar Biochem Physiol A 132:403–409

Soppela P, Turunen M, Heiskari U, Forbes B, Aikio P, Magga H, Sutinen M-L, Åhman B, Helle T, Nieminen M, Kyrö E, Lakkala K, Huttunen S, Uhlig C (2002a) Reindeer summer pastures and ultraviolet (UV) radiation: a research proposal. Proceedings of the 12th nordic conference on reindeer research, Kiruna, Sweden, 11–13 March 2002, Rangifer report no 6, p 100

Soppela P, Turunen M, Heiskari U, Forbes B, Aikio P, Magga H, Sutinen M-L, Åhman B, Helle T, Nieminen M, Kyrö E, Lakkala K, Huttunen S, Uhlig C (2002b) Reindeer summer pastures and ultraviolet (UV) radiation. 2nd AMAP International symposium on environmental pollution in the Arctic, 1–4 Oct 2002, Rovaniemi, Finland. Extended abstract P-C08, 3 p

Soppela P, Turunen M, Heiskari U, Forbes B, Aikio P, Magga H, Sutinen M-L, Åhman B, Helle T, Nieminen M, Kyrö E, Lakkala K, Huttunen S, Uhlig C (2003 c) Reindeer summer pastures and ultraviolet (UV) radiation. Proceedings of 11th Arctic ungulate conference, Saariselkä, Finland, 24–28 Aug 2003. Rangifer report no 7, p 30

Staaland H, Sæbo S (1993) Forage diversity and nutrient supply of reindeer. Rangifer 13:169–177

Storeheier PV, Mathiesen SD, Tyler NJC, Olsen MA (2002). Nutritive value of terricolous lichens for reindeer in winter. Lichenologist 34:247–257

Taalas P, Kaurola J, Kylling A (2000) The impact of greenhouse gases and halogenated species on future solar UV radiation doses. Geophys Res Lett 27:1127–1130

Taalas P, Kaurola J, Lindfors A (2002) Long-term ozone and UV estimates. In: Käyhkö J, Talve L (eds) Understanding the global system – the Finnish perspective. Finnish Global Change Research Programme FIGARE, University of Turku, 2002, pp 137–145

Tegelberg R (2002) Impact of elevated ultraviolet-B radiation on three northern deciduous woody plants. Dr Phil thesis, University of Joensuu

Tilley JMA, Terry RA(1963) A two-stage technique for the in vitro digestion for forage crops. J Br Grassland Soc 18:104–111

Turunen M, Heller W, Stich S, Sandermann H, Sutinen M-L, Norokorpi Y (1999a) The effects of UV exclusion on the soluble phenolics of young Scots pine seedlings in the subarctic. Environ Poll 106:219–228

Turunen M, Latola K (2005) UV-B radiation and acclimation in timberline plants. Environ Poll 137:390–403

Turunen M, Sutinen M-L, Derome K, Norokorpi Y, Heller W, Stich S, Sandermann H (1999b) Effects of UV exclusion on the soluble phenolics of subarctic forest plants. SPPS XIX Congress. Scandinavian Society for Plant Physiology, 21–23 June, 1999, Joensuu, Finland

Turunen M, Sutinen M-L, Derome K, Norokorpi Y, Lakkala K (2002) Effects of solar UV radiation on birch and pine seedlings in the subarctic. Polar Record 38(206):233–240

Utsi THA (1998) Digestive strategies in reindeer in winter. Dr Sci thesis, University of Tromsø, Norway

Von Post L (1937) The geographical survey of Irish bogs. Ir Nat J 6:210–227

Wallsten J (2003) *In vitro* and *in vivo* digestibility of lichens and silage for reindeer. Sveriges lantbruksuniversitet, Institution för husdjursgenetik, examensarbete 240, 25 pp

Walker MD (1996) Community baseline measurements for ITEX studies. In: Molau U, Mølgaard P (eds) International Tundra experiment. Danish Polar Centre, Copenhagen, pp 39–41

Warenberg K, Danell Ö, Gaare E, Nieminen M (1997) Porolaidunten kasvillisuus. Nordisk Organ för Renforskning (NOR) and A/S Landbruksförlaget, Helsinki

Weih M, Johanson U, Gwynn-Jones D (1998) Growth and nitrogen utilization in seedlings of mountain birch (*Betula pubescens* ssp. *tortuosa*) as affected by ultraviolet radiation (UV-A and UV-B) under laboratory and outdoor conditions. Trees 12:201–207

11 Consequences of Grazing on Soil Physical and Mechanical Properties in Forest and Tundra Environments

S. PETH and R. HORN

11.1 Introduction

Increasing grazing pressure as a result of increased reindeer numbers and pasture reduction due to the impact of forestry and hydropower industries has raised the issue of overgrazing in Fennoscandian landscapes. However, the term overgrazing as usually used can be misleading in that responsibilities are not correctly apportioned to the various forms of land use (Burgess 1999; Aikio 2001). Vegetation undergoes significant change not only from browsing but also from trampling. Such change, as seen in satellite imagery, is most obvious along fences in mountainous tundra where trampling during summer grazing decreases the lichen cover over large areas (Chap. 8, Evans 1994).

Soil plays an important role in ecosystem productivity and stability since it is the source of plants' nutrients and water. Evans (1994) and Löffler (1999) pointed out that grazing reindeer not only alter the vegetation pattern but also, due to erosion processes, affect the soil by reducing the humus layer and amount of finer soil particles. In heavily grazed areas this often results in a patchy vegetation canopy with sometimes >30 % bare soil exposed to the atmosphere. The loss of fine particles is in general accompanied by the export of mineral-bound soil nutrients. Nutrient cycling is, moreover, affected indirectly by a change in the microclimate due to disturbance of the vegetation cover by grazing (Broll 2000; Stark 2002). Whether the feedback of a changing microclimate on the ecosystem is positive or negative depends on how soil temperature and soil moisture changes, as both have strong influence on soil microbial activity, and thus on turnover rates (Svensson et al. 1975). Hinneri et al. (1975) argued that soil fertility, and hence the productivity of northern ecosystems, is determined not only by nutrient pools but also by the physical parameters of soil, such as moisture, aeration. and temperature. In other words, ecosystem productivity is to a great extent dependent on both soil structure and soil thermal and moisture regime.

Ecological Studies, Vol. 184
B.C. Forbes, M. Bölter, L. Müller-Wille, J. Hukkinen, F. Müller, N. Gunslay, Y. Konstatinov (Eds.)
Reindeer Management in Northernmost Europe

Animals not only alter vegetation by mechanically disturbing plants while trampling but also by exerting pressure on the ground, thereby increasing soil stresses. If soil stresses exceed the stability/strength of a soil, then compaction and change in soil structure occur. This structural change frequently goes along with a reduction in saturated hydraulic conductivity and infiltration (Gifford and Hawkins 1978; Willat and Pullar 1983), increasing the risk of soil erosion by surface water runoff and ultimately limiting plant growth (Greenwood and McKenzie 2001). This chapter presents some of the results on the effect of grazing/trampling on soil physical properties in relation to alteration of both soil surface properties and soil structure. These changes in turn may influence soil chemical properties (Chap. 12) and microbial processes (Chap. 13). The impact of reindeer husbandry compared with that of common forestry practices will be discussed.

11.2 Study Areas and Sampling Sites

Two areas were selected for the investigations. One is located in the subarctic tundra at the border between Finland and Norway, approximately 35 km southeast of Kautokeino, and referred to as the Näkkälä study area. The second study area, consisting of several individual locations, is situated within the boreal forest zone approximately 80 km north of Sodankylä, and is hereafter referred to as the Vuotso study area, Lappi Paliskunta (Chap. 1).

In the Näkkälä study area investigations have been concentrated on the effect of different grazing systems by comparing winter (only) grazed pastures on the Norwegian side of the border fence with pastures on the Finish side that have been grazed all year for over 30 years. Since the border fence was moved sometime in the past by a couple of hundred meters from south to north, an intermediate grazing intensity of about 10 years all-year grazing can be distinguished. This has enabled us to study a gradient of grazing intensity across the border fence. To study the effect of trampling on soil structural changes we also collected samples from both a trample path and in the grazed area within the >30-years-grazed pasture.

In the Vuotso study area the investigations were focused on the mechanical disturbance of soil structure resulting from different trampling intensities and their consequences for hydraulic functions and soil aeration. Two forest sites have been selected with soils that have developed from different parent materials (glacial till and glacifluvial sand) and samples have been taken from heavily trampled animal tracks, moderately trampled grazed spots, and untrampled controls. A wheeling track of a forestry harvester close to one of the forest sites has been chosen to compare the effects of reindeer trampling on soil structure with those to be expected from forestry practice. Furthermore, a corral has been examined to assess the effect of very high trampling

Table 11.1. Labels, diagnostic soil descriptions, and site characteristics of the investigated soil profiles

Profile-No.	Soil type[a]	Horizon	Upper/lower boundary[b] [cm]	Texture	Fabric[c]	Site characteristics
VP1-U[d]	Haplic podzol	L	+8			Not trampled
	Parent material:	O	+6			ungrazed control
	Glacial till	E	4	Sand	sg	
		Bs1	20	Sand	sg	
		Bs2	35	Sand	sg	
		BC	50			
		C	55 +			
VP1-G[d]	Haplic podzol	L/O	+2			Moderate trampling
	Parent material:	E	2	Sandy loam	pla	intensity
	Glacial till	Bs1	15	Sandy loam	sg	"grazed"
		Bs2	30			
		C	40 +			
VP1-T[d]	Haplic podzol	L/O	+1			High trampling
	Parent material:	E	3	Loamy sand	pla	intensity
	Glacial till	Bs1	13	Loamy sand	sg	"trample path"
		Bs2	30	Loamy sand		
		C	40 +			
VP2-I[d]	Haplic podzol	AE	2	Sand	pla	High trampling
	Parent material:	E	6	Sand	pla	intensity
	Glacifluvial sand	Bs1	23	Sand	pla...10 cm	"Coral inside"
		Bs2	33	Sand	sg	
		C	40 +	Sand	sg	

Table 11.1. (*Continued*)

Profile-No.	Soil type[a]	Horizon	Upper/lower boundary[b] [cm]	Texture	Fabric[c]	Site characteristics
VP2-II[d]	Haplic podzol	L	+4			Not trampled
	Parent material:	C	+3.5			"Coral outside"
	Glacifluvial sand	E	4	Sand	sg	
		Bs	22	Loamy Sand	sg	
		C	40 +	Sand	sg/lay	
VP3-U[d]	Haplic podzol	L	+8			Not trampled
	Parent material:	O	+2			"ungrazed control"
	Glacifluvial sand	E	2	Sand	sg	
		Bs1	10	Sand	sg	
		Bs2	15	Sand	sg	
		C	30 +	Sand	sg/lay	
VP3-G[d]	Haplic podzol	L/O	+0.5			Moderate trampling
	Parent material:	E	3	Sand	sg	intensity
	Glacifluvial sand	Bs	17	Sand	sg	"grazed"
		C	40 +	Sand	sg/lay	
VP3-T[d]	Haplic podzol	E	1	Sand	sg	High trampling
	Parent material:	Bs	8	Sand	sg	intensity
	Glacifluvial sand	C	30 +	Sand	sg/lay	"trample path"
VP4-F[d]	Haplic podzol	AE	10	Sandy loam	pla	Wheeling track of
	Parent material:	Bs1	20	Sandy loam	pla	forestry harvester
	Glacial till	Bs2	40 +	Sandy loam		

NP-G[e]	Haplic podzol	L/O	+4			Moderate trampling
	Parent material:	E	1		sg	intensity
	Glacial till	Bs	20	Loamy sand	sg	"grazed"
		C	25 +		sg	
NP-T[e]	Haplic podzol	AE	2		sg	High trampling
	Parent material:	Bs	20	Loamy sand	sg	intensity
	Glacial till	C	25 +		sg	"trample path"
MP-L[e]	Haplic podzol	L	+3			Over 30 years only
	Parent material:	O	+2.5			winter grazed
	Glacial till	E	4	Loamy sand	sg	undisturbed lichen
		Bs	20	Loamy sand	sg	cover (~ 7 cm thick)
		C	30 +	Loamy sand	sg	
MP-M[e]	Haplic podzol	L	+4			About 10 years summer
	Parent material:	O	+3.5			and winter grazed
	Glacial till	E	3	Loamy sand	sg	moderate disturbed
		Bs	10	Sandy loam	sg	lichen cover (~ 1 cm thick)
		C	30 +	Sandy loam	sg	
MP-H[e]	Haplic podzol	O	+2.5			Over 30 years summer and
	Parent material:	AE	1	Loamy sand	sg	winter grazed
	Glacial till	E	4	Loamy sand	sg	heavily disturbed lichen
		Bs	22	Sandy loam	sg	cover / absent lichen cover
		C	30 +	Silt loam	sg	lichen cover (~ 1 cm thick)

[a] FAO. ISRIC and ISSS 1998: World reference base for soil resources. Rome
[b] Numbers for litter horizons indicate upper boundaries; numbers for mineral horizons indicate lower boundaries
[c] sg = single grained; pla = platy; lay = layered
[d] Location: Vuotso
[e] Location: Näkkälä

intensity on soil physical properties. Site characteristics, soil types and profile descriptions are summarized in Table 11.1.

11.3 Materials and Methods

11.3.1 Assessment of Soil Structure and Pore Functioning

To assess changes in soil structure and pore functions related to mechanical stresses imposed by reindeer or forestry machines we have determined saturated and unsaturated hydraulic conductivities, pore size distributions (derived from water retention curves), and soil bulk densities. Undisturbed soil samples have been taken at two or three depths with the aid of standard steel cylinders varying in height and diameter depending on the parameter measured. Pore-size distributions and saturated hydraulic conductivities have been determined on 4 cm high cylinders with diameters of 6 cm. Water retention curves were obtained by means of a tension-plate assembly with suctions of –1, –3, –6, –15, and –30 kPa, and saturated hydraulic conductivities were measured with the aid of a falling-head permeameter (Hartge 1966). At some sampling sites, horizontal and vertical samples were taken to investigate anisotropy effects for saturated hydraulic conductivities. Unsaturated hydraulic conductivities [$K(\Psi)$] were determined on 6 cm high cylinders with diameters of 10 cm by simultaneously measuring volumetric water contents (TDR) and matric potentials at two depth while the initially saturated samples dried out by evaporation (Plagge 1991). Hydraulic conductivity functions were subsequently obtained by means of a nonlinear least-squares parameter fit using the RETC code of van Genuchten et al. (1991), applying the van Genuchten-Mualem model (van Genuchten 1980). When sensors could not be installed in the sample because of stones, unsaturated hydraulic conductivity functions were predicted by estimating the parameters α and n from the corresponding water retention curves. Changes in soil structure have also been assessed using X-ray computed tomography as a non-destructive technique. To do this, undisturbed soil cores (height 20–30 cm, ∅ 20 cm) have been scanned with a Siemens Somatom Plus-CT (Siemens, Forchheim, Germany) operated at an energy level of 120 keV at a resolution of 0.25 mm in the horizontal direction and 2.0 mm in the vertical direction. The two-dimensional CT images allowed a visual characterization of soil structure as well as the calculation of soil bulk density in 2 mm sections (Rogasik et al. 2003).

11.3.2 Assessment of Soil Stability

To assess the resistance of the soil structure to the increase in soil stresses due to the application of mechanical loads, so-called pre-compression stresses were determined. Pre-compression stress values were derived from compression curves of confined compression tests (oedometer tests) by the Casagrande method (Casagrande 1936). Samples (height 3 cm, ∅ 10 cm) were equilibrated to a common matric potential of –6 kPa before compression. In addition, cyclic loading tests were conducted to investigate the effect of repeated loading, unloading, and reloading on soil deformation as a function of soil texture. For this, sand, sandy loam, and loam were homogenized and molded into standard steel cylinders (height 3 cm, Ø 10 cm) with a loading frame to produce a standard initial soil bulk density of 1.60 g cm^{-3}. For sand, a second series with a slightly lower soil bulk density of ~1.50 g cm^{-3} was produced. After preparation, samples were also adjusted to a reference matric potential of –6 kPa (suction plate) and subsequently repeatedly loaded for 30 seconds with an axial load of 40 kPa in a standard oedometer device with up to 100 cycles. Between each loading step samples were unloaded for 30 seconds.

11.3.3 Assessment of Heat and Water Flow

To investigate the effect of vegetation disturbances by trampling on water infiltration and soil temperature, three monitoring profiles (MP-L, MP-M, and MP-H; Table 11.1) were installed in the Näkkälä study area along the grazing gradient across the border fence for the vegetation periods 2002 and 2003. Volumetric water contents were measured using ECH_2O Sensors (dielectric method; Decagon Devices, Pullman, WA, USA) at three depths. Soil-specific calibration of the sensors was carried out in the laboratory after their removal from the monitoring profiles, the influence of soil textures and soil bulk densities on output signals taken into account.

Soil temperatures were recorded using PT 100 temperature sensors (LKM Electronic, Geraberg, Germany) at four depths (plus surface). All measurements were collected with the aid of Delta T Loggers (Delta-T Devices, Burwell, UK) at 30 min intervals. Thermal conductivities were estimated from the field temperature measurements with the aid of an analytical solution of the diffusivity equation by determining the amplitude damping and the phase shift of the temperature wave at different soil depths (van Wijk and de Vries 1966). The practical application of this method is described in more detail in Peth (2004). Daily energy fluxes between soil and atmosphere have been estimated from the stationary heat flux equation by calculating the heat flux density for small time increments (30 min intervals) and summing up the incremental heat fluxes to obtain a net daily heat flux.

11.4 Results

11.4.1 Changes in Soil Structure and Conductivity Functions

Water retention curves and hydraulic conductivity functions for different trampling intensities at the Vuotso area site VP1 and the wheeling track of a forestry machine at site VP4-F are shown in Fig. 11.1. The distinct change in the shape of the pF curve indicates a change in pore size distribution at the trampled sites compared to the undisturbed site. At the undisturbed control total pore volume is very high (~60 Vol%) and drops by approximately 10 Vol% at the two trampled sites (VP1-G and VP1-T). Above pF 1.0 (matric potential = –1 kPa) the curve of the undisturbed control decreases steeply in contrast to the two trampled sites, indicating a high fraction of macropores at untrampled site VP1-U. As matric suction is further increased (higher pF values) water retention is higher at the two trampled sites than at the undisturbed control suggesting, on the other hand, that sites VP1-G and VP1-T have a higher amount of large mesopores than VP1-U.

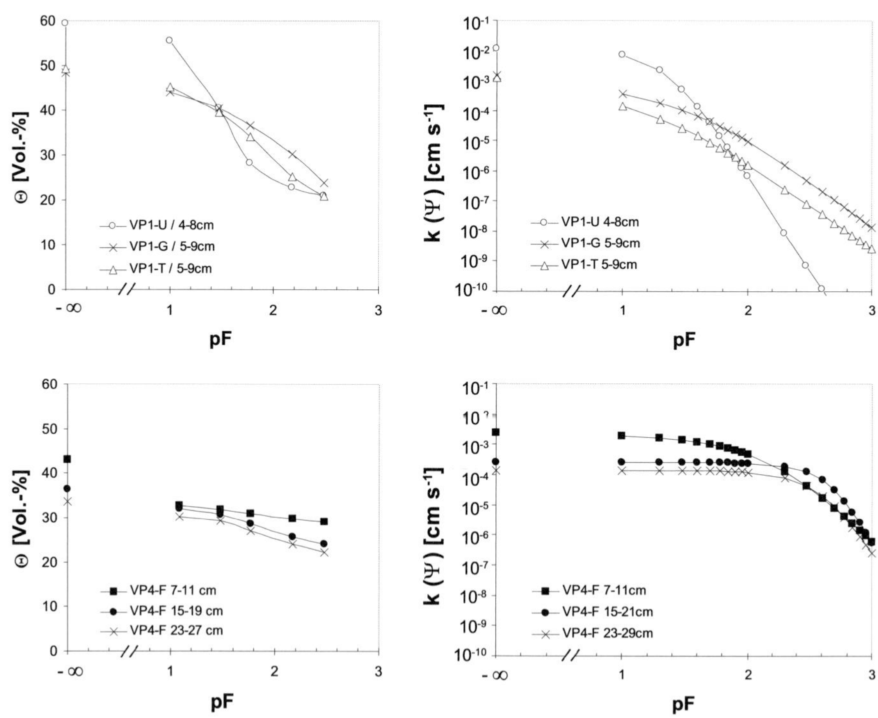

Fig. 11.1. Water retention curves (*left*) and unsaturated hydraulic conductivity functions (*right*) for different trampling intensities at site VP1 (*top*) and for the wheeling track at site VP4-F (*bottom*)

Calculated fractions of different pore size classes show that the reduction in total pore volume at the two trampled sites is accommodated by a decrease of macropores by 15–18 Vol% while the amount of large mesopores is increased by 5–6 Vol% (Table 11.2). This change in pore size distribution shows a noticeable effect on the hydraulic conductivity function (Fig. 11.1). At saturation (pF = $-\infty$) hydraulic conductivity is at the two trampled sites approximately one magnitude of order lower than at the undisturbed control (see also Table 11.3). However, as matric potentials become more negative, the hydraulic conductivity function decreases less steeply for the two trampled sites than for the undisturbed control. This is related to the dewatering of macropores at VP1-U which then are no longer available for the transport of water while the higher amount of smaller pores at site VP1-G and VP1-T are still water filled and thus water conducting. At suctions in the range of field capacity (pF 1.8–pF 2.5) unsaturated hydraulic conductivity is lower for the undisturbed control than for the two trampled sites.

At site VP4-F water retention curves are more flattened in comparison to site VP1. Total pore volume is lower than at site VP1, which is located in the vicinity, and decreases with increasing soil depth to less than 35 Vol% while soil bulk density increases. The number of macropores and large mesopores is further reduced compared to site VP1 (Table 11.2). As a result of this decrease in total pore volume, saturated hydraulic conductivity is further reduced showing the lowest values of ~10^{-4} cm s^{-1} at a depth of 23–29 cm (Fig. 11.1). Since most of the yet smaller pores are only dewatered at matric potentials lower than –30 kPa (pF = 2.5) unsaturated hydraulic conductivity only declines steeply at suctions below pF ~2.5. At the upper soil horizon of site VP4-F slightly higher numbers of macropores have been determined as in the subsoil. This is also reflected by the decline in the unsaturated hydraulic conductivity function at lower pF values (Fig. 11.1). Measurements of vertical and horizontal saturated hydraulic conductivities at site VP4-F showed a strong anisotropy, with lower values in the vertical than in the horizontal direction. Both vertical saturated hydraulic conductivity as well as the anisotropy in conductivity decreased with increasing soil depth at site VP4-F (Table 11.3).

Soil bulk densities and volume fractions of different pore size classes for all investigated soil profiles are summarised in Table 11.2. It can be seen that at the more sandy site (VP3) different trampling intensities showed a less pronounced effect on changes in pore-size distribution, and the fraction of macropores decreased only slightly. The number of macropores even increased at the heavily trampled site inside the corral (VP1-I) compared to the undisturbed control outside the corral (VP2-II), which might partly be related to the slightly higher amount of silt in the subsoil horizons of the undisturbed control. This textural effect on pore size distribution (primary pores) might also be true for site VP1. But differences in soil bulk density were found to be very high and CT images showed a well-developed system of sec-

Table 11.2. Soil bulk density (d_B) and pore size classes: TPV = total pore volume, MaP = macropores, LMP = large mesopores, SMP = small mesopores, FP = fine pores. CV = coefficient of variation

Profile No.		Depth [cm]	dB [g/cm³]	CV (d_B) [%]	TPV [%]	MaP > 50 µm	LMP 10–50 µm	SMP < 10 µm	FP < 0.2 µm
VP1-U	Depth 1	+4–0	0.15	76.4	89.1	59.9	4.7	24.5	–
	Depth 2	4–8	1.08	10.3	59.3	31.1	7.2	12.6	8.3
VP1-G	Depth 1	+1–-3	1.02	13.1	56.3	9.3	14.5	32.5	–
	Depth 2	5–9	1.36	3.1	48.5	12.0	12.6	13.7	10.1
VP1-T	Depth 1	+1–-3	0.60	27.1	74.5	34.0	14.2	26.3	–
	Depth 2	5–9	1.35	5.5	49.2	15.2	13.1	20.9	–
VP2-I	Depth 1	0–4	0.96	19.1	63.7	27.4	5.7	22.8	7.9
	Depth 2	6–10	1.16	9.7	56.3	30.3	8.2	10.8	7.1
VP2-II	Depth 1	+2–-2	0.81	21.5	59.6	10.1	23.9	25.7	–
	Depth 2	6–10	1.21	5.7	54.3	5.1	24.9	17.6	6.7
VP3-U	Depth 1	2–6	1.39	4	47.6	31.2	3.6	9.8	3.0
	Depth 2	8–12	1.36	3	48.6	36.3	2.8	4.6	4.9
	Depth 3	13–17	1.53	2.5	42.2	31.8	4.3	0.5	5.7
VP3-G	Depth 1	0–4	1.34	7.1	49.6	29.2	11.0	6.7	2.7
	Depth 2	5–9	1.40	4.2	47.1	30.2	8.4	3.3	5.3
	Depth 3	10–14	1.43	2.8	46.1	30.6	8.0	2.1	5.4
VP3-T	Depth 1	0–4	1.44	2.9	45.8	27.3	5.3	6.0	7.2
	Depth 2	6–10	1.49	4.1	43.6	31.0	4.0	2.3	6.3
	Depth 3	12–16	1.51	2.2	43.2	31.6	4.4	2.0	5.3
VP4-F	Depth 1	7–11	1.51	5.3	43.1	12.3	1.8	18.2	10.8
	Depth 2	15–19	1.69	6.8	36.3	7.7	4.5	12.5	11.6
	Depth 3	23–27	1.76	3.4	33.6	6.6	4.9	9.2	13.0
NP-G	Depth 1	3–7	1.49	4.9	43.7	16.2	14.9	6.6	6.1
	Depth 2	9–13	1.52	7.4	42.8	18.8	11.2	7.1	5.6
NP-T	Depth 1	3–7	1.52	5	42.5	16.8	14.3	5.2	6.2
	Depth 2	9–13	1.51	4.7	42.8	18.0	11.8	7.2	5.8
MP-L	Depth 1	3–7	1.50	4.4	43.2	19.6	12.3	7.7	3.6
	Depth 2	10–14	1.41	6.3	47.0	23.9	13.2	5.2	4.7
MP-M	Depth 1	1–5	1.36	5.8	48.5	26.9	8.6	8.9	4.1
	Depth 2	8–12	1.50	2.4	43.3	21.2	8.1	9.5	4.6
MP-H	Depth 1	3–7	1.64	3.3	38.0	19.7	7.4	8.0	2.9
	Depth 2	9–13	1.62	2.4	38.7	17.4	7.6	10.0	3.6

Table 11.3. Saturated hydraulic conductivities (K_f) of the investigated soil profiles. Direction: v = vertical, h = horizontal, n = number of samples, SD = standard deviation, CV = coefficient of variation

Profile No.		Direction v/h	Depth [cm]	n	geom. mean K_f [10^{-3} cm s^{-1}]	mean Log–K_f [cm s^{-1}]	SD Log–K_f [cm s^{-1}]	CV [%]	Conf. Limits $\alpha = 0.05$ Log–K_f [cm s^{-1}]
VP1-U	Depth 1	v	+4–0	4	17.7	–1.75	0.14	7.9	{–1.94;–1.56}
		h	–	–	–	–	–	–	––
	Depth 2	v	4–8	4	11.1	–1.95	0.11	5.6	{–2.10;–1.80}
		h	5–11	4	11.9	–1.92	0.17	8.7	{–2.15;–1.69}
VP1-G	Depth 1	v	+1-–3	4	4.4	–2.36	0.23	9.7	{–2.68;–2.04}
		h	0–6	4	10.4	–1.98	0.29	14.7	{–2.38;–1.58}
	Depth 2	v	5–9	4	1.48	–2.83	0.5	17.6	{–3.52;–2.14}
		h	6–12	4	1.61	–2.79	0.37	13.2	{–3.30;–2.28}
VP1-T	Depth 1	v	+1-–3	4	14.8	–1.83	0.11	6.3	{–1.99;–1.67}
		h	0–6	4	1.99	–2.7	0.31	11.6	{–3.13;–2.27}
	Depth 2	v	5–9	4	1.25	–2.9	0.2	6.7	{–3.17;–2.63}
		h	6–12	2	1.82	–2.74	0.37	13.5	{–3.86;–1.62}
VP2-I	Depth 1	v	0–4	4	2.46	–2.61	0.12	4.5	{–2.77;–2.45}
		h	–	–	–	–	–	–	––
	Depth 2	v	6–10	4	5.57	–2.25	0.06	2.8	{–2.34;–2.16}
		h	7–13	4	8.23	–2.08	0.12	5.7	{–2.25;–1.91}
VP2-II	Depth 1	v	+2-–2	4	8.59	–2.07	0.07	3.4	{–2.17;–1.97}
		h	–	–	–	–	–	–	––
	Depth 2	v	6–10	4	3.79	–2.42	0.08	3.3	{–2.53;–2.31}
		h	7–13	4	4.72	–2.33	0.13	5.6	{–2.51;–2.15}

Table 11.3. (*Continued*)

Profile No.		Direction v/h	Depth [cm]	n	geom. mean Kf [10^{-3} cm s^{-1}]	mean Log–Kf [cm s^{-1}]	SD Log–Kf [cm s^{-1}]	CV [%]	Conf. Limits $\alpha = 0.05$ Log–Kf [cm s^{-1}]
VP3-T	Depth 1	v	0–4	5	9.29	−2.03	0.05	2.7	{−2.09;−1.97}
		h	1–7	5	16.7	−1.78	0.15	8.2	{−1.95;−1.61}
	Depth 2	v	6–10	5	13.5	−1.87	0.04	2	{−1.91;−1.83}
		h	–	–	–	–	–	–	––
	Depth 3	v	12–16	5	11.4	−1.94	0.05	2.4	{−1.99;−1.89}
		h	10–16	5	11.4	−1.94	0.02	0.8	{−1.96;−1.92}
VP3-G	Depth 1	v	0–4	5	8.63	−2.06	0.07	3.2	{−2.14;−1.98}
		h	1–7	5	14.3	−1.84	0.08	4.1	{−1.93;−1.75}
	Depth 2	v	5–9	5	14.5	−1.84	0.03	1.5	{−1.87;−1.81}
		h	–	–	–	–	–	–	––
	Depth 3	v	10–14	5	11.5	−1.94	0.06	3.2	{−2.01;−1.87}
		h	10–15	5	13.1	−1.88	0.05	2.6	{−1.94;−1.82}
VP3-U	Depth 1	v	2–6	5	14.2	−1.85	0.04	2.3	{−1.90;−1.80}
		h	2–8	5	22.9	−1.64	0.1	5.9	{−1.75;−1.53}
	Depth 2	v	8–12	5	19.4	−1.71	0.04	2.3	{−1.76;−1.66}
		h	–	–	–	–	–	–	––
	Depth 3	v	13–17	5	16.6	−1.78	0.04	2.4	{−1.83;−1.73}
		h	12–18	5	19.8	−1.7	0.04	2.1	{−1.74;−1.66}
VP4-F	Depth 1	v	7–11	5	2.39	−2.62	0.65	24.7	{−3.36;−1.88}
		h	–	–	–	–	–	–	––
	Depth 2	v	15–19	5	0.25	−3.61	0.55	15.2	{−4.24;−2.98}
		h	15–21	5	0.96	−3.02	0.27	8.8	{−3.33;−2.71}
	Depth 3	v	23–27	5	0.14	−3.87	0.58	14.9	{−4.53;−3.21}
		h	23–29	5	0.62	−3.2	0.46	14.4	{−3.73;−2.67}

NP-T	Depth 1	v	3–7	5	3.75	−2.43	0.22	9.1	{−2.68;−2.18}
	Depth 2	v	9–13	5	1.98	−2.7	0.14	5.1	{−2.86;−2.54}
NP-G	Depth 1	v	3–7	5	4.22	−2.37	0.31	13.1	{−2.73;−2.01}
	Depth 2	v	9–13	5	1.44	−2.84	0.14	5.1	{−3.01;−2.67}
MP-L	Depth 1	v	3–7	5	2.71	−2.57	0.81	31.5	{−3.50;−1.64}
	Depth 2	v	10–14	5	2.13	−2.67	0.22	8.1	{−2.92;−2.42}
MP-M	Depth 1	v	1–5	5	6.92	−2.16	0.26	12.1	{−2.46;−1.86}
	Depth 2	v	8–12	5	0.77	−3.11	0.47	15.1	{−3.65;−2.57}
MP-H	Depth 1	v	3–7	5	1.99	−2.7	0.22	8.1	{−2.95;−2.45}
	Depth 2	v	9–13	5	1.56	−2.81	0.25	8.7	{−3.09;−2.53}

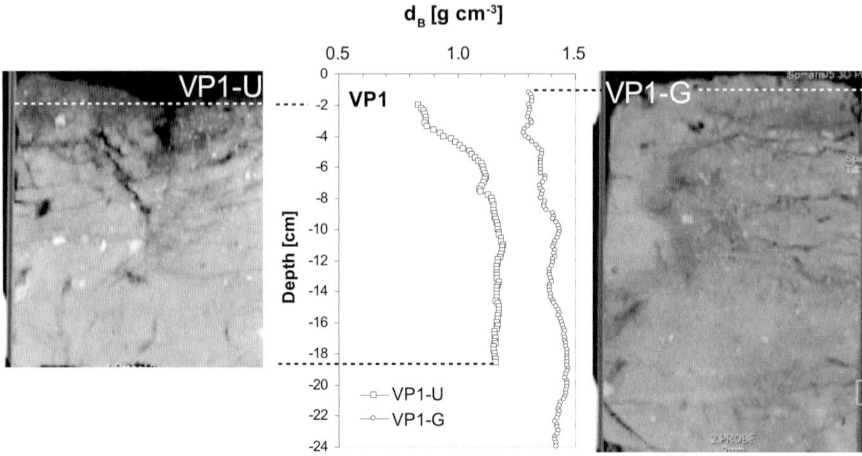

Fig. 11.2. CT Images showing the development of a system of secondary macropores at the ungrazed site VP1-U (*left*) in comparison to the grazed site VP1-G (*right*) where less macropores occur. The diagram in the middle shows the depth distribution of soil bulk density at the two sites derived from the CT scans

ondary macropores at the undisturbed control site VP3-U, which findings were missing at the grazed site VP1-G (Fig. 11.2).

The tundra sites in the Näkkälä study area did not show a clear trend in the change of pore size distribution in relation to trampling intensity or management practice (Table 11.2).

Measurements on saturated hydraulic conductivities of the investigated sites show, in general, high to very high values (AG Boden 1996) except for site VP4-F where saturated hydraulic conductivity is high in the topsoil layer but medium to low in the subsoil layers (Table 11.3). In the Vuotso study area saturated hydraulic conductivity was reduced with increasing trampling intensity at sites VP1 and VP3 but still remained very high due to the generally coarse textured soils. The very high trampling intensity inside the corral at site VP2 did not show a negative effect on saturated hydraulic conductivity.

Similar to observations of the pore size distribution, no clear correlation between management practice/trampling intensity and saturated hydraulic conductivity could be found in the Näkkälä study area.

11.4.2 Stress-Strain Relationships of the Investigated Soils

11.4.2.1 Stability of Soil Structure

Pre-compression stresses on the investigated soils range from 40 to 70 kPa with a few exceptions (Fig. 11.3) and are in general considered to be low to very low (DIN-V-19688 2002). At site VP1 pre-compression stresses seemed to

Fig. 11.3. Pre-compression stresses of the investigated soils. *Error bars* indicate confidence limits ($\alpha = 0.05$)

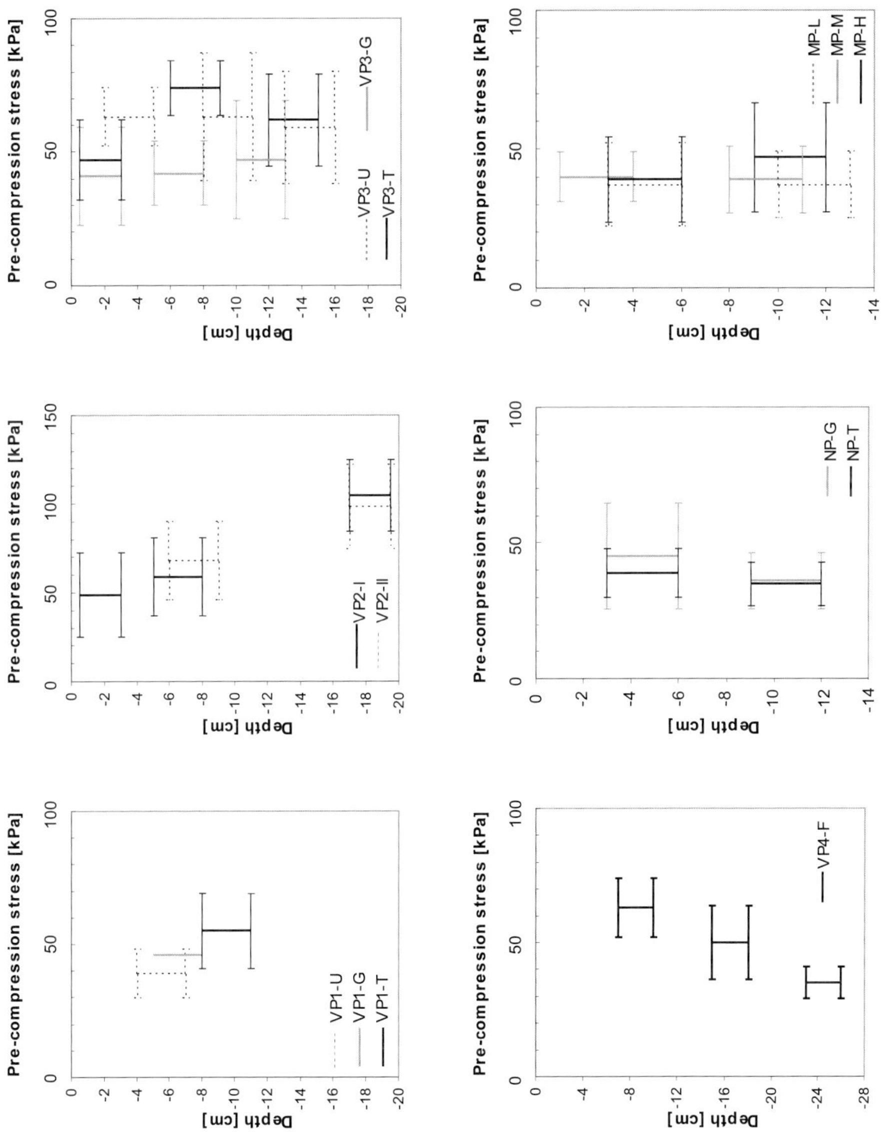

increase slightly with increasing trampling intensity, however, pre-compression stresses at very high trampling intensity inside the corral (site VP2-I) were found to be in the same order of magnitude as at the undisturbed control site outside the corral (VP2-II). At this site the pre-compression stresses increase with increasing soil depth, which corresponds to an increase in the amount of stones observed from the CT images (data not shown). The pre-compression stress on the trample path at site VP3 shows a significant increase in the second soil horizon compared to the topsoil, whereas the moderately trampled grazed site (VP3-G) as well as the undisturbed control (VP3-U) both were found to have a relatively uniform distribution of pre-compression stresses with depth. Remarkable is the relatively high pre-compression stress of the undisturbed control (VP3-U) where in the topsoil it was found even to exceed those at the heavily trampled site (VP3-T). At the wheeling track of site VP4-F, pre-compression stresses decreased as soil depth increased, reaching very low values at a depth of 23–26 cm.

Pre-compression stresses in the Näkkälä study area were somewhat lower than in the Vuotso area but seemed not to be influenced by management practices.

11.4.2.2 Soil Deformation Under Cyclic Loading

The cyclic compression tests showed that even if pre-compression stresses are not exceeded, plastic soil deformation occurs following a logarithmic trend with increasing number of load cycles (Fig. 11.4). The strength of the soil to resist cyclic loading strongly depends on soil texture. While sand and loamy sand are relatively insensitive to cyclic loading, loam reacts rather sensitively to repeated loading, as can be seen from the higher slope of the regression line in Fig. 11.4.

The mean of the slopes of all log-transformed datasets, henceforth termed "Z", are shown in Fig. 11.5. The figure indicates that the finer the soil texture the more sensitive is the soil to cyclic loading. This is also true when the initial soil bulk density decreases, as can be seen for the two different initial soil bulk densities for sand, where sensitivity doubles when soil bulk density is lowered from 1.6 to 1.5 g cm^{-3}. When comparing the change in total porosity of the initial load to the subsequent loads, it becomes obvious that fine-textured soils like loams are prone to further significant changes in pore volume when cyclic loading occurs (Table 11.4).

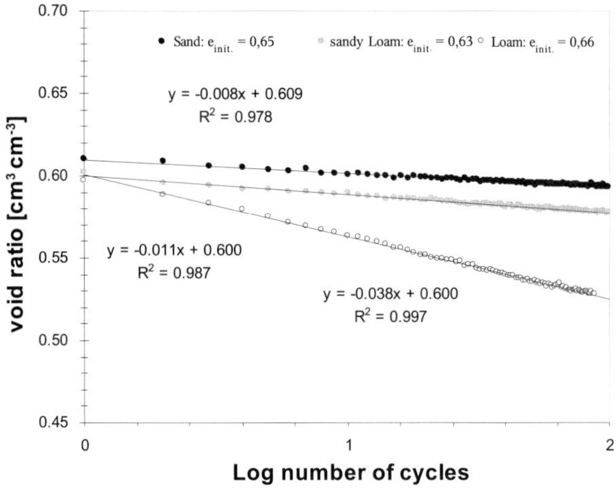

Fig. 11.4. Change in void ratio (cm^3 per cm^{-3}) as a function of number of load cycles for *Sand*, *sandy Loam*, and *Loam*. e_{init} = initial void ratio

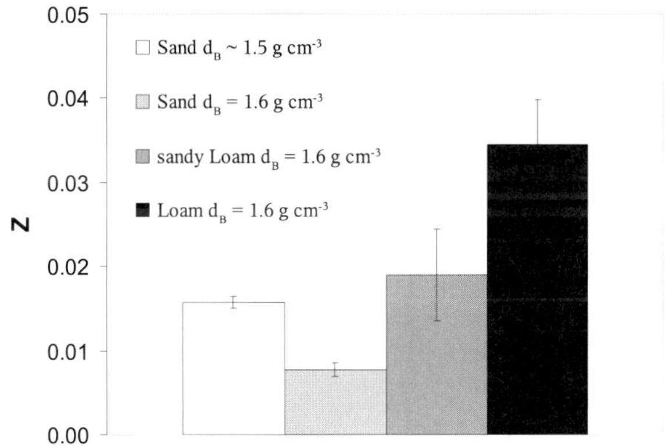

Fig. 11.5. Mean slope *Z* of the log-transformed cyclic compression curves. *Error bars* indicate confidence limits (α = 0.05)

Table 11.4. Pore volumes of samples before, after initial, and after subsequent loadings. n_0 = initial pore volume, MaP = fraction of macropores before loading, n_1 = pore volume after first load application, n_{100} = pore volume after 100 load applications, Δn_{0-1} = change in pore volume due to first loading event, Δn_{1-100} = change in pore volume due to subsequent loading events

Soil texture	n_0	MaP > 50 µm	n_1	n_{100}	Δn_{0-1}	Δn_{1-100}
	[Vol%]					
Sand	39.4	31.1	37.8	37.2	1.6	0.6
loamy Sand	35.5	15.3	34.1	33.1	1.4	1.0
Loam	39.8	7.8	37.5	34.4	2.3	3.1

11.4.3 Changes in Microclimate and Moisture Regime

11.4.3.1 Exchange of Heat Between Atmosphere and Pedosphere

The influence of the lichen cover on the soil's thermal household can clearly be seen from Fig. 11.6 (for details on vegetation characteristics along the grazing gradient, refer to Chap. 9). When the lichen cover is heavily disturbed, soils show, especially in topsoil layers, a stronger daily temperature fluctuation throughout the vegetation period. Differences are most pronounced during June and July and become smaller when mean air temperatures begin to decrease again in August. With increasing soil depth, differences in daily temperatures are dampened but may still be noticed at a depth of 30 cm. Temperature amplitudes of the undisturbed and moderately disturbed lichen cover only show small differences in the topmost soil layer and are virtually the same at depths below 10 cm. Daily mean temperatures engender a slightly different behavior. Highest mean soil temperatures are found when the lichen

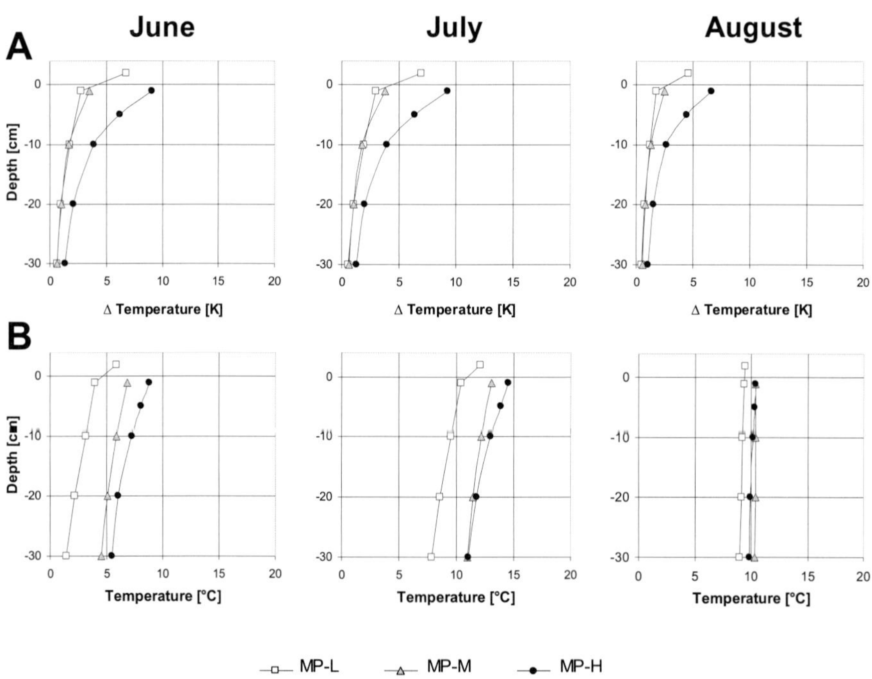

Fig. 11.6. Monthly averages of diurnal soil temperature amplitudes (*A*) in ΔK and mean soil temperatures (*B*) in the soil during the vegetation period 2003 for different disturbed lichen covers. The uppermost value of *MP-L* (undisturbed lichen cover) reflects the situation within the lichen cover, *MP-M* = moderately disturbed lichen cover, *MP-H* = heavily disturbed (absent) lichen cover

cover is heavily disturbed, but moderately disturbed lichen cover provokes noticeably higher soil temperatures than associated with undisturbed lichen cover. Mean soil temperatures indicate that soils warm up more quickly at the beginning of the vegetation period with increasing disturbance of lichen cover. During August soil temperatures start to decrease again at the sites with moderately and heavily disturbed lichen cover, while soil temperatures in the lower soil layers underneath an undisturbed lichen cover are still increasing due to a phase lag of the temperature wave.

Lichens obviously contribute strongly to insulation of the soil. The insulation effect can be seen on the temperature profile of a day with a relatively high temperature change (Fig. 11.7). Temperature within the lichen cover at noon strongly exceeds air temperature measured at a height of 2 m above ground. During the vegetation period in 2002 temperatures of more than 35 °C were measured in the lichen thallus. However, although the temperature gradient between the lichen cover and the underlying soil layer is very high, the heat is not conducted into the soil profile, as is indicated by the lower soil temperatures within the organic topsoil layer, but rather is retained within the lichen thallus causing its temperature to increase. In contrast to this, when the lichen cover is disturbed, ground surface temperatures are much lower and soil temperatures are higher. At night, on the other hand, heat flux from the soil into the lichen cover and hence toward the atmosphere is also reduced, as

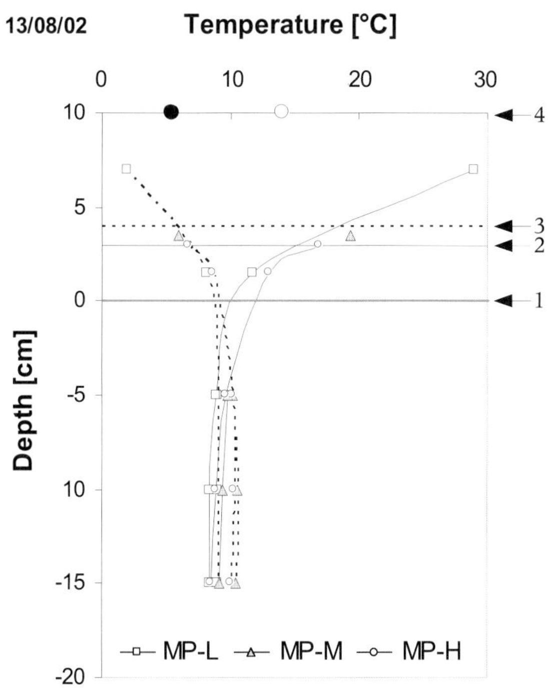

Fig. 11.7. Soil temperature profiles for different disturbed lichen covers at midnight (*dashed lines*) and at noon (*solid lines*) on a day in August. *MP-L* = undisturbed lichen cover, *MP-M* = moderately disturbed lichen cover, *MP-H* = heavily disturbed (absent) lichen cover. *1* = boundary between organic topsoil and mineralic soil horizon, *2* = ground surface (completely disturbed lichen cover), *3* = surface of moderately disturbed lichen cover, *4* = surface of undisturbed lichen cover, • = air temperature 2 m above ground at midnight, *o* = air temperature 2 m above ground at noon

shown by the very low temperatures within the undisturbed lichen cover at midnight – which occasionally reaches values well below 0 °C in early morning (in this case –3 °C at 0400).

The calculated net daily energy fluxes show that during the measurement period the energy flux where the lichen cover is heavily disturbed is significantly higher than where the lichen cover is moderately disturbed (i.e., partly damaged by trampling) or undisturbed; the latter two situations having approximately the same energy flux. This is true when the energy flux is directed into the soil profile, i.e., when soil is heated up, as well as during cooling periods when the soil loses energy to the atmosphere, as was the case at the beginning of July 2003 (Fig. 11.8). In total, the cumulative energy flux into the soil during the vegetation period is almost twice as high when the lichen cover is heavily disturbed than when it is only moderately so or undisturbed.

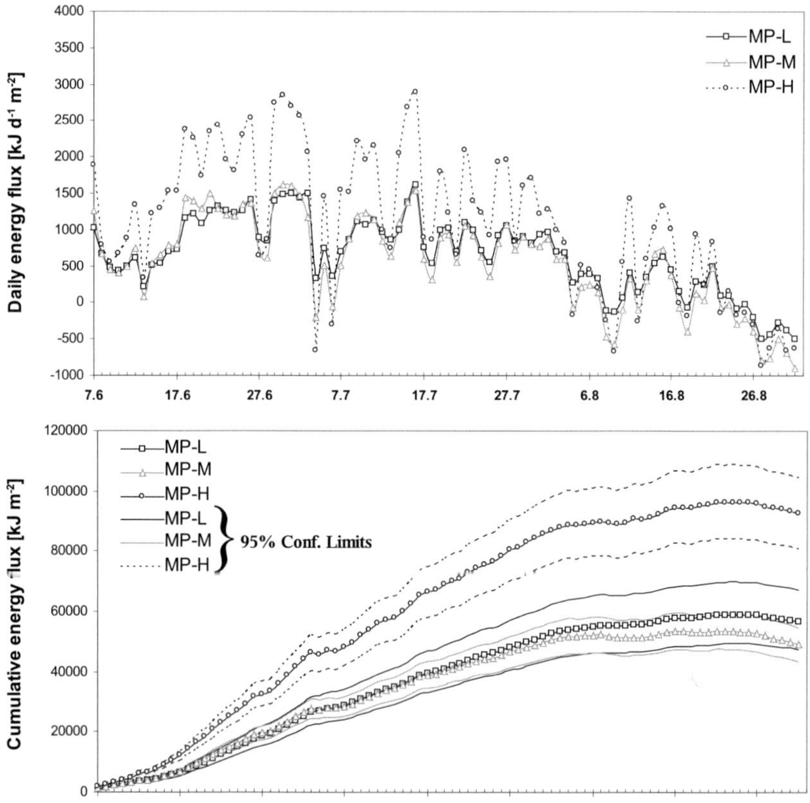

Fig. 11.8. Daily net (*top*) and cumulative energy fluxes (*bottom*) for different disturbed lichen covers during the vegetation period 2003. *MP-L*, *MP-M*, and *MP-H* as before. Note: Energy fluxes before 7.6 have not been considered in this graph. Therefore, the curves in the lower diagram have to be considered as shifted on the y-axes according the energy flux before 7.6

11.4.3.2 Exchange of Water between Atmosphere and Pedosphere

Lichen cover not only influences the exchange of heat but also has an effect on the infiltration of water into the soil after rainfall. Figure 11.9 shows the change in volumetric water content measured at a depth of 2 cm, and indicates that infiltration of water is higher where the lichen canopy is undisturbed. Differences in volumetric water content depend on the strength and duration of rainfall events, and may reach up to 10 Vol% between a totally disturbed and an intact lichen cover (Fig. 11.9, right). Two or three days after rainfall has ceased and the infiltration front has moved deeper into the soil profile, volumetric water content approaches field capacity.

Apart from the infiltration of water into the soil there is also evaporation at the soil surface and hence a water flux from the soil toward the atmosphere. The loss of water due to evaporation is also influenced by the lichen canopy, as shown in Fig. 11.10. Daily temperature fluctuations are in good agreement with fluctuations in volumetric water contents. Evaporation exceeds the rate of capillary rise in periods when soil temperature is increasing, thus reducing the water content while during periods when temperature decreases water is replenished in the topsoil layer probably mainly due to capillary rise from deeper soil layers and to a minor extent possibly also from condensation of water vapor of the atmosphere. In any event, loss of water, mostly from evaporation, is highest where lichen cover is heavily disturbed and lowest where soil is covered by an undisturbed lichen canopy.

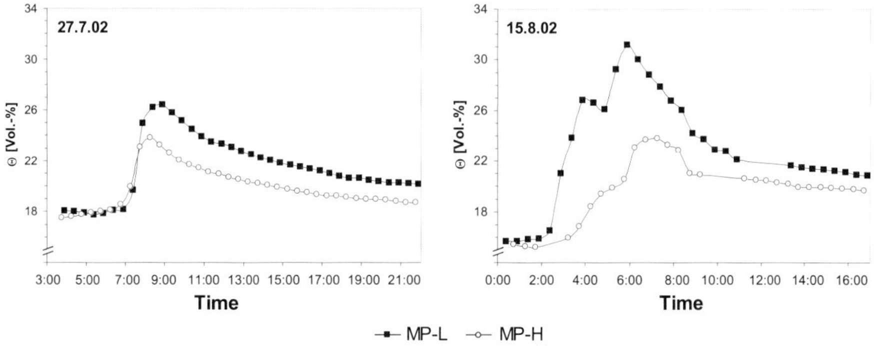

Fig 11.9. Change of volumetric water contents measured in a depth of 2 cm after rainfall beneath the undisturbed lichen cover (*MP-L*) and the heavily disturbed lichen cover (*MP-H*)

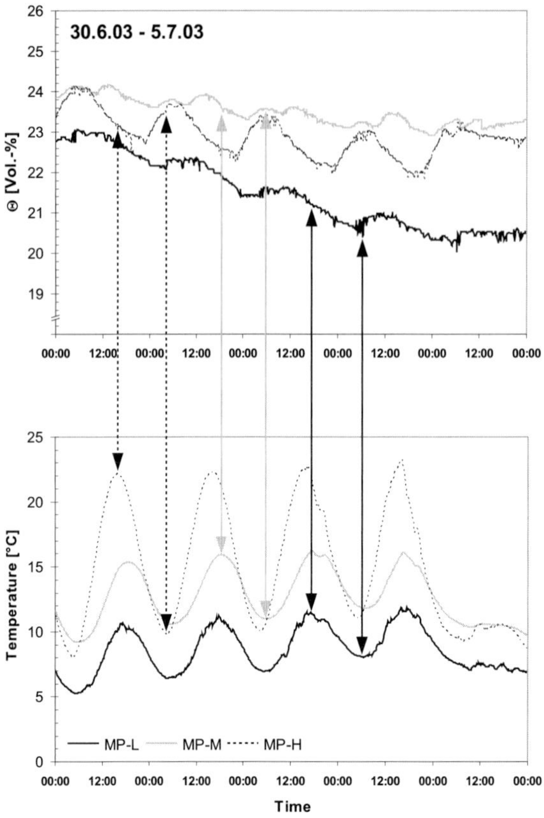

Fig. 11.10. Change in volumetric water content (*top*) in comparison to changes in soil temperature at a depth of 2 cm in the daily course for different disturbed lichen covers. *MP-L* = undisturbed lichen cover, *MP-M* = moderately disturbed lichen cover, *MP-H* = heavily disturbed (absent) lichen cover

11.5 Discussion

11.5.1 Changes in Soil Structure and Functioning as a Result of Mechanical Loading

In general soils of the study areas possess a low mechanical stability. If for the investigated soils mechanical stresses exceed 40 to 70 kPa, further plastic soil deformation is likely to occur. Assuming ground contact pressures for reindeer of approximately 30 to 40 kPa (estimated by mean summer body weight over hoof contact area) and for an ordinary forestry machine of about 80 kPa, soils may just withstand mechanical forces by reindeer, but not those imposed by forestry equipment. This assumption is confirmed by the strong reduction of pore volume and hydraulic conductivity underneath the wheeling track. During soil deformation by wheeling, soil structure is destroyed by homogenization, resulting in even lower soil stability as indicated by the very low precompression stresses in the subsoil layers.

Higher pre-compression stresses in the topmost soil layer are due to the initial re-aggregation of the soil by shrinkage after kneading has occurred. This is underlined by the higher vertical saturated hydraulic conductivity in the topsoil, which indicates the formation of a new secondary pore system, while subsoil layers, where re-aggregation has not yet started, show lower stability combined with lower saturated hydraulic conductivity. Air capacity in the subsoil at site VP4-F is only medium and much lower compared to the other sites, where air capacities are classified as high to very high. This is likewise true for saturated hydraulic conductivities, suggesting that under conditions encountered in the investigation areas the use of heavy machinery in forestry practice may be considered to cause a higher disturbance of pore functioning (aeration, water infiltration) and hence a higher limitation on plant growth than is to be expected from reindeer. Although reindeer trampling may also result in soil compaction (site VP1 and VP3) the coarse-textured soils of the study areas were not adversely affected by this compaction.

According to the theory of critical state soil mechanics, plastic soil deformation only occurs when pre-compression stresses are exceeded (O´Sullivan and Robertson 1996). However, cyclic loading tests have shown that, even if pre-compression stresses are not exceeded, some incremental plastic soil deformation may occur. The deformation that might be expected from the repeated mechanical loading of the soil depends, apart from soil structure, strongly on soil texture. Soil textures encountered in the investigation area were mainly coarse-grained sand to sandy loam, and therefore less sensitive to cyclic loading. In areas where finer soil textures prevail, especially where soils contain a higher clay fraction, repeated loading due to heavy trampling could significantly reduce pore volume. Bearing in mind that fine textured soils, in contrast to coarse textured soils, have a much lower fraction of primary macropores, and that volumetric soil deformation is accommodated predominantly by the destruction of secondary macropores, heavy trampling may have a negative impact on the conductivity of water and air and thus on plant growth.

11.5.2 Changes in Soil Thermal Regime as a Result of Changing Lichen Cover

Lichen canopies are known to be good insulators (Kershaw and Rouse 1971; Rouse and Kershaw 1971). Kershaw and Field (1975) showed that temperature gradients that develop within a canopy of *Cladina alpestris* are considerably greater than other canopy profiles examined, underlining the effectiveness of this particular species as a thermal insulator. Consequently, changes in lichen cover that results from trampling during summer grazing alters the soils microclimate (Stark et al. 2000). The reduction of the lichen cover increases the heat flux between soil and atmosphere, leading to higher average soil temperatures in the summer and also lower average soil temperatures in the winter (Broll 2000; Olofsson et al. 2002). When the lichen canopy is disturbed soils tend to warm up more quickly, especially at the beginning of the vegetation period, which could have a positive feedback on plant growth, since germination may start earlier and thus lengthen the vegetation period. Field observations revealed that at the beginning of June 2003 the soil underneath the undisturbed lichen cover was still frozen at a depth of 30 cm, whereas no ice was encountered at that depth at the sites with moderately and heavily disturbed lichen cover. As a consequence of the delayed thawing and impeded seepage when the lichen cover is undisturbed, soil water content remains for a longer time near saturation at the beginning of the vegetation period. This may lead to a reduction of soil aeration until macropores are emptied as drainage proceeds.

On the other hand infiltration of water into the soil is decreased and evaporation increased when the lichen canopy is disturbed, which may in dry summers with low groundwater tables reduce plant available water and lead to water stress for plants. Horn (1994) showed that the production of abscissic acid (ABA) in xylem sap as an indicator of shortage of plant-available water in the soil is not only, as usually assumed, related to the water stored in pores <2 μm (pF > 4.2), but also depends on the minimum water potential value that plants have ever encountered in the soil. Hence, water stress may occur at pF values well below the actual wilting point, i.e., drying of soil to a water potential that has not been reached before possibly leads to an impeded stomatal conductance. Plants often avoid stress due to moisture deficits by enhanced root growth, thus increasing water-drawing power and therefore their chance for survival (Hillel 1998). This might be one reason for the strong increase in root production in tundra plant communities, as has been reported by Olofsson et al. (2002) in heavily grazed drier continental sites.

Soil temperature and moisture are abiotic factors that control microbial activity and thus turnover rates (Svensson et al. 1975). A change in microclimate may increase mineralization and nutrient availability when the lichen canopy is removed, resulting from higher soil temperatures as long as water is not the limiting factor for microbial turnover. The change in microclimate is assumed to be one of several mechanisms (e.g., shift of vegetation into more decomposable plant species, fertilization by urine and feces) by which herbivores have a positive impact on plant production (Olofsson et al. 2002). However, the often-postulated transition of a less productive moss-rich heath tundra into a more productive graminoid-dominated tundra vegetation (Olofsson et al. 2001) is considered to be a slow process that might take several decades (Olofsson et al. 2002). In this case one may argue that under nutrient-poor conditions, as is frequently encountered in the continental tundra heath, nutrients are released from soil pools in a manner – namely, by leaching – which allows neither full use for primary production nor nutrient availability for the long term (see also Chap. 12). Apart from this, the importance of the lichen cover, in view of the expected global warming, needs further discussion since then soil temperatures will increase even more when it is disturbed.

11.6 Conclusion

Impacts of grazing on soil structure are considered to be less prominent compared to impacts of heavy forestry machinery. High trampling intensities, however, in areas where fine-textured soil prevails may increase the risk for unfavourable soil deformation leading to deterioration of soil functions. This could reduce plant production and increase the risk for erosion, especially along reindeer paths and close to border fences. However, the more dominant impact of grazing on a soil's physical properties is assumed to be related to the change of lichen cover caused by summer grazing, since a reduction in the lichen canopy significantly alters the soil's thermal and water household. Although positive effects for plant growth may be connected with increased soil temperatures, resulting in increased mineralization rates, nutrient pools could in this process also be reduced by leaching. This might lead to a drop in plant nutrition, consequently reducing pasture productivity in the long term. Lichens are considered to play an important role with respect to global warming. They should be protected from heavy summer trampling not only to preserve them as one of the most important winter diets for reindeer but also to insulate the soil from future atmospheric temperature increases.

References

AG-Boden (1996) Bodenkundliche Kartieranleitung, 4th edn. Schweitzbart´sche Verlagsbuchhandlung, Stuttgart

Aikio P (2001) The prospects of Sami reindeer herding in Finland. In: Kankaanpää S, Sara E, Rybkin A, Suomi K, Magga J, Turi JM, Aatsinki S, Valkeapää L (eds) The 2nd world reindeer herders´ congress, pp 36–40

Broll G (2000) Influence of overgrazing by reindeer on soil organic matter and soil microclimate of well-drained soils in the Finnish subarctic. In: Lal R, Kimble JM, Stewart BA (eds) Global climate change and cold regions ecosystems. Lewis Publ, London, pp 163–172

Burgess P (1999) Human Environmental Interactions in Upper Lapland, Finland. Arctic Centre Rep 27. Arctic Centre University of Lapland, Rovaniemi

Casagrande A (1936) Characteristics of cohesionless soils affecting the stability of slopes and earth fills. Soc Civil Engrs 23:13–32

DIN-V-19688 (2002) Ermittlung der mechanischen Belastbarkeit von Böden aus der Vorbelastung. Handbuch der Bodenuntersuchung, vol 6. Beuth, Berlin

Evans R (1994) Impacts of reindeer grazing on soils and vegetation in Finnmark. Norut information technology, rep 2033/02-94

Gifford GP, Hawkins RH (1978) Hydrologic impact of grazing and infiltration: a critical review. Water Resourc Res 14:305–313

Greenwood KL, McKenzie BM (2001) Grazing effects on soil physical properties and the consequences for pastures: a review. Austr J Exp Agricult 41:1231–1250

Hillel D (1998) Environmental soil physics. Academic Press, London

Hinneri S, Sonesson M, Veum AK (1975) Soils of Fennoscandian IBP tundra ecosystems. In: Wielgolaski FE (eds) Fennoscandian Tundra ecosystems, vol 1. Springer, Berlin Heidelberg New York, pp 31–40

Hartge KH (1966) Ein Haubenpermeameter zum schnellen Durchmessen zahlreicher Stechzylinderproben. Z Kulturtechn Flurbereinig 7:155–163

Horn R (1994) The effect of aggregation of soils on water, gas, and heat transport. In: Schulze E-D (ed) Flux control in biological systems: from enzymes to populations and ecosystems. Academic Press, London, pp 335–361

Kershaw KA, Field GF (1975) Studies on lichen-dominated systems. XV. The temperature and humidity profiles in a *Cladina alpestris* mat. Can J Bot 53:2614–2620

Kershaw KA, Rouse WR (1971) Studies on lichen-dominated systems. I. The water relations of *Cladonia alpestris* in spruce-lichen woodland in northern Ontario. Can J Bot 49:1389–1399

Löffler J (1999) Nutzungspotentiale und Nutzungsgrenzen in den norwegischen Hochgebirgslandschaften. In: Jung G (ed) Norwegen eine Naturlandschaft? Ökologie und nachhaltige Nutzung, vol 4. Bibliotheks und Informationssystem der Carl von Ossietzky Universität Oldenburg, Oldenburg, pp 163–188

O´Sullivan MF, Robertson EAG (1996) Critical state parameters from intact samples of two agricultural topsoils. Soil Till Res 39:161–173

Olofsson J, Kitti H, Stark S, Oksanen L (2001) Effects of summer grazing by reindeer on composition of vegetation, productivity and nitrogen cycling. Ecography 24:13–24

Olofsson J, Stark S, Oksanen L (2002) Herbivore influence on ecosystem processes in tundra plant communities. In: Stark S (ed) Reindeer grazing and soil nutrient cycling in boreal and tundra ecosystems, A382. Department of Biology, University of Oulu, Oulu, pp 1–12

Peth S (2004) Bodenphysikalische Untersuchungen zur Trittbelastung von Böden bei der Rentierweidewirtschaft an borealen Wald- und subarktisch-alpinen Tundrenstandorten – Auswirkungen auf thermische, hydraulische und mechanische Bodeneigenschaften. PhD Thesis, Institut für Pflanzenernährung und Bodenkunde, Christian-Albrechts-Universität zu Kiel, Kiel, 160 pp

Plagge R (1991) Bestimmung der ungesättigten hydraulischen Leitfähigkeit im Boden. PhD Thesis, Institut für Ökologie, TU Berlin, Berlin, 152 pp

Rogasik H, Onasch I, Brunotte J, Jegou D, Wendroth O (2003) Assessment of soil structure using X-ray computed tomography. In: Mees F, Swennen R, van Geet M, Jacobs P (eds) Applications of X-ray computed tomography in the geosciences, vol 215. Geological Society of London, London, pp 151–165

Rouse WR, Kershaw KA (1971) Studies on lichen-dominated systems. II. The growth pattern of *Cladonia alpestris* and *Cladonia rangiferina*. Can J Bot 49:1401–1410

Stark S (2002) Reindeer grazing and soil nutrient cycling in boreal and Tundra ecosystems. PhD Thesis, Department of Biology, Univ Oulu, Oulu, 31 pp

Stark S, Wardle DA, Ohtonen R, Helle T, Yeates GW (2000) The effect of reindeer grazing on decomposition, mineralisation and soil biota in a dry oligotrophic Scots pine forest. Oikos 90:301–310

Svensson BH, Veum AK, Kjelvik S (1975) Carbon losses from Tundra soils. In: Wielgolaski FE (ed) Fennoscandian Tundra ecosystems: plants and microorganisms, vol 1. Springer, Berlin Heidelberg New York, pp 279–286

Van Genuchten MT (1980) A closed-form equation for predicting the hydraulic conductivity of unsaturated soils. Soil Sci Soc Am J 44:892–898

Van Genuchten MT, Leij FJ, Yates SR (1991) The RETC code for quantifying the hydraulic functions of unsaturated soils. US Salinity Laboratory / US Department of Agriculture, Riverside

Van Wijk WR, de Vries DA (1966) Periodic temperature variation in a homogeneous soil. In: Van Wijk WR (ed) Physics of plant environment. North-Holland Publ, Amsterdam, pp 102–143

Willat ST, Pullar DM (1983) Changes in soil physical properties under grazed pastures. Austr J Soil Res 22:343–348

12 Changes in Organic Horizon Soil Properties Due to Reindeer Herding and Changing Management

C. UHLIG and A. ZINK

12.1 Introduction

Mammalian herbivores can generally affect soil nutrients and nutrient cycling by their defecation and trampling (Fig. 12.1). The direct consumption of plants by herbivores normally enhances nutrient cycling, as urine and feces create an organic soil nutrient pool where nutrient release tends to be faster than nutrient release from litter (e.g., McKendrick et al. 1980; Zimov et al. 1995). Over time, increased nutrient turnover by grazing herbivores can improve soil fertility and thus support more fertile and grazing-tolerant plant species. This is true, for example, in the alpine/arctic environment where grazing and browsing can change both tall-herb and dwarf-shrub heath communities into ones of grass domination (Zimov et al. 1995; Olofsson et al. 2001). However, results from Uhlig et al. (2004) indicate that reindeer grazing decreases soil fertility of lichen-dominated mountain birch forests in Fennoscandia, and signs of soil erosion due to reindeer herding are repeatedly reported (Helle and Aspi 1983; Foster 1985; Väre et al. 1995; Evans 1996; Johansen and Karlsen 1998; Uhlig et al. 2004).

Cryptogams, e.g., fruticose lichens, are important in the structure of tundra and taiga ecosystems, as exemplified by cover values, biomass, mineral content; significance for soil, thermal, and water household, and effect on other ecosystem components (Longton 1992; Chap. 11). Usnic acid and other sparingly mobile lichen compounds are, for example, suggested as contributing significantly to podzolation and profile development in alpine tundra soils (Dawson et al. 1984). Thus, fruticose lichens influence pedogenesis on mineral soils by contributing organic matter, through their impact on nutrient cycling, and by their stabilizing effect on soil temperature and moisture regimes (Rouse and Kershaw 1971; Edward and Miller 1977; Chap. 11) Furthermore, it is well recognized that trampling and grazing of lichens and plants by reindeers today are major factors controlling vegetation cover and

Ecological Studies, Vol. 184
B.C. Forbes, M. Bölter, L. Müller-Wille, J. Hukkinen, F. Müller, N. Gunslay, Y. Konstatinov (Eds.)
Reindeer Management in Northernmost Europe

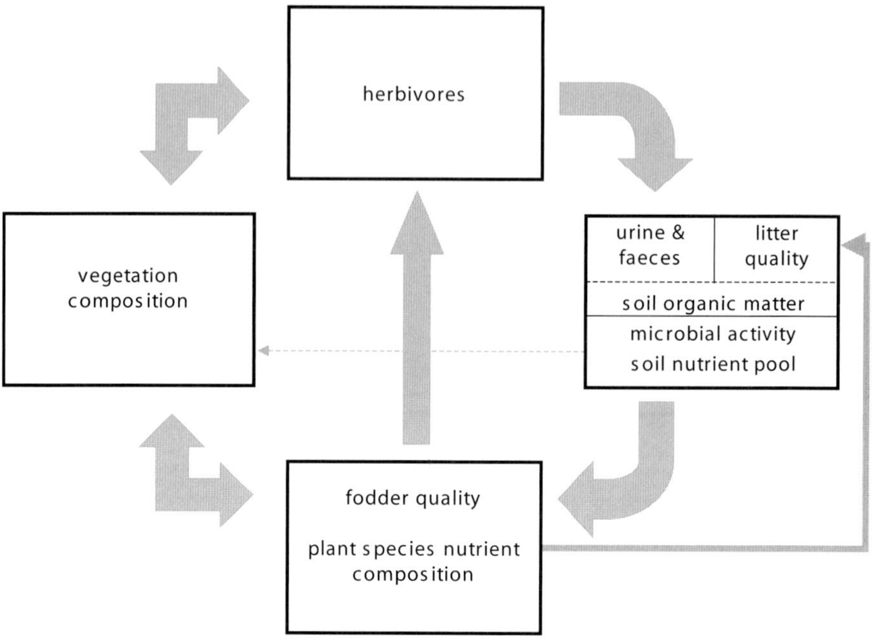

Fig. 12.1. A simplified flow scheme illustrating interactions between herbivores, nutrient cycling, fodder quality, and composition of vegetation. For further details see text

composition all over the Northern Fennoscandia (Oksanen 1978; Helle and Aspi 1983; Väre et al. 1995; Stark et al. 2000; Kumpula 2001; Bråthen and Oksanen 2001).

Particularly, lichen-dominated vegetation types are strongly affected, and Richardson and Young (1977) emphasized the importance of reindeer population density for the effect on lichen flora. Polunin (1936), Pegau (1970), and Johansen and Karlson (1998) have reported that reindeer grazing for the most part reduces the thickness and cover of fruticose lichen species, while the abundance and species numbers of minor lichens, partly those participating in early succession, has been found to increase (Oksanen 1986). In final consequence, overgrazing by reindeer may terminate all vegetation in an area and thus facilitate the erosion of the organic layer and expose the bare mineral soil (Foster 1985; Väre et al. 1996; Evans 1996; Olofsson et al. 2001). Similar influences of wild caribou herds on vegetation have been reported in North America – i.e., caribou grazing reduces lichen cover and growth of deciduous shrubs and bryophytes, and increases coverage by non-forage lichens and bare soil (Klein 1987; Manseau et al. 1996). According to Ahti and Oksanen (1990) and Vilchek (1997), depletion of lichen cover in tundra due to reindeer herding is also widespread in Russia. The impact of grazing on soil physical properties is assumed to be related to the change of lichen cover (by summer grazing) since reduction of the lichen canopy significantly alters the soil's thermal and water household.

Based on the reported impact of mammalian herbivores on cryptogams, plants, and soils, it can be expected that reindeer husbandry will lead to significant changes in soil properties, especially those of the uppermost O horizon (Chap. 11, 13). The aims of our investigation were first to investigate the effects of reindeer grazing on physical and chemical properties of O horizons, and second to compare these impacts under different reindeer management systems, namely, those of winter grazing pastures in Norway and of year-around-grazing pastures in Finland.

12.2 Materials and Methods

12.2.1 Selection of Study Area

In general, it is rather difficult to study the impact of a single herbivore species on an ecosystem. However, reindeer densities in the Finnmarksvidda region are at least 10–50 times higher than in most places outside Fennoscandia, e.g., North America (Batzli et al. 1980) or Russia/Siberia (Chernov and Matveyeva 1997), and reindeer may have become regionally the main mammalian herbivore. This investigation was performed at Jauristunturit, Finnmarksvidda, at about 550 m asl at the border between Finland and Norway (68°50' N, 23°50'E; Fig. 12.2). The climate of the nearest weather station (Siččajavri, ca. 400 m asl) is characterized by a mean annual air temperature of –3.1 °C and an annual precipitation of 366 mm (DNMI 1999). On both sides of the fenced national border, *Betula nana* and *Cladina* types dominate the heath vegetation (Olofsson et al. 2001; Chap. 8), which by area and usage represents an important habitat for reindeer husbandry in both nations. Management in Finland is of year-around pastures, while pastures in Norway are used only in winter (Chap. 7). Today, both reindeer management regimes are permanently separated by a fence that was erected in the early 1950s. The investigated area also includes pastures, which due to relocation of the reindeer fence during the 1990s, changed from Norwegian winter to Finnish year-around pastures (Fig. 12.2). In the following results from sample sites within the area where the fence was moved are referred to as N/Fin.

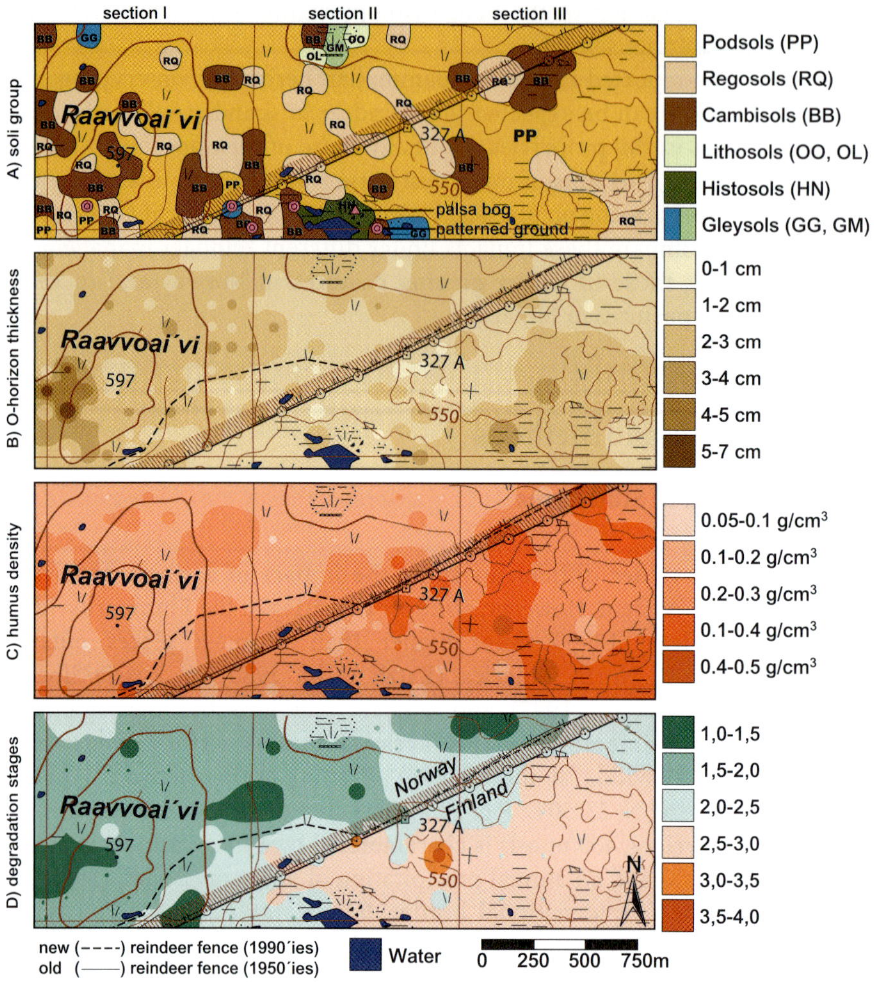

Fig. 12.2a–d. The regional appearance of: **a** soil groups, **b** average thicknesses of O horizons, **c** average humus densities, and **d** dominant degradation stages of lichen dominated dwarf shrub tundra in the investigated area in Jauristunturit, Finnmarksvidda. Background map: topographic map 1:200,000 Finland sheet: 281412+09

12.2.2 The concept of Degradation/Succession Stages

Jauristunturit, as many other tundra landscapes, represents a mosaic of different arctic ecosystems, reaching from dry and windblown lichen ridges to wet peatlands. Any studies of changes in soil properties in such a patchy landscapes needs to ensure collection of data from comparable sample sites according to essential abiotic factors, e.g., quaternary geology, topography, and winter conditions. In general, different vegetation types mirror the sum of site-specific abiotic conditions to a reasonable degree (and vice versa). However, pastures with comparable vegetation types are not grazed entirely equally by migrating herbivores, and small-scale variations in grazing pressure normally occurs, even within one management regime.

In general, reported pattern for impacts of reindeer grazing on lichen pastures (Polunin 1936; Pegau 1970; Richardson and Young 1977; Oksanen 1978; Helle and Aspi 1983; Oksanen 1986; Manseau et al. 1996; Johansen and Karlson 1998; Uhlig et al. 2004) are rather similar, but quite unlike those found for the post-fire succession of many lichen-dominated boreal and arctic ecosystems, which floristically are characterized by the succession of cryptogams (Ahti 1977): (1) Bare soil stage; (2) Crustose lichen stage; (3) Cup lichen stage; (4) First reindeer lichen stage, (5) Second reindeer lichen stage. Sites in stage 2 are often first colonized by mosses and crustose lichens, mainly *Polytrichum* spp. and *Lecidea* spp., that stabilize the surface. Stage 3 is characterized by the establishment of cup lichens, e.g., *Cladonia cornuta* and *C. gracilis*, while larger fruticose lichens, generally *Cladonia* spp., become distinctly dominant in stages 4 and 5.

Ahti (1977) distinguished between two reindeer lichen stages. The younger stage 4 is dominated by *Cladina mitis*, *C. arbuscula*, *C. rangiferina* and *Cladonia uncialis*, the older stage 5 by *Cladina stellaris*. Based on results on the impact of reindeer grazing on lichen grounds, post-fire succession stages described by Ahti (1977), and field observations from the investigated area, four different degradation/succession stages (DS) for the Fenno-Scandinavian lichen-dominated *Betula nana* tundra ecosystem were characterized (Table 12.1).

In the following, the concept of DSs was used as a strategy for the selection of comparable sample sites in Jauristunturit. First a 3 km^2 area covering about equal parts in Finland and Norway was selected (Fig. 12.2). Then a rectangular grid with 100 m (200 m for sector III) distances between grid points was determined to systematically indicate potential sample sites. Along a 100 m long transect between two neighboring grid points DSs were determined every fifth meter and their average was used to select the DS of the approaching sample site. DS in vegetation types along transects other than lichen-dominated heath, e.g., peat lands, were not recorded since the DS model does not apply for them. However, for presentation purposes DSs were interpolated over the whole study area, excepting lakes, and presented in distinct classes (Fig. 12.2). It was assumed that the vegetation differences among the different degradation stages were mainly caused by different reindeer-grazing activities.

Table 12.1. Characterization of degradation/succession stages (DS) of lichen dominated *Betula nana* tundra in Fennoscandia

Degradation/ succession stages (DS)	I	II	III	IV
Succession stage (Ahti 1977)	1st reindeer lichen stage	2nd reindeer lichen stage	Cup lichen stage	Crustose lichen and bare soil stage
Dominant lichens	Fruticose lichens (FL)		Cup, navel, and crustose lichens	
Average height of fruticose lichens	12 to 4 cm	4 to1 cm	1 to 0 cm	0 cm
Characteristic cryptogams	*Cladina stellaris, Cetraria nivalis, Cetraria islandica, Cladina mitis, Cladina rangiferina, Cladonia uncialis*	*Cladina stellaris, Cetraria nivalis, Cetraria islandica, Cladina mitis, Cladina rangiferina, Cladina arbuscula, Cladonia uncialis*	*Cladonia cornuta, Cladonia gracilis, Cladonia crispata, Stereocaulon* spp., *Polytrichum* spp.	*Lecidea oligotropha, L. uliginosa, L. granulosa, Polytrichum* spp.
Other characteristics	>50 % FL cover	50 to 25 % FL cover, or FL cover >50 % but clear signs for recent grazing of lichen carpet	25 to 5 % FL cover, vegetation free bare patches with exposed O horizon. Plant stems and roots partly apparent	<5 % FL cover, Thin O horizon with patches of bare mineral soil in between. Plant stems and roots partly apparent

12.2.3 Soil Description, Sampling, and Analysis

From a total of about 230 GPS-recorded sample sites within the investigated area (Fig. 12.2) detailed descriptions of soil profiles were carried out according to AG-Boden (1996), including soil texture, type, and color. Further, type soil profiles for each DS were carefully excavated and thoroughly described by horizons according to FAO (1990) subsequent to a detailed vegetation analysis (Table 12.2). At each sample site the total depth of the organic horizon (O horizon) was calculated as the mean of five individual measurements from subsample sites in between *B. nana* canopies. For humus density determinations and chemical analyses organic matter from the entire O horizon (Ol, Of, and Oh) was sampled at three different subsample sites using 100 cm^3 steel cylinders. Before combining soil samples of each sample site in polyethylene bags, depths of the organic layer within cylinders were measured for determination of sample volume.

At the Chemical Analysis Laboratory of the Holt Research Centre, weights of humus samples were measured before and after drying at 35 °C for 48 h, and humus density was calculated by dividing the dry weight of the samples by the volume. Prior to chemical analysis, fragments >2 mm were removed from all samples by using a sieve. Soil samples were than analyzed for $pH_{(H20)}$, Kjeldahl-N (Bremner 1960), organic carbon (loss of ignition), total (HNO_3 soluble), and plant-available Ca, Mg, K, and P (Egnér and Riehm 1960). Analy-

Table 12.2. Vegetation composition and cover (100 × 100 cm) according to Braun-Blanquet (1932) directly above the representative pedons (Table 12.3) of the degradation stages. Cover ranges: "5" = 75 to 100 %; "4" = 50 to 75 %; "3" = 25 to 50 %; "2" = 5 to 25 %; "1" = 1 to 5 %; "+" = < 1 % but frequently found; "R" = rare; – = absent

Species	DS I	DS II	DS III	DS IV
Cryptogams				
Cladina stellaris	5	3	R	R
Dicranum spp.	1	2	–	–
Polytrichum spp.	1	1	1	2
Cladina rangiformis	1	+	–	–
Cetraria islandica	+	1	1	+
Cladonia uncialis	+	1	–	+
Cladina mitis	+	+	–	–
Cetraria nivalis	–	3	1	1
Lecidea spp.	–	1	2	4
Cladonia spp.	–	1	–	3
Stereocaulon spp.	–	+	1	+
Cladonia aphylla	–	+	–	–
Alectoria nigricans	–	–	–	1
Vascular plants[a]				
Vaccinium vitis-idaea	1	1	1	2
Festuca ovina	1	1	–	R
Carex biglowii	R	–	+	1
Empetrum nigrum	–	+	–	1
Deschampsia flexuosa	–	–	1	–

[a] *B. nana* is not included since vegetation analyses were performed in between *B. nana* canopies (see also section on Materials and Methods)

sis of filtered solutions were performed using a Perkin-Elmer Optima 3300 DV (Perkin-Elmer Instruments, Norwalk, CT, USA). For technical reasons analyses of pH, Kjeldahl-N, and plant available Ca (Ca_{AL}), Mg (Mg_{AL}), K (K_{AL}) and P (P_{AL}), were performed from soil samples representing sections I and II, while total Ca (Ca_{tot}), Mg (Mg_{tot}), K (K_{tot}), P (P_{tot}), and C-org were only analyzed for section I (Fig. 12.2). However, based on correlations between plant nutrients arrived from section I, results for Ca_{tot}, Mg_{tot}, K_{tot}, $N_{Kjeldahl}$, and C-org for section II were calculated according to following equations: $Ca_{tot} = 2.27$ $Ca_{AL} + 15.56$ ($r^2 = 0.9701$); $Mg_{tot} = 1.145$ $Mg_{AL} + 65.29$ ($r^2 = 0.9417$); $K_{tot} = 1.02$ $K_{AL} + 75.53$ ($r^2 = 0.9423$); $P_{tot} = 535.11$ $N_{Kjeldahl} + 50.32$ ($r^2 = 0.8061$).

12.2.4 Statistical Analysis

Normality tests showed that the investigated soil chemical parameters are normally distributed, in general. Thus, analysis of variance was performed on all data using the GLM procedure of SAS (version 6.12). Chemical properties of the different soil horizons with different degradation stages were compared using the different locations as replicates. The individual means were compared using the Tukey's Studentized Range (HSD) test. Significance was assigned to a test if $P \leq 0.05$.

12.3 Results

Podzols (Fig. 12.2a; Table 12.3), which had developed on coarse-ground morainic till deposited by glaciers of the Pleistocene glaciations (Olsen et al. 1996), was the dominant soil type (63 %), while regosols and cambisols covered about 19 % and 14 % of the investigated area. According to DS (Fig. 12.3) about 41 % of the area was classified as DS II or better, about 58 % between DS II and DS III, and less than 1 % scored less than DS III. However, DSs were not equally distributed within the area (Figure 12.2d). Finnish year-around-grazed pastures were not graded better than DS II, winter-grazed pastures in Norway not worse than DS III.

With an average depth of 2.3 cm, O horizons in Norway were about 0.6 cm thicker than those in Finland (Fig. 12.4), while average humus densities in Finland significantly ($P<0.05$) exceeded those detected in Norway (Fig. 12.2b, c). The mass of O horizon humus appeared to be higher in Norway, although the differences were not significant. With increased DSs, average depth of the O horizon decreased from 2.7 cm (DS I) to 2.0 cm (DS II) and 1.6 cm (DS III), while soil density showed the opposite trend (Fig. 12.4). Calculated humus mass per area from DS II and DS III was about 15 % and 25 % lower, respectively, than those of DS I. Results form the chemical analysis showed that N

Table 12.3. Selected soil properties for representative pedons of the four degradation stages (DS I-IV) at Jauristunturit. The capital letters O, A, E, B, and C represent the master horizons of the soil

DS and horizons	Horizon thickness (cm)	Color (moist)	Volume weight kg l^{-1}	pH (H_2O)	N Kjeldahl %	C org. %	CEC meq	Σ bases[a] 100 g^{-1}	BS %	Ca mg kg^{-1} total	Ca mg kg^{-1} AL[b]	Mg mg kg^{-1} total	Mg mg kg^{-1} AL[b]	P mg kg^{-1} total	P mg kg^{-1} AL[b]	K mg kg^{-1} total	K mg kg^{-1} AL[b]
DS I Norway																	
Ol	+3.0																
Of	+2.5	5YR 2.5/2	0.17	4.0	1.60	54	111	13.6	12	2400	102	480	30	1000	14	680	40
E	4.0	10YR 5/2	1.44	4.6	b.d.l.	0.6	3.2	0.12	3.9	130	5.6	140	1.4	40	0.4	380	1.0
Bhs	11	5YR 3/3	1.69	5.0	0.10	2.2	6.7	0.12	1.7	440	7.8	860	1.4	250	3.5	530	0.8
BsC	20	10YR 4/6	1.81	5.0	0.05	1.2	4.8	0.08	1.6	390	6.4	1000	1.2	160	1.0	560	0.6
C	>20	2.5Y6/2	1.94	5.5	b.d.l.	0.4	2.0	0.05	2.4	770	8.9	750	1.1	250	1.9	460	0.3
DS II Finland																	
Ol	+3.5																
Of	+3.0	5YR 2.5/2	0.18	3.9	1.42	54	117	13.0	11	2200	77	600	34	950	13	680	42
E	3.0	10YR6/2	1.63	4.7	b.d.l.	0.6	3.0	0.10	3.4	270	5.5	290	1.4	50	1.0	400	0.9
BsC	10	7.5YR4/4	1.93	5.0	b.d.l.	1.1	4.7	0.08	1.6	670	6.6	740	1.3	310	6.0	510	0.7
C	>10	2.5Y6/2	1.83	5.5	b.d.l.	0.6	2.7	0.06	2.1	840	10	960	1.1	270	2.7	630	0.4
DS III Finland																	
Of (in patches)	+1.0	7.5YR4/2	0.24	4.2	1.25	42	86	8.50	10	1600	68	280	15	760	6.9	590	31
Ahe	1.0	7.5YR5/0	0.66	4.2	0.62	16	47	1.60	3.3	540	18	140	4.2	360	1.2	570	5.6
E	3.0	7.5YR6/2	1.53	4.8	b.d.l.	0.6	3.3	0.07	2.1	130	5.8	300	1.2	60	0.4	440	1.1
Bs	22	7.5YR4/4	1.85	5.3	b.d.l.	1.3	3.7	0.07	1.8	580	6.2	810	1.1	250	0.5	570	0.7
C	>22	5Y5/2	1.68	5.5	b.d.l.	0.5	2.8	0.10	3.6	1060	14	1270	1.0	300	3.3	730	0.9
DS IV Finland																	
Ahe	1.0	10YR4/1	0.75	4.2	0.53	12	31	2.10	6.7	400	19	180	5.3	500	4.1	570	11
E	5.0	7.5YR6/2	1.44	4.7	0.06	1.0	4.0	0.20	4.9	180	7.0	240	1.7	140	1.2	430	2.3
Bs	15	7.5YR3/4	1.73	5.1	0.08	1.9	5.9	0.07	1.2	570	6.3	710	1.1	240	1.3	520	1.1
BsCv	30	2.5Y5/6	1.89	5.4	b.d.l.	0.8	2.8	0.06	2.0	830	7.7	860	1.1	260	0.8	550	0.7
C	>30	2.5Y5/4	1.78	5.4	b.d.l.	0.6	2.7	0.08	3.0	960	9.1	1070	1.0	300	2.0	640	0.7

[a]Less H^+, does not include Na. [b]AL= Ammonium-lactate extracted nutrients: considered plant available; CEC = cation exchange capacity; BS = base saturation; b.d.l.. = below detection limit (<0.05 %)

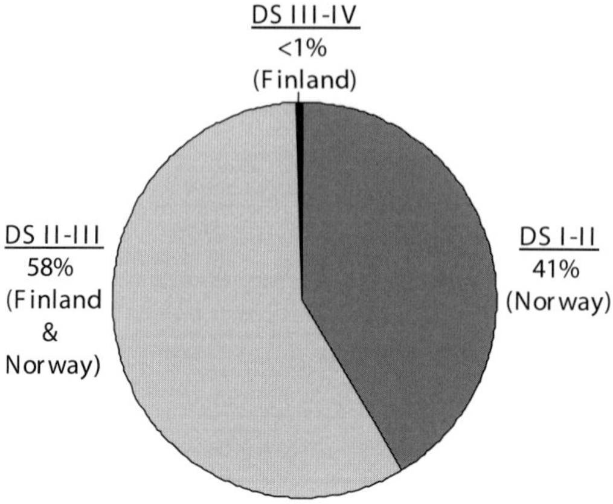

Fig. 12.3. Percentage coverage of degradation stages (DS) within the investigated area. Note that DSs below II were limited to Norway, while DSs above II were only detected in Finland

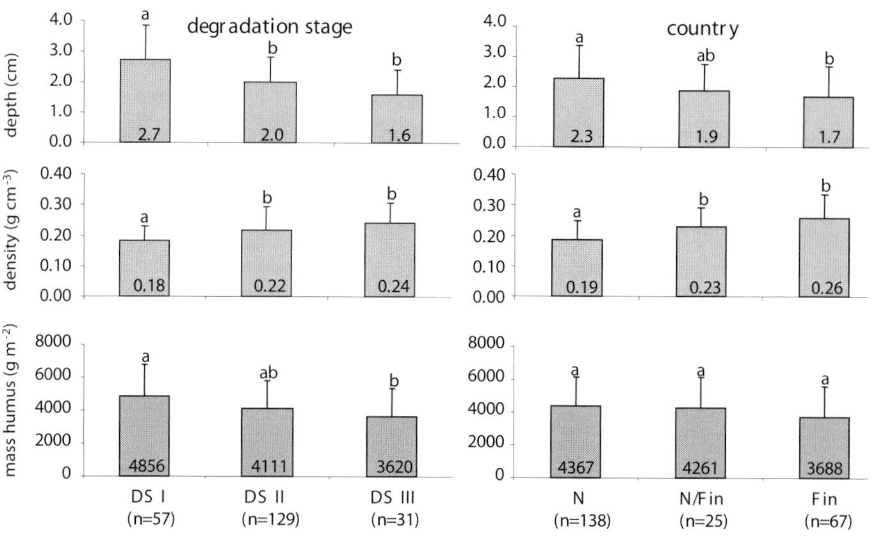

Fig. 12.4. Average depth (cm) of O horizon, humus density (g/cm^3), and humus mass (g/m^2) according to degradation stage and country. *N* = Norway, *Fin* = Finland, and *N/F* = area where the reindeer fence had been moved during the 1990s (see Fig. 12.2). Comparisons significant at the $P<0.05$ level are indicated by *different letters*. Note: the O layer beneath *B. nana* canopies is not considered (see Materials and Methods)

concentrations in O horizons of year-around pastures in Finland were about 10 % less than those of winter pastures in Norway (Fig. 12.5).

Furthermore, pastures in Finland had significant less amounts of total and plant-available Mg and Ca and total P (Table 12.4). The comparison of calculated amounts of plant nutrients per area stored in the O horizon from both grazing regimes revealed that Norwegian winter pastures generally exceeded

Fig. 12.5. Average O horizon N concentrations according to country and degradation stage. *N* = Norway, *Fin* = Finland. Note: the O-layer beneath *B. nana* canopies is not considered (see Materials and Methods). Comparisons significant at the *P*<0.05 level are indicated by *different letters*

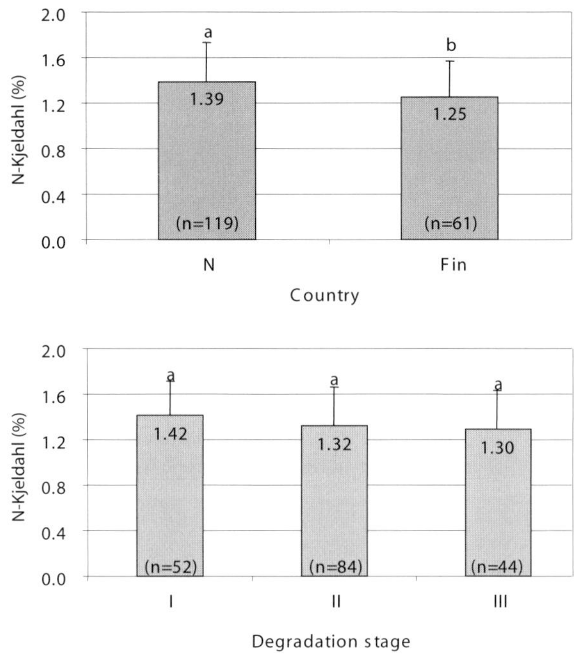

Table 12.4. Concentrations of total (tot) and plant-available (AL) Ca, Mg, K, and P in O horizons from the winter pastures in Norway and year-around pastures in Finland in Jauristunturit, and according to degradation stage (DS). Comparisons significant at the *P*<0.05 level are indicated by different letters

	Ca_{tot} mg kg^{-1}	Ca_{AL} mg kg^{-1}	Mg_{tot} mg kg^{-1}	Mg_{AL} mg kg^{-1}	K_{tot} mg kg^{-1}	K_{AL} mg kg^{-1}	P_{tot} mg kg^{-1}	P_{AL} mg kg^{-1}
Norway	1,740^a	763^a	332^a	231^a	504^a	397^a	788^a	261^a
(n=119)	(±830)	(±360)	(±133)	(±112)	(±148)	(±102)	(±208)	(±64)
Finland	1,456^b	627^b	276^b	182^b	503^a	395^a	727^b	251^a
(n=61)	(±636)	(±281)	(±73)	(±64)	(±114)	(±105)	(±173)	(±82)
DS I	1,878^a	824^a	339^a	237^a	517^a	405^a	791^a	260^a
(n=52)	(±924)	(±404)	(±131)	(±111)	(±174)	(±100)	(±205)	(±59)
DS II	1,558^a	677^a	310^a	212^a	498^a	394^a	753^a	249^a
(n=84)	(±739)	(±322)	(±124)	(±102)	(±111)	(±103)	(±186)	(±67)
DS III	1,532^a	665^a	288^a	193^a	501^a	389^a	765^a	271^a
(n=44)	(±607)	(±264)	(±87)	(±80)	(±114)	(±104)	(±183)	(±88)

Table 12.5. Differences between the calculated amounts of N and total (tot) and plant-available (AL) amounts of plant nutrients stored in the O horizon from winter pastures in Norway, year-around pastures in Finland, and in accord with DSs. Note: the O layer beneath *B. nana* canopies is not considered (see Materials and Methods). Comparisons significant at the $P<0.05$ level are indicated by different letters

	N g m^{-2}	Ca_{tot} g m^{-2}	Ca_{AL} g m^{-2}	Mg_{tot} g m^{-2}	Mg_{AL} g m^{-2}	K_{tot} g m^{-2}	K_{AL} g m^{-2}	P_{tot} g m^{-2}	P_{AL} g m^{-2}
Norway	60^{a}	7.5^{a}	3.3^{a}	1.5^{a}	1.0^{a}	2.2^{a}	1.7^{a}	3.4^{a}	1.1^{a}
(n=121)	(±30)	(±6.0)	(±2.6)	(±0.9)	(±0.6)	(±1.0)	(±0.8)	(±1.6)	(±0.6)
Finland	50^{b}	5.3^{b}	2.3^{b}	1.1^{b}	0.7^{b}	2.0^{a}	1.6^{a}	2.9^{a}	1.0^{a}
(n=59)	(±30)	(±2.8)	(±1.2)	(±0.5)	(±0.3)	(±1.1)	(±0.9)	(±1.7)	(±0.6)
DS I	69^{a}	9.6^{a}	4.2^{a}	1.7^{a}	1.2^{a}	2.5^{a}	2.0^{a}	3.9^{a}	1.3^{a}
(n=52)	(±35)	(±8.1)	(±3.6)	(±1.0)	(±0.8)	(±1.3)	(±1.1)	(±1.9)	(±0.7)
DS II	53^{b}	5.9^{b}	2.6^{b}	1.2^{b}	0.8^{b}	2.0^{b}	1.6^{b}	3.0^{b}	1.0^{b}
(n=84)	(±26)	(±3.0)	(±1.3)	(±0.7)	(±0.4)	(±0.8)	(±0.7)	(±1.4)	(±0.4)
DS III	49^{b}	5.3^{b}	2.3^{b}	1.0^{b}	0.7^{b}	1.8^{b}	1.4^{b}	2.9^{b}	1.1^{a,b}
(n=44)	(±29)	(±2.5)	(±1.1)	(±0.5)	(±0.3)	(±1.0)	(±0.8)	(±1.7)	(±0.6)

those of the Finnish year-around pastures (Table 12.5). With about 50 g m^{-2} N contents, O horizon of year-around pastures contained approximately 17% less N than winter pastures, while amounts of Ca and Mg were decreased by approximately 30% (Table 12.5). There were indications that nutrient contents in O horizons decreased with increasing DS (Table 12.4). However, no significant differences were detected. But calculated storage of N, total Ca, Mg, K, P, and plant-available cations of DS I were significantly higher ($P<0.05$) than those of DS II and DS III (Table 12.5).

12.4 Discussion

12.4.1 Quantitative Changes

In accordance with the distribution of DSs between countries (Figs. 12.2 and 12.3), winter pastures in Norway were less intensively grazed than year-around pastures in Finland. Findings are in good agreement with remote-sensing data (Chap. 8) and reflect differences in animal densities. The relatively strong reduction in fruticose lichen cover in Finland is partly due to its consumption during winter, but may be primarily a result of trampling during the snow free seasons (Oksanen 1978; Bayfield et al. 1981; Käyhkö and Pellikka 1994; Väre et al. 1996). Thus, while lichens beneath deep snow are afforded protection against trampling by reindeer (Miller 1976, cited in Crittenden 1999), frequent recurrence of summer trampling creates a disturbed habitat from which lichens will rapidly become eliminated. Along with the degradation of the fruticose lichen cover, results indicate that reindeer trampling increases humus density of the O horizon, partly at the expense of humus depth (Figs. 12.2 b, c; Fig. 12.4). In accord with the concept of DS, humus depth decreased by approximately 40%, while humus density increased by about 25%. Average humus densities of 0.18 g cm^{-3} and 0.19 g cm^{-3} for DS I and Norwegian pastures (Fig. 12.4) are somewhat higher than 0.13 g cm^{-3}, the value given by Tedrow (1968) as the natural soil density of the O horizon for tundra soils. Beside the reduction in lichen cover and humus depth and the increase in humus density, reindeer husbandry also seems to reduce humus mass (Fig. 12.4). It was found that humus mass per area, likewise in accordance with the DS conception, declined from 4,800 g/m^2 (DS I) to 3,600 g m^2 (DS III). Differences in humus mass, though not significant ($P>0.05$), were also detected between Norwegian and Finish pastures (Fig. 12.4). Though deterioration of the organic layer due to Fennoscandian reindeer husbandry has been repeatedly observed (Helle and Aspi 1983; Foster 1985; Väre et al. 1995, 1996; Evans 1996; Johansen and Karlsen 1998; Olofsson et al. 2001; Uhlig et al. 2004) and similar effects of wild caribou herds on soils have also been reported from North America (Klein 1987; Manseau et al.

1996), this is, to our knowledge, the first time losses have been quantified. However, since O horizons beneath *B. nana* canopies were not taken into account (cf. Sect. 12.2), the possibility cannot be excluded that portions of the apparently vanished organic matter were redistributed to deeper locations.

12.4.2 Erosion

Losses of organic matter from sites with disturbed lichen cover could also partly be due to erosion. Results from Fahnestock et al. (2000) indicate that the redistribution of litter by wind and snow during winter is an important mechanism of nutrient transfer across the arctic landscape. At sites with continuous vegetation cover, redistribution of litter may be limited to aboveground plant debris. However, once vegetation cover is disrupted, either by grazing, trampling, or digging, soil organic matter may also become exposed to forces of erosion, as for example deflation processes (Thannheiser 1977). Evans (1996) reports that many of the descriptions of erosion in northern Scandinavia refer to the importance of wind erosion. Incidences of erosion due to reindeer grazing has been noted previously in Finnmark (Lyftingsmo 1965; Väre et al. 1995; Johansen and Karlsen 1998).

12.4.3 Microbial Activity

In ungrazed systems lichens usually contribute directly to undecomposed organic matter (Longton 1992). In grazed systems, however, litter mass derived from lichens is reduced due to their consumption, in our case by reindeer. Kumpula (Chap. 8) for example found that lichen biomass in Finnish year-around pastures is significantly lower than that in Norwegian year-around pastures. Changes in litter quality and quality can be expected to cause changes in microbial activity and thus nutrient cycling. Sendstad (1981) found in a field experiment on Svalbard that the artificial removal of lichens resulted in a 50 % decrease in soil respiration, along with a significant reduction in soil organic matter. For the site we investigated in Jauristunturit, it is reported in this volume (Chap. 13) that respiration rates of organic soil samples from the year-around-grazed area in Finland were lower than in samples from the winter-grazed Norwegian area. In contrast, results from Olofsson et al. (2001) indicate that reindeer grazing increases the activity of decomposer microbes.

12.4.4 Qualitative Changes

Our results clearly show that the quantitative deteriorating of the O horizon is associated with qualitative losses of essential plant nutrients. P and N concentrations in the O horizon from Finnish pastures were approximately 10 % lower than in Norwegian pastures (Table 12.4, Fig. 12.5), while concentrations of total and plant available Ca and Mg were reduced by approximately 20 % (Table 12.4). A decline of nutrient concentrations at comparable magnitudes were also detected with increasing DS's (Table 12.4, Fig. 12.5), though differences were not significant. These results are in full conformity with Väre et al. (1996) who reported a significant decline in total P with increased grazing, and Sendstad (1981) who found that removal of lichen cover resulted in reduced concentrations of soil N and P. Levels of total and plant-available P, Ca, Mg, and K in the O horizon beneath ample lichen cover corresponded well with results reported from mountain birch forests in Finland (Wielgolaski 2001) and Norway (Uhlig et al. 2004). The findings that plant-available Ca and Mg in the organic layer are reduced with increased grazing and decreased lichen cover are generally in good agreement with results obtained from a North Finnish pine forest (Väre et al. 1996) and Norwegian mountain birch forests (Uhlig et al. 2004), although, with reductions of 30–50 % of plant available Ca and Mg, losses reported in those two studies were about two times those found here. The significant differences in concentrations of P, Ca, Mg, and N within the O horizon between countries and among DSs indicate that considerable quantitative and qualitative changes in the nutrient composition of the humus have taken place.

Nutrient mass calculations per area show that quantities of nutrients within DS I O horizons exceeded those of DS III by about 20 g N, 4 g Ca, 1.0 g P and 0.7 g K m^{-2} (Table 12.5). O horizons of Norwegian winter pasture exceeded those of Finnish year-around pasture by about 10 g N, 2.2 g Ca and 0.4 g Mg m^{-2} (Table 12.5); thus differences between managing regimes were less distinct. Results clearly show that reindeer husbandry on lichen dominated dwarf shrub tundra leads to a significant loss of essential plant nutrients from the O horizon, and that nutrient losses go along with the degradation of lichen mats. Lichens require a similar range of elements to other plants, and through growth and decomposition they inevitably participate in nutrient cycling. Even if rates of lichen decomposition in Arctic communities appear generally to be low when compared to angisperms (Rosswall et al. 1975), they clearly may dominate the process in many tundra communities. Especially fruticose lichens' mineral content may play an important part in nutrient cycling in tundra ecosystems. The pattern of nutrient release from decomposing *Cladina* litter shows that particularly Ca is rapidly lost, presumably through a leaching process, followed by a slower release of Mg (Moore 1984). Consequently, these macronutrients rapidly return to the soil and plant root systems, although the magnitude of winter losses may effect the availability of these nutrients to plants (Moore 1984).

12.4.5 Pathways of Nutrient Loss

Winter leaching also has to be considered as an important mechanism for nutrient loss from reindeer pastures. The direct consumption of plants by herbivores normally enhances nutrient cycling, as urine and feces create an organic soil nutrient pool in which nutrient release tends to be faster than from litter (McKendrick et al. 1980; Holland et al. 1992). Over time, this increased nutrient turnover by grazing herbivores can improve soil fertility. However, increases in soil fertility presumes that herbivore-released nutrients add to the soil nutrient pool, and that this pool is accessible for microorganism and plants (Fig. 12.1). Not all nutrients in urine and feces released at below zero temperatures in the snow during the winter season may enter the soil nutrient pool. It is reasonable to assume that a considerable proportion remains in the snow above the frozen soil until the final snowmelt in spring. During the first period of final snowmelt many tundra soils are commonly still frozen and surface run-off frequently occurs even on well-drained soils. Unfrozen soils are infiltrated by relatively high amounts of melt water, and losses of winter mineralized nutrients may occur especially during this period. Streams draining Alaskan tundra show large amounts of inorganic- and organic-bound nutrients during snowmelt (Peterson et al. 1992; Whalen and Cornwell 1985), indicating that terrestrial ecosystems do not retain nutrients tightly early in the season. Thus, it can be generally expected that nutrient cycling in winter pastures will be less complete than in summer pastures.

The retention of nutrients in soils is highly dependent on the cation exchange capacity (CEC) of the soil, which thus is a key function for soil fertility. Results from Uhlig et al. (2004) emphasize that the CEC of soils developed from morainic material in Finnmarksvidda is highly determined by the soil's organic matter content. Loss of organic matter would consequently increase leaching and thus decrease soil fertility. It is well known that lichens can be major contributors of organic matter to soils, and thus may increase exchangeable cations (Filion and Payette 1989). Additionally, lichens promote trapping of wind-blown organic and inorganic material (Longton 1992). The CEC_{pot} (pH 7.0) of fruticose lichens (mainly *Cladina stellaris*) was measured at 670 $mmol_c\ kg^{-1}$ (unpublished results), which is approximately 60 % of the CEC_{pot} of the organic matter at comparable sites (Uhlig et al. 2004). At these sample sites in Finnmarksvidda Uhlig et al. (2004) found that the organic layer beneath reduced lichen cover had about a 20 % lower CEC and a 30–50 % lower concentration of plant-available Ca and Mg than that beneath ample lichen cover.

12.4.6 Soil Fertility and Pasture Productivity

McKendrick et al. (1980) found that elevated concentration of soil nutrients, particularly N and P, are associated with activities of mammalian herbivores, and that plants growing within areas of elevated soil nutrients tend to have higher nutrient concentrations in their tissue. A similar impact of reindeer grazing on tundra ecosystems has been suggested by Zimov et al. (1995) in that reindeer increase soil fertility and thus initiate the vegetational change from nutrient-poor dwarf shrub communities to nutrient-rich grasslands. However, for the continental area of Finnmarksvidda no significant increase in soil fertility, nutrient concentrations of plants, or measures of graminoid cover could be detected (Olafsson et al. 2001; Uhlig et al. 2004. In contrast, it seems that reindeer grazing leads to a decrease in soil fertility, and the more the grazing, the greater the decrease in soil fertility (Tables 12.4 and 12.5; Figs. 12.4 and 12.5). Since growth of low arctic plant species is particularly limited by N and P supply (Shaver and Chapin 1980; Kummerow et al. 1987), any reduction in soil N and P, as found here (Tables 12.4 and 12.4; Fig. 12.5), is likely to decrease plant and thus pasture productivity.

12.5 Conclusions

Our results showed a significant decrease in humus depth with increasing DS, while humus density increased. Humus mass per area decreased from approximately 4,800 g m^{-2} (DS I) to 4,100 g m^{-2} (DS II) and 3,600 g m^{-2} (DS III). Along with these qualitative changes concomitant with DSs, total calculated amounts (in g m^{-2}) of the essential plant nutrients N, P, Ca, Mg, and K stored in the O horizon were reduced by about 30 to 45 %. Comparison of the two management regimes showed that lichen cover of Norwegian winter pastures was less degraded than those of Finnish year-around pastures. Norwegian pastures had a higher average humus depth and humus mass per area, but lower soil density. Furthermore, nutrient concentrations of N, total P, Ca, Mg, and plant-available Ca and Mg were about 10 to 20 % higher in Norway. Total amounts of N in winter pastures were about 17 % higher, while amounts of Ca and Mg exceeded those of year-around pastures by approximately 30 %. As for differences in organic matter properties between the reindeer-grazing regimes, an obviously greater deterioration of O horizons was found where year-around grazing husbandry was practiced. Thus, results clearly indicate that degradation of fruticose lichen carpets by reindeer negatively influence organic matter quantity and quality in the soil beneath. Consequently, cautious reindeer management is required to maintain the productivity of fruticose-lichen-dominated dwarf shrub tundra in Finnmarksvidda.

References

AG-Boden (1996) Bodenkundlliche Kartieranleitung, 4th edn. Schweitzbart'sche Verlagsbuchhandlung, Stuttgart

Ahti T (1977) Lichens of the boreal coniferous forest zone. In: Seaward MRD (ed) Lichen ecology. Academic Press, London, pp 145-181

Ahti T, Oksanen J (1990) Epigeic lichen communities of taiga and tundra regions. Vegetatio 86:39-70

Batzli GO, White RG, MacLean SF, Pitelka FA, Collier BD (1980) The herbivore-based trophic system. In: Brown J, Miller RG, Tieszen LL, Bunnell FL (eds) An arctic ecosystem. The coastal tundra at Barrow, Alaska. US/IBP Synthesis Series, Dowden, Hutchinson and Ross, PA, pp 335-410

Bayfield NG, Urquhart UH, Cooper SM (1981) Susceptibility of four species of *Cladonia* to disturbance by trampling in the Cairngorm Mountains, Scotland. J Appl Ecol 18:303-310

Bråthen KA, Oksanen J (2001) Reindeer reduce biomass of preferred plant species. J Veg Sci 12:473-480

Braun-Blanquet J (1932) Plant sociology; the study of plant communities. McGraw-Hill, New York, 438 pp

Bremner JM (1960) Determination of nitrogen by the Kjeldahl method. J Agric Sci 55:11-33

Chernov YI, Matveyeva NV (1997) Arctic ecosystems in Russia. In: Wielgolaski FE (ed) Ecosystems of the World. Elsevier, Amsterdam

Crittenden PD (1999) Aspects of the ecology of mat-forming lichens. Rangifer 20(2/3):127-139

Dawson HJ, Hrutfiord BF, Ugolini FC (1984) Mobility of lichen compounds from *Cladonia mitis* in arctic soils. Soil Sci 138(1):40-45

DNMI (1999) Normaler 1961-1990. Lufttemperatur og nedbør. Foreløpig liste pr 23.3.1999, DNMI Klima

Edward N, Miller PC (1977) Validation of a model on the effect of tundra vegetation on soil temperature. Arct Alp Res 9:89-104

Egnér HH, Riehm WR (1960) Untersuchungen über die chemische Boden-Analyse als Grundlage für die Beurteilung des Nährstoffzustandes der Böden. Royal Agricult High School Ann 26:199-215

Evans R (1996) Some impacts of overgrazing by reindeer in Finnmark, Norway Rangifer 16(1):3-19

Fahnestock JT, Povirk KL, Welker JM (2000) Ecological significance of litter redistribution by wind and snow in arctic landscapes. Ecography 23:623-631

FAO (1990) Guidelines for soil description, 3rd rev ed. FAO, Rome

Foster DR (1985) Vegetation development following fire in *Picea mariana* (black spruce)-*Pleurozium* forests in south-eastern Labrador, Canada. J Ecol 73:517-534

Filion L, Payette S (1989) Subarctic lichen polygons and soil development along a colonization gradient on eolian sands. Arct Alp Res 21:175-84

Helle T, Aspi J (1983) Effects of winter grazing by reindeer on vegetation. Oikos 40:337-343

Holland EA, Partoin WJ, Detling JK, Coppock DL (1992) Physiological responses of plant populations to herbivory and their consequences for ecosystem nutrient flow. Am Nat 140:685-706

Johansen B, Karlsen SR (1998) Endringer i lavdekket på Finnmarksvidda 1987-1996 basert på Landsat 5/TM data. NORUT IT Rapport, IT475/1-98, 12 pp

Käyhkö J, Pellikka P (1994) Remote sensing of the impact of reindeer grazing on vegetation in northern Fennoscandia using SPOT XS data. Polar Res 13:115-124

Klein DR (1987) Vegetation recovery patterns following overgrazing by reindeer on St. Matthew Island. J Range Manage 40(4):336-338

Kummerow JJN, Mills JN, Ellis BA, Hastings SJ, Kummerow A (1987) Downslope fertilizer movement in arctic tussock tundra. Holarct Ecol 10:312-319

Kumpula J (2001) Productivity of semi-domesticated reindeer (*Rangifer t. tarandus* L.) stock and carrying capacity of pastures in Finland during 1960-1990`s. Thesis Department of Biology, University of Oulu. Acta Universitatis Ouluensis A 375

Longton RE (1992) The role of bryophytes and lichens in terrestrial ecosystems. In: Bates JW, Farmer AM (eds) Bryophytes and lichens in a changing environment. Clarendon Press, Oxford, pp 32-76

Lyftingsmo E (1965) Oversyn over Fjellbeite i Finnmark. Norsk Fjellbeite Bind 15, 367 pp

Manseau M, Huot J, Crete M (1996) Effects of summer grazing by caribou on composition and productivity of vegetation: community and landscape level. J Ecol 84:495-502

McKendrick JD, Batzli O, Everett KR, Swanson JC (1980) Some effects of mammalian herbivores and fertilization on tundra soils and vegetation. Arct Alp Res 12(4):565-578

Moore TR (1984) Litter decomposition in a subarctic spruce-lichen woodland, eastern Canada. Ecology 65:299-308

Ogner G, Haugen A, Opem M, Sørlie M (1975) Kjemisk analyseprogram ved Norsk institutt for skogsforskning. Meddr Norsk Inst Skogforsk 32:207-232

Ogner G, Haugen A, Opem M, Sørlie M (1977) Kjemisk analyseprogram ved Norsk institutt for skogsforskning. Meddr Norsk Inst Skogforsk 33 [Suppl 1]:87-101

Oksanen J (1978) Lichen grounds of Finnmarksvidda, northern Norway, in relation to summer and winter grazing by reindeer. Kevo Subarctic Res Station 14:64-71

Oksanen J (1986) Succession, dominance and diversity in lichen-rich pine forest vegetation in Finland. Holarct Ecol 9:261-266

Olofsson J, Kitti H, Rautiainen P, Stark S, Oksanen L (2001) Effects of summer grazing by reindeer on composition of vegetation, productivity and nitrogen cycling. Ecography 24:13-24

Olsen L, Reite A, Riiber K, Sørensen E (1996) Finnmark County, map of quaternary geology, scale 1:500,000 with description. Geological Survey of Norway

Pegau RE (1970) Effect of reindeer trampling and grazing on lichens. J Range Manage 23:95-97

Peterson BJ, Corliss TL, Kriet K, Hobbie JE (1992) Nitrogen and phosphorous concentrations and export for the upper Kuparuk River on the North Slope of Alaska in 1980. Hydrobiol 240:61-69

Polunin N (1936) Plant succession in Norwegian Lapland. Journal of Ecology 24:372-391

Richardson DHS, Young CM (1977) Lichen and vertebrates. In: Seaward MRD (ed) Lichen ecology. Academic Press, London, pp 121-144

Rosswall T, Veum AK, Kärenlampi L (1975) Plant litter decomposition at Fennoscandian tundra sites. In: Wielgolaski FE (ed) Fennoscandia tundra ecosystems, part 1. Plants and microorganisms. Springer, Berlin Heidelberg New York, pp 268-278

Rouse WR, Kershaw KA (1971) The effects of burning on the heat and water regime of lichen-dominated subarctic surfaces. Arct Alp Res 3:291-304

Sendstad E (1981) Soil ecology of lichen heath at Spitzbergen, Svalbard: effects of artificial removal of lichen plant cover. J Range Manage 34:442-445

Shaver GR, Chapin FS (1980) Response of fertilization by various plant growth forms in an Alaskan Tundra: nutrient accumulation and growth. Ecology 61(3):662-675

Stark S, Wardle DA, Ohtonen R, Helle T, Yeates GW (2000) The effect of reindeer grazing on decomposition, mineralization and soil biota in a dry oligotrophic Scots pine forest. Oikos 90:301-310

Tedrow J CF (1968) Pedogenic gradients of the polar regions. J Soil Sci 19(1):197-204
Thannheiser D (1977) Subarctic birch forests in Norwegian Lapland. Naturaliste Can.104:151-156
Uhlig C, Sveistrup T, Schjelderup I (2004) Impacts of reindeer grazing on soil properties on Finnmarksvidda, Northern Norway. Rangifer Special Issue 15:83-91
Väre H, Ohtonen R, Oksanen J (1995) Effects of reindeer grazing on vegetation in dry *Pinus sylvestris* forests. J Veg Sci 6:523-530
Väre H, Ohtonen R, Mikkola K (1996) The effect and extent of heavy grazing by reindeer in oligotrophic pine heath in northeastern Fennoscandia. Ecography 19:245-253
Vilchek GE (1997) Arctic ecosystem stability and disturbance: a western Siberian case history. In: Crawford RMM (ed) Disturbance and recovery in Arctic lands. Kluwer Academic, Dordrecht, pp 179-189
Whalen SC, Cornwell JC (1985) Nitrogen, phosphorous and organic carbon cycling in an arctic lake. Can J Fish Aquat Sci 42:797-808
Wielgolaski FE (2001) Nordic mountain birch ecosystems. In: Jeffers JNR (ed) Man and the biosphere series, vol 27. UNESCO, Paris, The Parthenon Publishing Group
Zimov SA, Chuprynin VI, Oreshko AP, Chapin FS, Reynolds JF, Chapin MC (1995) Steppe-Tundra transition - a herbivore-driven biome shift at the end of the Pleistocene. Am Nat 146:765-794

13 Changing Microbial Ecology with Changes in Grazing and its Management

M. Bölter and R. Möller

13.1 Introduction

Management of tundra systems has been but poorly investigated, especially with regard to agricultural use. The growth of cereals and other crops is basically limited at these high latitudes by prevailing low temperatures, which do not provide enough energy for appropriate tillage (Lauer and Rafiqpoor 2002). A much better agricultural use of a tundra environment is found in grazing. Indeed, reindeer herding in the landscape of northern Fennoscandia has been carried out for a long time (Chap. 2). But now these landscapes and soils are being affected by new forms of land use, prominently forestry and tourism, among others (Chap. 16).

The RENMAN project has focused on various influences of reindeer on man and nature (Chap. 1). Recent studies in Fennoscandia have shown various effects of land use changes on vegetation, soil properties, and soil biology (e.g., Grellmann 2001; Kumpula 2001; Eilertsen 2002; Stark 2002; den Herder 2003). In our study we compared the effects of grazing by means of microbiological characteristics, i.e., by changes in community and soil respiration in forests and on fjeld used in different ways by reindeer herding.

13.1.1 Effects of Changes in Land Use

Intense grazing affects the soil habitat in different ways; for just three examples, it alters plant communities, affects fertilization, and introduces trampling. Vice versa, reindeer density in Norway has been significantly influenced by land development (Vistnes et al. 2001; Chap. 2). Extended grazing is regarded worldwide as a cause of land degradation, especially in areas with low plant and root density and only thin layers of organic material, which is typical for the forest and tundra soils in northern Scandinavia. Excessive

Ecological Studies, Vol. 184
B.C. Forbes, M. Bölter, L. Müller-Wille, J. Hukkinen, F. Müller, N. Gunslay, Y. Konstatinov (Eds.)
Reindeer Management in Northernmost Europe

grazing inevitably destroys vegetation, compacts soils (Chaps. 9, 11, 12), and causes imbalances in the land integrity. Special problems occur when fences confine herbivores and when their demands on forage exceed plant production rates. Further important effects of plants are their influences on water infiltration, evaporation, and soil stability, all important links in soil health of riparian areas and uplands (den Herder et al. 2003). In such areas, the time spans of grazing have strong effects on plant reproduction.

Herbivory is an important structuring force within arctic ecosystems. The individual effects, however, differ locally (Miller et al. 1980; Manseau et al. 1996; Virtanen et al. 1997), as do the magnitude of the processes involved (Augustine and McNaughton 1998). Effects on the environment from herding have undergone drastic changes during the last few decades (Kumpula 2001; Chap. 2) with local and regional alterations in management. There is an ongoing debate as to whether plant communities are affected either by grazing or by nutrient limitations, especially from the viewpoint of environmental carrying capacity. Studies in Fennoscandia have shown different responses to fertilization (Jonasson 1992; Parsson et al. 1994; Press et al. 1998). However, such populations under stable conditions can be regarded as balanced by growth and grazing (Oksanen 1993; Oksanen and Oksanen 2000). Plants promote healthy soils by providing organic matter, which in turn stimulates the microbial processes of mineralization.

Feedback mechanisms between grazing and plant nutrient availability as well as plant productivity are well known from different grazed ecosystems (McNaughton 1985; Chapin 1991; Insam et al. 1996). Investigations of Stark et al. (2002) show effects of grazing on vegetation and soil biota in Finland. Both positive and negative effects of grazing on soil biota can be monitored. An important aspect is change of the near-bottom microclimate when plant cover is removed. Grazing affects not only the plants on the ground (grasses, etc.) but also woody plants such as shrubs and trees in their growth and reproduction (den Herder and Niemelä 2003).

But not only man-made decisions provide the frames for these influences; natural impacts such as fodder availability and weather conditions are also controlling factors, and they reflect patterns of global climate change (Weladji and Holand 2003). Yet man-made influences indeed have to be taken into account, and these may act on the long scale in matters like habitat fragmentation and contamination, such as result from commercial logging and increased tourist affairs (Müller-Wille 1987; Nellemann et al. 2001).

Any impact on the landscape scale thus needs to be considered for its influence on the appropriate smaller scale. Changes in soil organisms can be due to the direct impact of local activities or stem from regional long-term policies (Kumpula and Colpaert 2003). It is often difficult to distinguish among such impacts, especially at the level of the lower organisms. An example is seen in soil biodiversity. Highly fluctuating temperatures and moisture levels pose considerable stress (Bölter 2004a,b), and the differentiation between a "normal" band width of environmental fluctuation and one derived from man-made conditions is difficult. On the other hand, changes in community structure may not be as dramatically affected as forecast. Effects may become visible only over the long haul, i.e., at time scales out of man's direct perception.

Other changes in environmental conditions are consequences of habitat fragmentation, or they come when secondary effects have become the driving force, as is the case with erosion, persistent vegetation damage, accumulation of pollutants, diminished foraging success, or reduced breeding success. These cumulative impacts are increasingly evident in Arctic and sub-Arctic regions. Influences also beyond local responsibility have far-reaching ramifications beyond their visible effects. An example is that the increasing atmospheric nitrogen deposition and N_2O emission due to cultivation efforts and cropping (in combination with increased grazing pressure) lead to changes in northern plant communities (Maljanen et al. 2003; van der Wal et al. 2003). This is an effect similar to increasing UV-B radiation on the growth, production, and phenology of shrubs and mosses (Phoenix et al. 2001; Chap. 10). The global trend of increasing temperature and precipitation has been analyzed by Jylhä et al. (2004). The changes in aboveground features have significant consequences for the microbiological processes of mineralization and nutrient turnover. Effects on plant communities of global change, especially the shift from lichen heaths to higher plant communities due to their ability to be more efficient competitors, have been described by Cornelissen et al. (2001). The shifts in plant communities, driven by direct or indirect effects are reflected in soils, soil organisms, and soil activity.

13.1.2 Soil Habitats in Northern Fennoscandia

Tundra environments cover an important part of the earth's surface; their share of the land surface is estimated to be 5.5 % (Brown et al. 1980). Soils in this area are young and still developing, and mostly nutrient-poor Podzols or Cambisols and Regosols (Schultz 2000). However, tundra soils are not a uniform habitat and differences in plant cover and soil quality are visible at low scales (Bölter and Blume 2002). Wet tundra regions are considered to be driving systems in affairs of global change and is suspected that they drive this process because of their high stocks of carbon, amounting to 14 % of the total soil C (Post et al. 1982; Gilmanov and Oechel 1995). Dry tundra regions on fjells are not nearly so involved in this matter.

Rieger (1974) describes the tundra as a patchwork of individual islands with distinct populations of plants and animals at scales of meters or below. Studies on interactions between landscape relief (Peterson and Billings 1980; Batten and Svoboda 1994), hydrology (Webber 1978), and soil conditions (Bliss 1981; Bliss et al. 1984; Gebauer et al. 1996; Chapin et al. 1988) show effects on vegetation patterns and individual soil variables even on scales of a few meters. Extent and duration of snow layers and their thickness depend on relief, and thus so do vegetation patterns (Press et al. 1988).

Similarly, drastic changes are evident in the boreal forests. The dominating species, Scots pine (*Pinus sylvestris* L.) and Norway spruce (*Picea abies* L.), produce an acidic environment in soil low in nutrients. These forests have a poorly developed understory, and over 80 % of the aboveground biomass is regarded as bound in trees (Havas and Kubin 1983). The boreal forests are known to play an important role in the global C budget (Apps et al. 1993; Dixon et al. 1994). Commercial forestry is carried out in the northern forests of Europe, northwest Russia, Siberia, Canada, and Alaska. Economic pressures to harvest virgin timber for pulp, paper, and wood products are increasing. The use of "improved" harvesting techniques – often with subsequent planting of foreign species – has led to the fragmentation of critical habitats. Clear-cutting in the vulnerable transition or taiga zone between the northern forests and the treeless tundra has had a severe effect on the biodiversity of forest ecosystems. Soil C decreases strongly after deforestation in boreal forests (Alaska), i.e., during land-use change. This decrease is probably more pronounced than in other biomes. Soil C disappears due to combustion, respiration, and leaching. Compensation takes place at longer time scales in mature meadows (Grünzweig et al. 2004).

Hence, it is likely that changes in land use affect not only soil cover but also soil-dwelling organisms, which relate directly or indirectly to inputs of all kinds of energy and material from the surface and from lateral transport. Such disturbances come from changes in land use: increased building construction, tourism activity, and/or increased farming and grazing. These disturbances are the result not only of feeding on plants, but also of side effects such as damage to fragile plant covers, soil compaction, or resource extraction. The cumulative ecological damage now taking place in the Arctic is cause for much concern. Habitat loss has ramifications reaching far beyond what is immediately visible. Continued habitat fragmentation in the Arctic has devastating effects on its plants and animals, including many migratory populations.

13.2 Locations, Sampling, and Analytical Methods

13.2.1 Areas of Investigation

The sampling locations are in the vicinity of Vuotso, which represents a forested area, and north of the village of Näkkälä, the main place to analyze effects of grazing on a dry tundra system. The locations analyzed in this study were: (a) a forest area near Vuotso (68°09' N, 26°36' E) and (b) a fjell tundra environment north of Näkkälä in the Jauristunturit area (68°49' N, 23°48' E; Fig. 13.1). The latter investigation site is characterized by a separation into a winter forage area in Norway and a summer and winter forage area in Finland, separated by a reindeer fence. Part of the fence was shifted northward 12 years ago, which revealed a "transition area" of vegetation and land use.

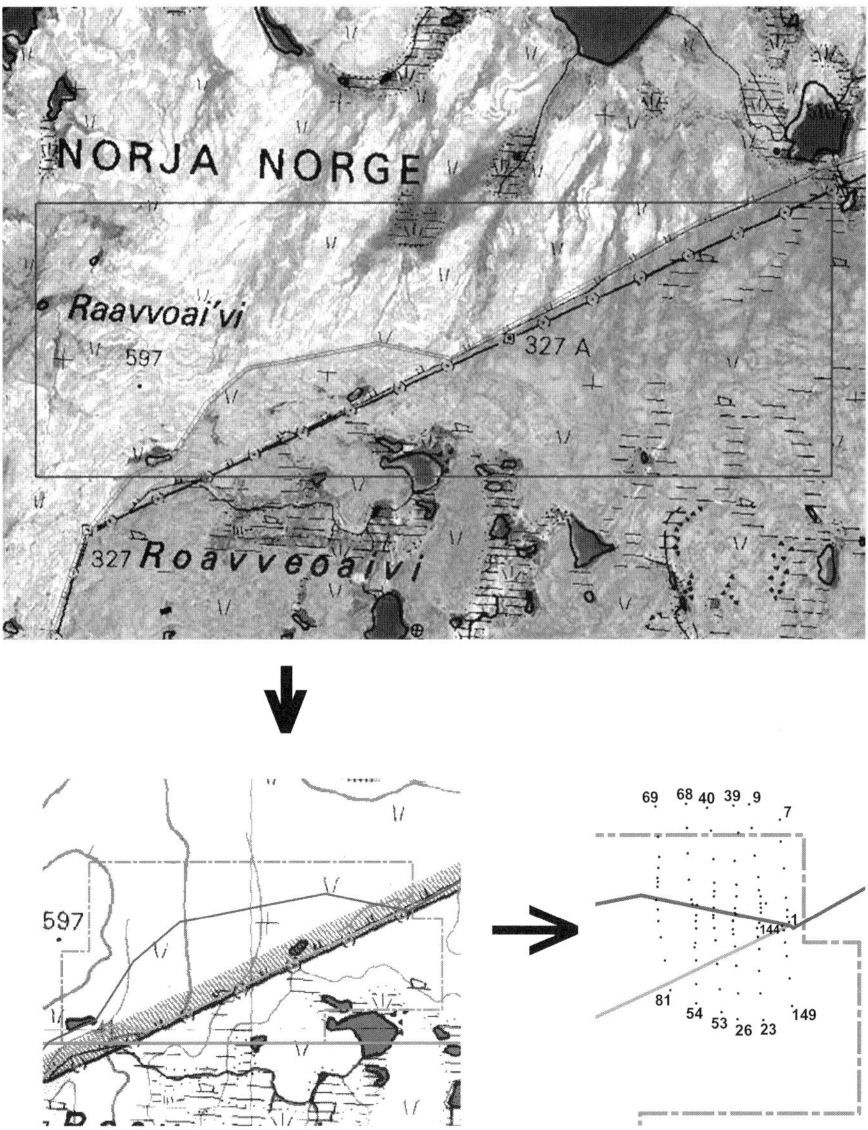

Fig. 13.1. Area of the sampling site at Jauristunturit and location of transects 1–6

13.2.2 Sampling

The forest sites were analyzed for vertical profiles (samples "VP–nn", collected in 2001 and part of 2002; for details see Chap. 11). Surface samples (0–1 cm) were collected with special respect to contamination by feces; these samples were collected over a wide range of sampling points along forest roads around Vuotso (samples V1–V10 collected in 2001, and samples V1–V20 collected in 2002). Details of the locations can be found in Chapters 11 and 12.

Sampling in the tundra environments concentrated on surface horizons and were collected during the field study in 2002 (for details see Chaps. 11 and 12).

1. Along six North-South transects across the three management areas (Norway, Transition, Finland) surface samples were collected (numbers T1–T149),
2. Samples from monitoring sites: MP-L (low grazing, Norway), MP-M (medium grazing, transition zone), MP-H (high grazing, Finland),
3. Samples N1–N13 were collected in all three areas with special regard to fecal sheddings.

13.2.3 Analyses

Samples were transported to the laboratory under air-dried conditions and analyzed for:

1. Bulk organic material by loss on ignition (LOI, 540°C, 6h)
2. Microbial counts and biomass (bacteria, fungi; Bölter et al. 2002; Schmidt and Bölter 2002);
3. DNA by DAPI reaction (Russel et al. 1975; Williamson and Fennel 1975; Teucher 1986)
4. Chlorophyll *a* and phaeopigments by acetone extraction and fluorometric measurement (Lorenzen 1967; Arar and Collins 1997);
5. Soil respiration (Sommerkorn et al. 1999; Bölter et al. 2003).

These analyses were performed under temperature and moisture controlled conditions in open flow chambers.

Soil samples V1–V10 (2001) and V1–20 (2002) were further analyzed for the abundance of enteric bacteria, e.g., *Escherichia coli* and other coliform bacteria, in two ways. On one hand, the spectra of size classes were used to indicate enteric bacteria, which are significantly larger than the bulk of soil bacteria. On the other hand, a genetic test was used to identify coliform bacteria by genetic probes and subsequent fluorescent microscopy. Test kits from Vermicon, München (www.vermicon.com) were used for this procedure.

As the individual data sets do not fulfill the statistical requirements for normal distribution, the data subsets are represented by minimum, maximum, and median values.

13.3 Results

13.3.1 Forest Environments (Vuotso Area)

The microbiological properties of the soil profiles in this region are given in Tables 13.1 and 13.2. Undisturbed soils show high levels of organic carbon in the uppermost horizon, which is depleted to low or lowest values in areas

Table 13.1. Microbiological parameters of soil profiles from sampling sites near Vuotso. LOI (loss on ignition: % of dwt); TBN (total bacterial number (10^9 n g^{-1} dwt); BBM (bacterial biomass: µg C g^{-1} dwt), and TBS (total bacterial surface: cm^2 g^{-1} dwt. Profiles: VP1-U: undisturbed area near location 1; VP2-I: high trampling intensity inside a former corral; VP2-II: non-trampled area outside this corral; VP3-U: non-trampled area in the forest; VP3-G: moderately trampled, grazed area; VP3-T: trampling path (for details and soil description see Table 11.1)

Profile	Horizon	Depth cm	LOI % dwt	TBN 10^9 g^{-1}	BBM C µg g^{-1}	TBS cm^2	Respiration CO_2 mg g^{-1} h^{-1}		
							5 °C	20 °C	35 °C
VP1-U	Of/Oh	+6–0	93.6	2.28	8.35	0.064	0.079	0.047	0.047
	Ae	0–4	4.9	0.22	0.47	0.056	0.035	0.033	0.028
	Bhs	4–7	9.4	0.23	0.58	0.058	0.038	0.036	0.033
	Bs	7–20	2.6	0.22	0.35	0.051	0.040	0.038	0.038
VP2-I	Ahe	1.5	16.3	0.61	1.80	0.061	0.037	0.024	0.009
	Of	2	63.4	1.74	3.97	0.057	0.041	0.036	0.022
	Ae	6	1.5	0.16	0.32	0.055	0.039	0.039	0.039
	Bhs	23	7.7	0.15	0.21	0.049	0.038	0.036	0.036
VP2-II	Of	+3.5	76.9	1.26	5.19	0.067	0.110	0.099	0.067
	Ae	4	1.5	0.26	0.73	0.060	0.042	0.039	0.038
	Bs	20	3.4	0.17	0.48	0.060	0.039	0.040	0.039
	C	40 +	1.1	0.06	0.060	0.044	0.043	0.044	0.041
VP3-U	L/Oh	+8	68.8	0.88	1.17	0.048	0.005	0.031	0.042
	Ae	2	2.5	0.04	0.05	0.046		0.003	0.054
	Bhs	10	3.4	0.07	0.12	0.050	0.002	0.002	0.005
	Bsw	15	2.1	0.04	0.06	0.050	0.001	0.002	0.006
	C	30	1.8	0.04	0.09	0.055		0.002	
VP3-G	Ae	3	0.7	0.13	0.19	0.050		0.004	0.014
	Aeh	5	24.2	0.18	0.37	0.055	0.005	0.008	0.014
	Bsh	17	1.7	0.08	0.14	0.052	0.007	0.015	0.022
VP3-T	Ae	1	8.3	0.04	0.05	0.047	0.007	0.004	0.008
	Bws	8	1.9	0.43	0.80	0.053	0.005	0.012	0.036
	Cw	30	1.9	0.06	0.09	0.049		0.004	0.011

Table 13.2. Contents of DNA. chlorophyll *a*., and phaeopigments (μg g^{-1} dwt) of sites near Vuotso (for site descriptions see Table 11.1)

Site	Depth cm	Horizon	DNA	Chlorophyll *a*	Phaeopigments
VP3-T	0–1	Ae	220.8	0.65	7.07
	1–8	Bws	113.1	0	0.31
	30+	Cw	33.8	0	0.02
VP3-G	0–3	Ae	57.9	1.23	23.16
	3–5	Aeh	815.1	0.48	3.26
	5–17	Bsh	15.1	0	0.17
VP3-U	+8–0	L/Of/Oh	1172.7	1.86	62.94
	0–2	Ae	4.2	0	0.17
	2–10	Bhs	110.3	0	2.69
	10–15	Bsw	79.3	0	0.03
	30+	C	24.4	0	0.04
VP4-F	0–2	Ah	407.2	1.61	0.4
	3–6	Bvs	103.9	0.81	12.47
	15–20	Bvs	124.3	0	0.02
	22–25	Bv	5.5	0	0.01
	30–33	Bv	0	0	0.01

where they are disturbed by trampling (e.g., samples VP3-G and VP3-T) and where just below the active plant cover the soil profile starts with the Ae horizon. These amounts of organic material are strongly reflected by the microbial community – i.e., its number, biomass, and respiratory activity – comparable to C-depleted horizons of other profiles. Respiratory activity (Table 13.1), however, was found at low levels at all sites, due to the probably low amounts of available carbon for metabolic processes. This becomes evident from experiments with increasing temperature; the results show only slight or even no related effects, indicating that no active metabolic processes are present.

Concomitant data on contents of DNA, chlorophyll *a*, and phaeopigments show also decreases with depth (Table 13.2). It becomes evident, that these distributions show some common behaviour. Chlorophyll, an indicator for active photosynthetic biomass, can be found only in the uppermost horizons. Phaeopigments, decay products of chlorophyll, are transported to deeper horizons, down to the B horizons. Profile VP3-U shows a good relation between DNA, phaeopigments, and bacterial biomass. Profile VP3-G bacterial biomass and DNA also coincide. Other profiles show only weak relationships between bacteria and DNA. Thus, conversions between DNA and biomass figures do not seem to be applicable generally.

A detailed study of the surface with respect to contamination by enteropathogenic microorganisms was conducted on 10 surface samples (2001), on the surfaces of the 6 profiles (2001, 2002), and on those samples collected during a wider survey in the Vuotso region (29 samples, 2002). These samples were also analyzed for respiratory activity (Table 13.3). Due to the

Table 13.3. Microbiological properties of surface samples from sites of the Vuotso region. Data shown are number of measurement (N), minimum (Min), and maximum (Max) value as well their medians

a Soil respiration (μg CO_2 h^{-1} g^{-1} dwt) at different temperatures

Sites	Param.	5 °C	15 °C	25 °C	35 °C
V 1-10	N	10	10	9	9
	Min	34	36	33	18
	Max	210	210	170	210
	Median	90.5	77.5	61	68
VPs	N	6	6	6	6
	Min	1.6	4.7	12.5	4.7
	Max	110	100	87	67
	Median	34	36	31	27
V1-20	N	31	31	31	31
	Min	0.2	0.6	0.8	0.9
	Max	18.6	60.6	51.1	65.4
	Median	1.6	3.4	4.7	5.7
All samples	Median	4.0	8.2	15.6	9.3

b LOI (loss on ignition: % of dwt); MCV (mean cell volume: μm^3 $cell^{-1}$); TBN (total bacterial number (10^9 n g^{-1} dwt); BBM (bacterial biomass: μg C g^{-1} dwt) and TBS (total bacterial surface: cm^2 g^{-1} dwt)

Sites	Param.	LOI	MCV	TBN	BBM	TBS
V 1-10	N	10	10	10	10	10
	Min	0.9	0.013	0.051	0.07	0.048
	Max	99.0	0.034	2.09	7.02	0.063
	Median	43.2	0.027	0.531	1.15	0.051
VPs	N	6	6	6	6	6
	Min	16.3	0.013	0.19	0.37	0.048
	Max	93.3	0.041	2.28	8.35	0.067
	Median	72.7	0.026	1.07	2.89	0.059
V1-20	N	29	29	29	29	29
	Min	1.4	0.010	0.0578	0.06	0.044
	Max	97.7	0.038	5.53	17.20	0.065
	Median	11.0	0.021	0.312	0.56	0.055
All samples	Median	19.5	0.021	0.391	0.89	0.055

very different locations, a wide range of soils was covered, described by the wide range of organic matter, expressed as LOI. The reaction to temperature is not clear with respect to individual samples or the three samples sets; just the total median value shows a tendency, which seems to reflect an optimum temperature of 25 °C.

The tests for *Escherichia coli* and other coliform bacteria in natural soil samples were negative or so low that indication of significant quantity could not be established in original samples without prior enrichment. Positive results for these organisms could only be found in feces, which were collected concomitantly with these soil samples (Chap. 14). The frequency of large cells has pointed to the possibility of occurrences of other soil bacteria, seemingly fecal bacteria, although their state of activity is not clear. A more detailed description of the bacterial communities is presented below.

13.3.2 Dry Tundra (Jauristunturit Area)

Table 13.4 summarizes the microbiological properties of the surface soils in this area. Most emphasis was put on the transect samples T1–T149, which were sampled in order to differentiate the land uses. From the ranges of the data sets, no significant differentiation can be made; they show wide overlapping. Nevertheless, the median values point to some tendencies. From this, it can be supposed that DNA values are highest at sites in Finland, which correlates with chlorophyll *a*, an indicator for active photosynthetic organisms. As soil samples were analyzed and visible plant material was removed, it can be concluded that soil algae are more abundant in lichen-depleted areas. Highest values for bacterial parameters are found in samples from Norway. This result is suspected to be due to higher amounts of low-molecular-weight substances and higher actual water content due to the dense lichen cover, which prevents evaporation and provides a more stable microclimate (Chap. 11).

Strong patchiness can be seen from the LOI data (Fig. 13.2). It should be noted that samples with high amounts of soil organic matter can be found next to samples with highest values of vegetation cover. Taking into account all six transects, a clear separation between Norwegian and Finnish samples cannot be recognized. Such separation only becomes evident from the full data set. Similarly, the bacterial biomass distribution is governed by strong patchiness (Fig. 13.3); numbers and biomass fluctuate within one order of magnitude, while mean cell volume and total bacterial surface show only low variability (Table 13.4). Further, direct relationships between LOI and BBM are not deducible from these data sets.

The data on soil respiration for the samples from the six transects show a positive response to temperature increase (Fig. 13.4, Tables 13.4 and 13.5). This can bee seen also for the surface samples from the reference points MP-L, MP-M, and MP-H, whereas samples N1–N13 show an optimum at 25 °C. The

Table 13.4. Microbiological properties of surface samples from sites of the Jauristunturit region. Data show number of measurement (N) minimum (Min) and maximum (Max) value as well their medians

a Soil respiration (µg CO_2 h^{-1} g^{-1} dwt) at different temperatures

Sites	Param.	5 °C	15 °C	25 °C	35 °C
T 1-149	N	83	82	83	83
	Min	0.56	0.52	4.58	3.25
	Max	50.8	47.7	70.9	63.2
	Median	4.81	5.37	18.2	29.2
MPs	N	3	3	3	3
	Min	0.96	0.31	1.54	10.8
	Max	21.5	21.5	23.8	33.4
	Median	1.54	5.76	9.59	24.9
N1-13	N	13	13	13	13
	Min	1.43	3.15	7.08	0.88
	Max	6.22	14.5	29.0	12.4
	Median	3.08	5.4	10.7	7.9
All samples	Median	4.01	6.7	15.7	23.7

b LOI (loss on ignition: % dwt); MCV (mean cell volume: µm^3 $cell^{-1}$); TBN (total bacterial number (10^9 g^{-1} dwt); BBM (bacterial biomass: µg C g^{-1} dwt) and TBS (total bacterial surface: cm^2 g^{-1} dwt)

Sites	Param.	LOI	MCV	TBN	BBM	TBS
T1-149	N	83	83	83	83	83
	Min	5.13	0.009	0.102	0.14	0.043
	Max	89.1	0.003	1.90	4.99	0.061
	Median	48.2	0.002	0.455	0.94	0.053
MPs	N	3	3	3	3	3
	Min	23.5	0.002	0.131	0.25	0.053
	Max	60.8	0.003	0.441	1.01	0.061
	Median	56.4	0.008	0.339	0.98	0.056
N1-13	N	13	13	13	13	13
	Min	7.7	0.001	0.115	0.32	0.046
	Max	98.0	0.004	0.986	2.25	0.063
	Median	19.2	0.002	0.409	0.67	0.054
All samples	Median	43.2	0.002	0.408	0.88	0.053

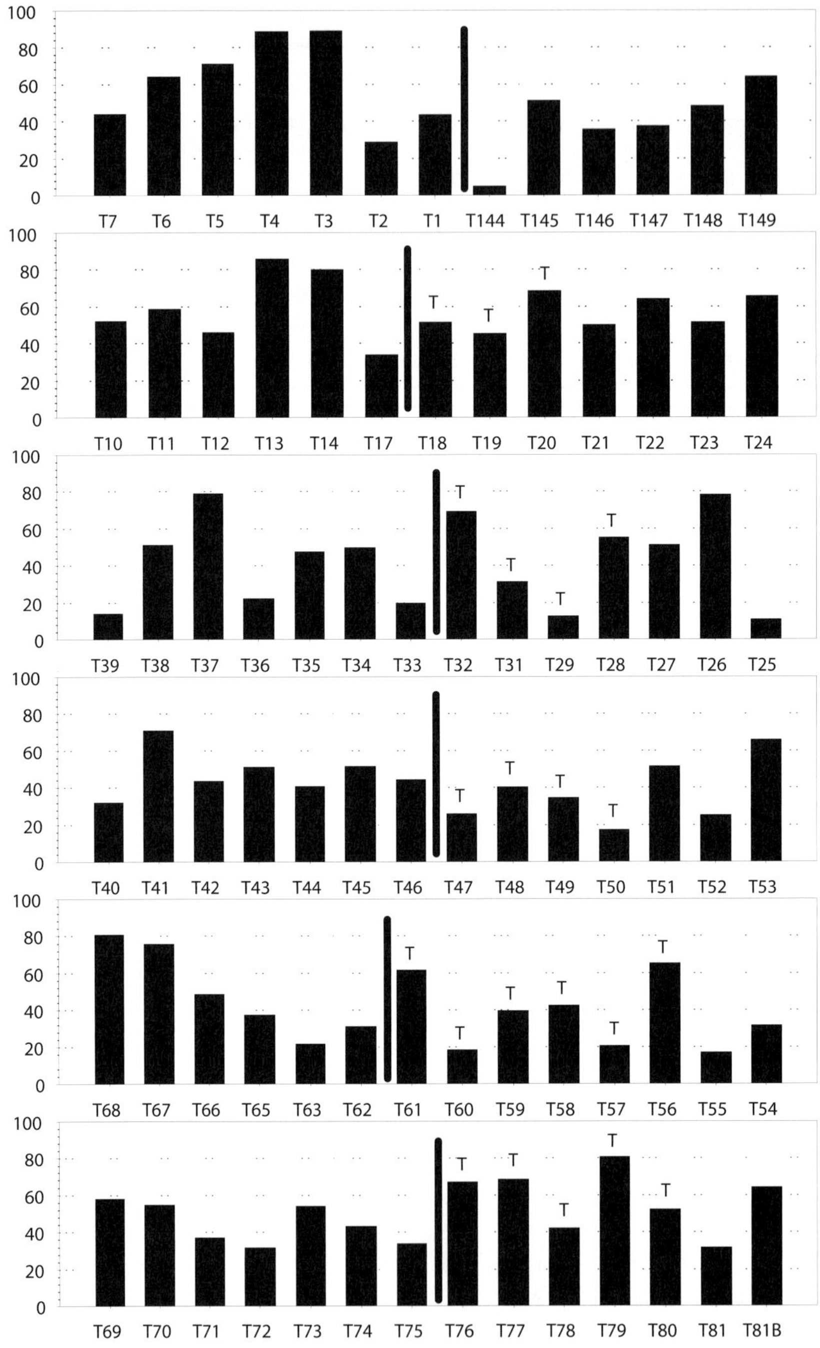

Fig. 13.2. Organic matter contents (loss on ignition: % dwt) of samples from the Jauristunturit transects. The *thin bars* indicate the position of the fence, Norwegian samples are displayed *left of the marker*, samples from the transition area are marked by a *T on top of the columns*

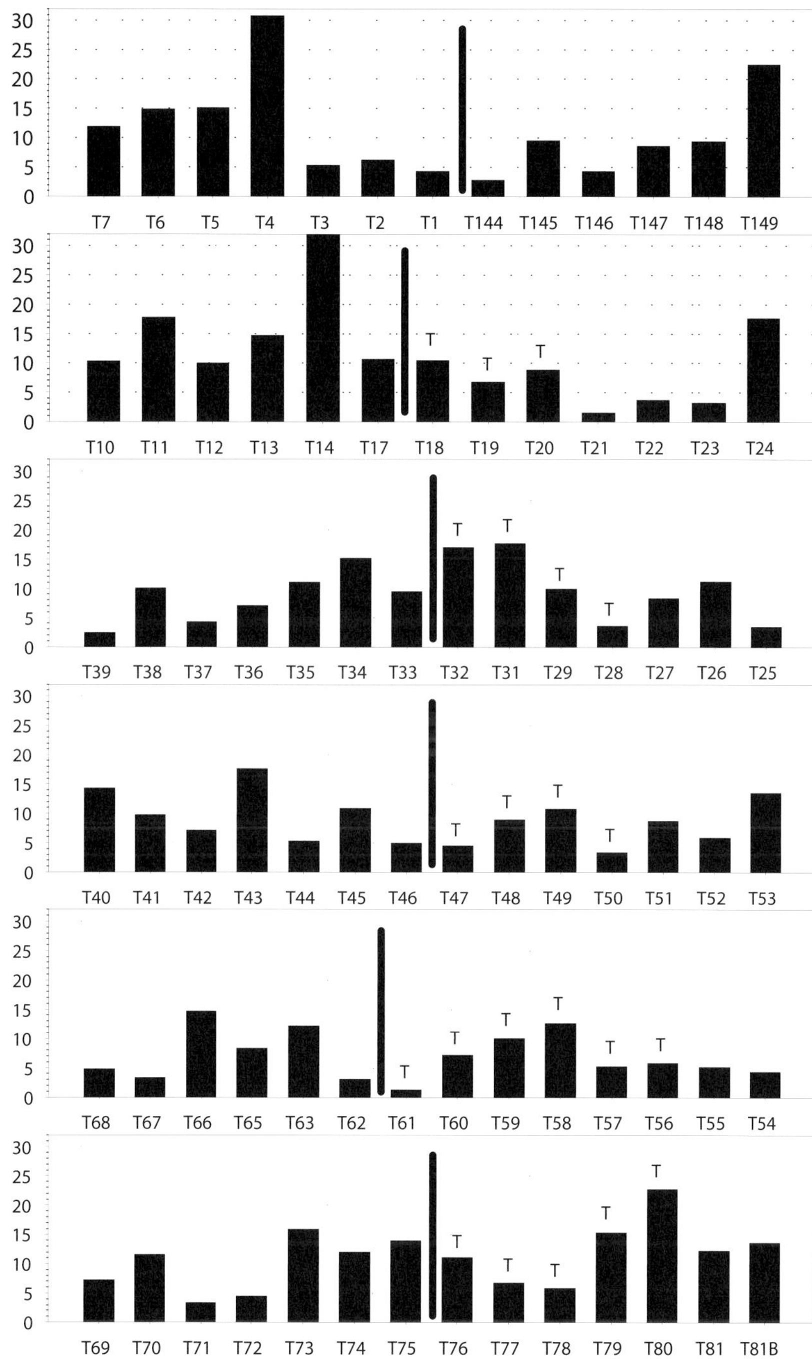

Fig. 13.3. Bacterial biomass (μg g^{-1} soil dwt) of samples from the Jauristunturit transects. The *thin bars* indicate the position of the fence, Norwegian samples are displayed *left of the marker*, samples from the transition area are marked by a *T on top of the columns*

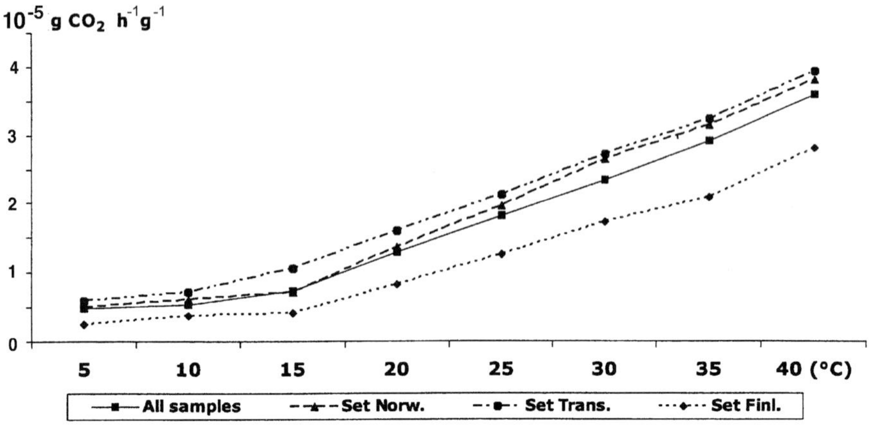

Fig. 13.4. Soil respiration (g CO_2 h^{-1} g^{-1} soil dwt) of sample sets from the transects in the Jauristunturit region

Table 13.5. Soil respiration of samples from the Jauristunturit – transects, calculated with respect to area (mg CO_2 h^{-1} m^{-2}). All data refer to soil depth of 0–5 cm (data for bulk density were provided by S. Peth). Given are number of samples (N), minimum (Min) and maximum (Max) values, and their medians related to different incubation temperatures

Sites	Param.	5 °C	15 °C	25 °C	35 °C
T 1-149	N	82	81	82	82
	Min	2.45	8.07	22.2	15.7
	Max	221	279	357	273
	Median	23.8	39.8	90.1	14.9
Samples No	N	40	40	40	40
	Min	3.33	9.00	22.3	30.7
	Max	203	215	256	286
	Median	26.2	38.9	99.2	155
Samples Tr	N	22	22	22	22
	Min	6.41	11.1	42.9	23.5
	Max	155	181	287	297
	Median	26.0	48.2	98.4	15.7
Samples Fi	N	20	19	20	20
	Min	2.45	8.07	22.2	15.7
	Max	221	279	357	273
	Median	14.4	21.3	71.4	124

Table 13.6. Differentiation of sites in the Jauristunturit region into the three sub-regions Norway, Transition, and Finland, with inclusive related sites MP-L, MP-M, MP-H, and N1-13 (for abbreviations see Tables 13.1 and 13.3)

Sites		LOI % dwt	DNA μg g^{-1}	Chl. *a* μg g^{-1}	Phaeop. μg g^{-1}	MCV μm^3	TBN 10^9 g^{-1}	BBM μg C g^{-1}	TBS cm^2 g^{-1}
Norw.	N	44	41	42	42	44	44	44	44
	Min	7.7	39.8	0.07	4.08	0.012	0.102	0.14	0.046
	Max	95.5	20,389	9.52	136.4	0.030	1.90	4.99	0.061
	Median	48.3	612.4	1.29	28.53	0.018	0.5	1.02	0.053
Trans.	N	26	21	23	23	26	26	26	26
	Min	9.3	61.5	0.15	2.71	0.009	0.12	0.22	0.043
	Max	80.9	49,989	8.01	95.5	0.030	1.20	2.27	0.061
	Median	42.4	759.0	1.14	16.86	0.020	0.385	0.80	0.055
Finl.	N	29	19	19	19	29	29	29	29
	Min	5.1	29	0.02	1.23	0.012	0.129	0.16	0.046
	Max	98.0	19,228	19.33	82.85	0.035	1.37	2.25	0.063
	Median	48.2	1,008	1.8	14.96	0.018	0.409	0.84	0.053

total data set shows fair ranges for the parameters describing the bacterial community. The differentiation into samples from Norway, Finland, and the transition zone is given in Tables 13.4 to 13.7. Here it becomes evident that samples from the higher-grazed area in Finland have lower respiration rates. This fact can also be considered to be due to the very low amount of available organic material. Concomitant medians of Q_{10} values for these data can be found between 1.21 and 2.0; low values are evident for low temperature spans. Individual values can be found at levels up to 3.4; the lowest value was 0.5.

13.3.3 Microbial Communities

Studies on the microbial community were performed with regard to shifts in size classes, which were assumed to indicate fecal pollution. The bacteria present in tundra environments, as known from other studies (see below), show figures for mean cell volumes of about 0.02–0.03 μm^3 and lengths of 0.3–1.5 μm. Hence, these values were used as base lines to indicate communities not influenced by enteric organisms. The samples used to test this assumption were derived from Vuotso (samples V1–V20) and Jauristunturit (samples N1–N13, and samples from the transect). The results of this study are presented in Table 13.7.

The overall picture of the samples analyzed ($n = 81$) meets this assumption and from this it can be concluded that there is no significant influence on the bacterial community by fecal shedding. However the maximum values (col-

Table 13.7. Data of the bacterial communities of all samples and sample groups from Vuotso (V1-20) and Jauristunturit regions Norway (No), Finland (Fi) the Transition area (Tr), and Sites N1-13 (for details see text). Percentages of cocci are related to total bacterial count (n %), total bacterial volume (vol %), and total bacterial surface (surf %)

		Mean length µm	Mean surface µm²	Mean volume µm³	Max. length µm	Cocci n %	vol %	surf %
All samples	N	81	81	79	80	81	79	79
	Min	0.32	0.32	0.02	0.94	7.0	6.8	19.7
	Max	0.67	0.86	0.19	3.13	88.8	64.3	69.0
	Med	0.45	0.49	0.03	1.72	66.9	39.6	44.5
V1-20	N	29	29	29	29	29	29	29
	Min	0.32	0.32	0.02	0.94	45.1	18.7	22.4
	Max	0.67	0.79	0.06	2.74	84.2	60.1	65.8
	Med	0.46	0.51	0.04	1.67	64.2	34.7	39.9
All samples transects (No, Tr, Fi, N1-13)	N	52	52	50	51	52	50	50
	Min	0.34	0.35	0.02	1.15	7.0	6.8	19.7
	Max	0.63	0.86	0.19	3.13	88.8	64.3	69.0
	Med	0.43	0.47	0.03	1.74	67.9	39.8	44.6
Samples No	N	25	25	25	25	25	25	25
	Min	0.34	0.35	0.02	1.15	53.7	21.0	26.5
	Max	0.56	0.64	0.05	3.13	83.3	64.3	66.0
	Med	0.43	0.46	0.03	1.66	69.1	41.1	54.8
Samples Tr	N	7	7	5	6	7	6	6
	Min	0.36	0.38	0.03	1.19	54.6	6.8	22.0
	Max	0.56	0.86	0.05	2.72	88.8	44.9	50.9
	Med	0.49	0.59	0.04	2.37	61.2	28.7	34.3
Samples Fi	N	7	7	7	7	7	6	6
	Min	0.34	0.35	0.02	1.16	63.2	41.6	47.7
	Max	0.52	0.83	0.19	2.09	86.2	63	69.0
	Med	0.39	0.41	0.03	1.7	77.2	48.9	55.7
Samples N1-13	N	13	13	13	13	13	13	13
	Min	0.38	0.39	0.03	1.49	7	15.8	19.7
	Max	0.63	0.75	0.06	2.85	75.5	45.2	51.6
	Med	0.47	0.52	0.04	1.92	62.1	30.9	36.6

umn "max. length") indicate that local effects cannot be excluded, since those large cells may well indicate the occurrence of fecal bacteria because they are much larger in size than others found. This is also deducible from the extremely low numbers of cocci-shaped bacteria in the samples – which organisms usually comprise more than 50 % of the total number or than 30 % of the total volume in such samples. Although the tests for *E. coli* and other coliform bacteria were negative for the sites at Vuotso, individual microscopic

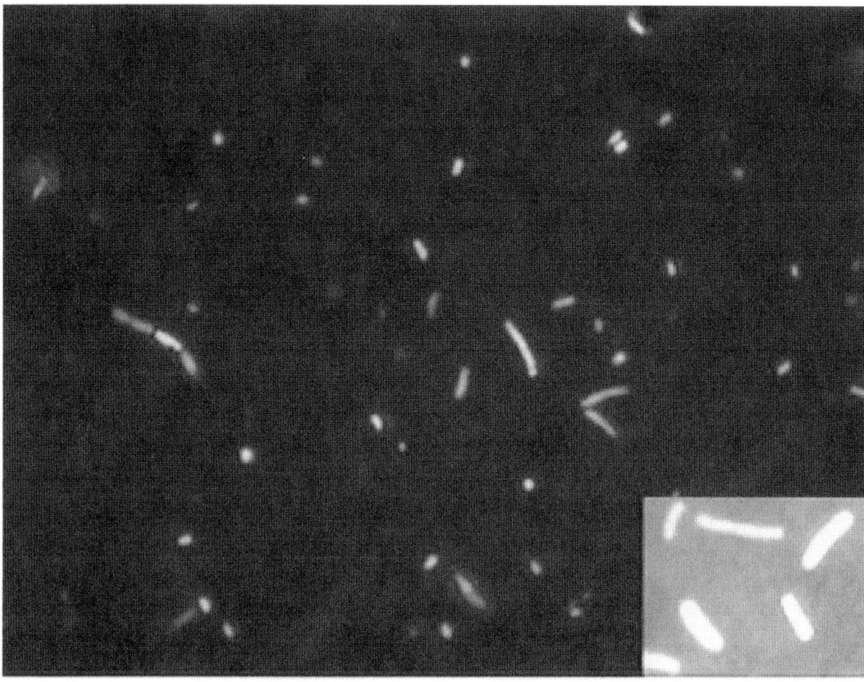

Fig. 13.5. Microscope image of a soil sample from Vuotso with large cells. The *insert* (*bottom right*) shows cells of *E. coli* from a fresh, pure culture for comparison. Length of the *insert*: 10 μm

pictures clearly show these large forms (Fig. 13.5); their presence is also reflected in elevated figures of the mean cell volume. Hence, it would seem that the size of enteric bacteria can be used as a non-specific indicator for such pollutants.

13.4 Discussion

Reindeer and other ruminants are described as "disturbance factors" (Persson et al. 2000; Suominen and Olofsson 2000; Edenius et al. 2002) as well as "ecosystem engineers" (Suominen et al. 1999; Suominen and Olofsson 2000; Persson 2003) and "key-stone species" (Angelstam et al. 2000). The herd size plays a crucial role in this respect and is under much discussion in Lapland. Reindeer grazing has been found to increase the proportion of graminoids in tundra heaths (Oksanen and Virtanen 1995; Manseau et al. 1996; Post and Klein 1996; Olofsson et al. 2001). In Finland, grazing by mammalian herbivores has had a decelerating effect on soil nutrient cycling and reduced soil respiration in both fertilized and unfertilized areas (Stark and Grellmann 2002). However, it is difficult to predict changes in vegetation patterns as they depend on the type of grazing and the primary plant cover (Adler et al. 2001). Reindeer are opportunistic feeders on various vegetations, varying locally and seasonally (White et al. 1981; Eilertsen 2002). However, in some regions they are highly selective and responsible for drastic shifts in plant communities. This is true for sites on the island of Spitsbergen, where reindeer have selectively removed several lichen species (Cooper and Wookey 2001; van der Wal et al. 2001a, b) as well as flowering plants (Cooper and Wookey 2003).

The data of this study have shown effects at different levels of reindeer herding on environments in northern Scandinavia. Changes in soil microbiology due to this land use are difficult to see because of different effects at different scales of the microenvironment and locally different soil conditions and plant cover. Moreover, local weather conditions can reveal stronger changes in microbiological properties than base line date extrapolated from long-term records. This is due to reactions of a microbial community, which has to react properly to wide ranges of environmental factors.

13.4.1 Effects of Climate and Soil Conditions on Soil Biota

The harsh climate imposes various stresses on the living world in the arctic region. There are frequent freeze–thaw cycles, dry–wet situations, and hard battles for nutrients and favorable habitats. Arctic plants, soil dwelling organisms, and soil microbes have to cope with all of these brutal environmental conditions. Extremes in temperature and uncertain moisture or nutrient availability favor a community with the ability to adapt to wide environmental ranges. Vegetation patterns and plant growth forms change drastically within distances of few meters due to differing small-scale patterns of topography and related variables of hydrology and soil chemistry (Bliss 1981; Bliss et al. 1984; Chapin et al. 1988; Gebauer et al. 1996). Lichens use the shelter of snow to protect from lowest temperatures and to start primary production in early spring at temperatures below 0 °C under the snow cover. Such extreme factors significantly affect parameters of soil biology and soil physics (Shaver and Chapin 1991; Johnson et al. 1996; Cheng et al. 1998) and could be appreciated in our study at Jauristunturit (Chaps. 11, 12). Fjells and northern forest ecosystems show generally a low nutrient status, low production rates and low decomposition activity (Heal 1981; Bliss et al. 1984; Binkley and Högberg 1997; Bhatti et al. 2002), although nutrient input by precipitation, especially nitrogen, has increased during the last decades (Ingerslev et al. 2001).

Moisture and micro-environmental habitats are the dominant factors in setting the frames for tundra ecosystems (Wynn-Williams 1990; Kennedy 1993; Broady 1996). Increasing temperatures in combination with stable precipitation in central parts of Scandinavia (IPCC 2001) may have an important effect on lichen growth. Actual soil water content were lower in the tundra soils, because the soils have only low clay and silt contents and thus a lower field capacity (Chap. 11). The years 2002 and 2003 were very dry and thus hampered microbial activity and growth. This aridity has to be considered regarding the low respiration rates in our study. Short or low precipitation events cannot compensate long-term stress from dryness, moreover, they lead to short dry–wet cycles. Although dry–wet cycles were found to have only little effect on community structure, bacterial diversity, and richness (Fierer et al. 2002), they probably affect total number and biomass.

While Table 13.3 shows that during growing season CO_2 efflux correlates well with soil temperature, it also reveals that the relationship between CO_2 production and temperature can be applied only when based on a large data set. However, data from individual samples show often no direct relationships or different trends. An optimal temperature appears to lie between 15 and 25 °C (Tables 13.1, 13.3, and 13.4).

But the relationship in total is even more complicated: For comparison, soils from a boreal coniferous forest show a temperature optimum between 7 and 12 °C (Huxman et al. 2003). Ecosystem respiration rate (R_e) increases with increasing temperature, but winter and summer soil respiration react differently to temperature. At 0 °C in winter there is a higher respiration proportion than in summer at 22 °C (Lipson et al. 2002). Temperature again controls ecosystem fluxes in spring and autumn; during spring an increase in temperature has a positive effect on net ecosystem CO_2 exchange (NEE), as temperature rises from 0 °C to a level optimum for photosynthesis, while R_e remains low. In summer, increasing temperatures have a negative effect on NEE because of inhibition in photosynthesis and increases in R_e. Further, root respiration has been variously calculated to be responsible for 90 % of total soil respiration (Dugas et al. 1999), 15–70 % (Norman et al. 1992), 50 % (Hanson et al. 1993), and 40 % (Kucera and Kirkham 1971).

The data which form the base for Fig. 13.4 also allows one to establish Q_{10} values for soil respiration. These were found to range from 0.04 to 3.78, and the mean values for the individual temperature steps show a general slowly decreasing trend from 2 to 0.56; the maximum height can be seen for the temperature step 10 to 20 °C (2.8). For soils from Abisko a mean was found at 2.4 with increasing values at low temperatures (Sjögersten and Wookey 2002). Highest shifts in Q_{10} values were found at temperatures below 0 °C (Mikan et al. 2002). A threshold for a respiration Q_{10}, i.e., activation energy, has been observed to occur at 7 °C (Sjögertsen and Wookey 2002). This relates to a threshold found by Nadelhoffer et al. (1991) around 9 °C. Median Q_{10} for 100 sites at Marion Island was seen to be about 2 for temperatures between 5 and 20 °C (Smith 2003), indicating a mixed community with different responses to temperature. Temperature relationships between Q_{10} and temperature for CO_2-evolutions were also found by Schleser (1982), Lloyd and Taylor (1994) and Widén (2001).

13.4.2 Effects of Grazing on Soil Biota

Grazing has significant influence on vegetation and thus the soil and soil biota, with important consequences for several ecosystem functions. It affects not only aboveground, but also belowground processes, i.e., nutrient mineralization and organic matter decomposition (Chapin 1980; Chiariello and Gulmon 1991). In forests, grazing can reduce microbial respiration but mineralization increases; in tundras, strong spatial variation of respiration and mineralization was found to depend on local characteristics (Väre et al. 1996; Austrheim and Eriksson 2001; Stark 2002; Stark et al. 2002; den Herder et al. 2003). Table 13.1 discloses the differences in microbial abundance and respiration, underlining this for forest soils. In a study in Norway, grazed areas were dominated by graminoids, whereas dwarf shrubs disappeared. Grazing stimulated plant production and N-cycling, and it can promote the change of vegetation patterns in tundra systems (Olofsson et al. 2001). Most important are removals of lichen and moss covers with their effect on soil temperature and water-holding capacity (Kershaw and Field 1975; van der Wal et al. 2001a, b).

Grazing can also affect soil structure via compaction, and thus changes in water penetration, porosity, aeration, and other soil physical parameters (Horn 1986; Vallamil et al. 2001; Chap. 11). As such, soil compaction, i.e., higher bulk density, caused by grazing of cattle in forest plantations of British Columbia is known to affect negatively the growth of young trees and the production of older plants by restricted root development, shifts in soil porosity and a lower nutrient availability (Broersma et al. 2000). Similar effects were noted by Zhang and Horn (1996) for sheep grazing on dike forelands in Germany. Main effects from grazing can be observed in connection with increased bulk density and penetration resistance, which limit root growth. However, these parameters need to exceed the corresponding critical values 1.4 g cm^{-3} and 2.500 kPa, respectively (Broersma et al. 2000). Although roots may develop strong growth pressures, up to 0.6 MPa (Clark et al. 2003); trampled soils can be too strong for many tundra plants when they contain high amounts of fine material. In our study at Jauristunturit, bulk density showed values between 0.94 and 1.5 g cm^{-3} (Chap. 11). Hence, direct effects of soil compaction (trampling) on soil biology could be neglected in this study. Hence, grazing and concomitant trampling may have additive or conflicting effects on soils. However, they become critical in fenced areas along the trampling paths on mineral soils of slopes (Evans 1996) and further explain the data from our study, which show low values for microbial activity on trampling paths.

Effects of grazing on soil C, N, and biomass are controversial (Bauer et al. 1987; Kieft 1994; Berg et al. 1997; Li and Chen 1998) and impacts can be due to management factors as well as local environmental conditions. Organic layers under undisturbed lichen cover had higher CEC and plant-available Ca and

Mg than sites with missing plant cover, where the organic layer was thin or gone. The organic layer is important for soil fertility and thus crucial to grazing (Chap. 12). Henry and Svoboda (1994) report from a high-arctic environment grazed by musk ox that grazed meadows show higher shoot densities than in non-grazed areas, and elevated contents in nitrogen, phosphorous, and organic matter. Owen (1980) hypothesized a mutually beneficial relationship between grazers and graminoids. This can be of special importance for impoverished environments, but such enhancement of grasses could not be observed during our study as a general feature.

Elevated concentrations of soil nutrients are associated with mammalian herbivores, with changes in vegetation from lichens to grasses. Deciduous plants are assumed to react positively to fertilization, evergreen plants are supposed to react negatively (Zimov et al. 1995; Chapin et al. 1996). Such effects can be forced by fertilization with N and P. Herbivores, and rodents, may induce significant local effects by nutrient transports on the long scale (McKendrick et al. 1980; Henry and Svoboda 1994). A study of Väre et al. (1996) showed a decrease in exchangeable nutrient stocks in grazed landscapes of 30–40 %, and microbial activity was significantly lower due to dry stands during summer. However, microbial biomass could not be directly related to respiration in Alaskan taiga forest systems (Vance and Chapin 2001). Our data from Jauristunturit show a lower respiration in disturbed than in untouched soils (Fig. 13.4), but individual data on soil organic matter and bacterial biomass are very variable (Figs. 13.2 and 13.3).

13.4.3 Effects on Microbial Communities

Soil microbial biomass is recognized as an indicator for environmental change (Jenkinson and Ladd 1981). It was found to be related generally to climatic conditions, as shown by Insam (1990) who showed a close correlation between C_{mic}/C_{org} vs. precipitation/evaporation in North America. Intensive forestry can alter decomposer communities as well as their functions at different trophic levels (Huhta 1976; Sundman et al. 1978; Binkley 1984; Butterfield 1999; Siira-Pietikäinen et al. 2003). Microbial communities change when fungi, including mycorrhiza and its feeding organisms, decrease (Bååth 1980; Pietikäinen and Fritze 1995; Hagerman et al. 1999), and bacterial biomass increases (Bååth et al. 1995). Abandonment of managed grasslands in the Alps has been found to cause changes in soil microbial communities, e.g., increase in fungal biomass and other ratios; bacterial biomass remaining stable, but the bacterial counts increasing (Zeller et al. 2001).

Our data on bacterial biomass and counts showed such effects for forest sites (Tables 13.1 and 13.2), but no significant differences between the grazed and non-grazed sites (Tables 13.5 and 13.6). Local heterogeneity reflects more the general pattern and the close relationship between plants and soil organisms. The numbers show levels comparable to other arctic tundra environments (Bölter and Kanda 1997; Bölter and Pfeiffer 1997). Soil algae and cyanobacteria, estimated by chlorophyll *a* and by microscopy, as well as fungi, were not significant contributors to the microbial community, a difference from other tundra sites (Bölter 2001; Schmidt and Bölter 2002). However, elevated contents of DNA (Table 13.2 and 13.6) refer to local patches of organisms not observed microscopically, but present.

The results of the survey on communities of surface soils with respect to size classes showed only partly the effect of an enteric pollution (Fig. 13.5, Table 13.7). The actual community thus does not seem to be influenced significantly by fecal indicators. This was confirmed by analyses performed with genetic probes for *Escherichia coli* and other coliform bacteria, although such organisms could be identified in original feces. The actual number in the soil environment was obviously strongly below the detection level. Thus, soils themselves are probably not reservoir for enteric bacteria, a result of low nutrients, dry conditions, and low temperatures (Bölter and Höller 1996). However, data obtained by cultivation (Chap. 14) describe their potential abundance (and also their low virulence).

13.4.4 Effects on Soil Respiration

In arctic tundra heaths microbial respiration and microbial C showed patterns different in grazed from ungrazed areas (Tables 13.5, 13.6). Nevertheless, low respiration data can be regarded as indicative for lowered levels of available C in soils. Mammalian grazers also affect the coupling between plant and microbes. Fecal nutrients are additional N sources outside the plant growth season and thus do not affect their growth (Stark and Grellmann 2002). Urine and fecal deposition were shown to have positive effects on nutrient cycling in soil (McKendrick et al. 1980; Ruess and McNaughton 1987; Frank and Groffmann 1998). Tracy and Frank (1998) hypothesize that microbial populations are favored by inputs of low weight soil organic matter from dung in moderately grazed grasslands, which leads to increased root growth and stimulates degradation processes.

Affects of reindeer grazing was found with positive consequences for soil respiration and microbial C development in the lichen and litter layer of the soil profile in a Scots pine forest. Grazing also was noted to stimulate nematode abundance, pointing toward a wide effect on soil biology, cycling of nutrients, and microclimate (Stark et al. 2000). Belowground effects of reindeer grazing were significant in relation to C mineralization and microbial activity; reductions occurred irrespective of moisture content. An effect may be the reduction of labile dissolved organic carbon compounds because of destroying roots by trampling and thus hampering exudate release. There was no consistent effect on N mineralization (Stark et al. 2003). Data from this study, however, showed a remarkably lower respiration in the Finnish area of Jauristunturit (Fig. 13.4, Table 13.5), where grazing has occurred for a long time and shifts in plant (and root) abundance have taken place.

It has been shown in alpine meadows in Tibet that grazing by sheep alters soil respiration intensity. Light grazing (2.5 sheep ha^{-1}) had an efflux of 2.04 kg CO_2 m^{-2} y^{-1}, the heavier-grazed site (5.3 sheep ha^{-1}) released only 1.53 kg CO_2 m^{-2} y^{-1}; and Q_{10} values at the low-grazed site were 3.22 compared to 2.75 at the high-grazed site (Cao et al. 2004). Grazing on prairie environments in the US also shows an increase in soil CO_2 flux (Frank et al. 2002). Reflecting on these data, it might be supposed that soil respiration was probably enhanced by grazing in nutrient-rich tundra heaths, but not in continental tundra.

Grazing was also shown to increase substrate-induced respiration (SIR), carbon availability index (CAI) (Cheng et al. 1998), and the qCO_2 (Ohtonen 1994) in suboceanic tundra heaths. Our date from Jauristunturit (Table 13.5) shows that there is much agreement with literature from elsewhere, as cited in this chapter. However, adequate comparison cannot be made with all data from other sources. Individual methods, soils, seasons, and other factors often differ too much.

The data in our study, it should be noted, reveals that respiration has led to only small losses of soil carbon. Even with the stimulations of elevated moisture and addition of glucose, soil respiration yielded only minor positive effects. This is probably due to an already starving community, the result of dryness and low amounts of non-carbon nutrients, especially nitrogen. Extrapolations to a seasonal or yearly budget cannot be done from these data, as microbial respiration occurs down to −18 °C and 80 % of the CO_2 could be trapped in the soil between 0 and −9 °C; soil respiration in winter can make up to 40 % of the yearly budget (Elberling and Brandt 2003).

13.5 Conclusion

Changes in plant diversity create environmental heterogeneity at different spatial scales. It has been stated by Weber et al. (1998) that management should account for spatial grazing aspects. Consequently, a management task is to consider vegetation changes and their effects on soil biota, i.e., on soil fertility. Degradation of vegetation occurs mainly along the fences in Finnmark, which results in a great extend of bare soil exposure, followed by erosion on slopes and drumlins. Soil denudation occurs at 1–3 mm y^{-1}. High amounts of dung and disappearance of lichens by grazing and trampling are other factors (Evans 1996). The trampling associated with large increases in caribou herds in Quebec and Labrador have affected the grazing grounds, their plant composition and biomass. Lichens and vascular plants show limited resilience to summer and winter browsing, consequently large areas are barren or produce a lower quality diet. Effects have been observed in animal health and calf production (Manseau et al. 1999) and should be considered in future management strategies.

Acknowledgements. We are grateful to the organizers of the logistics at Jauristunturit, namely, Anti Näkkälä, as well as to B. Burkhard and S. Peth, who carried out samplings in 2002 and 2003 at Vuotso and in the sampling area at Jauristunturit. Further samples were provided by N. Kemper. Mrs. A. Schelz kindly performed the measurement of pigments and DNA.

References

Adler PB, Raff DA, Lauenroth WK (2001) The effect of grazing on the spatial heterogeneity of vegetation. Oecologia 128:465–479

Angelstam P, Wikberg P-E, Danilov P, Faber WE, Nygrén K (2000) Effects of moose density on timber quality and biodiversity restoration in Sweden, Finland and Russian Karelia. Alces 36:133–145

Apps MJ, Kurz WA, Luxmoore RJ, Nilsson LO, Sedjo RJ, Schmidt R, Simpson LG, Vinson T (1993) The changing role of circumpolar boreal forests and tundra in global C cycle. Water Air Soil Pollut 70:39–53

Arar EJ, Collins GB (1997) *In vitro* determination of chlorophyll *a* and pheophytin *a* in marine and freshwater algae by fluorescence. Revision 1.2, September 1997. National Exposure Research Laboratory, Office of Research and Development US Environmental Protection Agency, Cincinnati, pp 445.0-1–445.0-17

Augustin DJ, McNaughton SJ (1998) Ungulate effects on the functional species composition of plant communities: herbivore selectivity and plant tolerance. J Wildlife Manage 62:1165–1183

Austrheim G, Eriksson O (2001) Plant species diversity and grazing in the Scandinavian mountains – patterns and processes at different spatial scales. Ecography 24:683–695

Bååth E (1980) Soil fungal biomass after clear-cutting of a pine forest in central Sweden. Soil Biol Biochem 12:495–500

Bååth E, Frostegard Å, Pennanen T, Fritze H (1995) Microbial community structure and pH response in relation to soil organic matter quality in wood-ashed fertilized, clearcut or burned coniferous forest soils. Soil Biol Biochem 27:229–240

Batten DS, Svoboda J (1994) Plant communities on the uplands in the vicinity of the Alexandra Fiord Lowland. In: Svoboda J, Friedman B (ed) Ecology of a polar oasis, Alexandra Fiord, Ellesmere Island. Captus Univ Press, Toronto, pp 97–110

Bauer A, Cole CV, Black AL (1987) Soil property comparisons in virgin grasslands between grazed and nongrazed management systems. Soil Sci Soc Am J 51:176–182

Berg WA, Bradford JA, Sims PL (1997) Lon-term soil nitrogen and vegetation change on sandhill rangeland. J Rang Manage 50:482–486

Bhatti JS, Apps MJ, Jiang H (2002) Influence of nutrients, disturbances and site conditions on stocks along a boreal forest transect in central Canada. Plant Soil 242:1–14

Binkley D (1984) Does forest removal increase rates of decomposition and nitrogen release? For Ecol Manage 8:229–233

Binkley D, Högberg P (1997) Does atmospheric deposition of nitrogen threaten Swedish forests? For Ecol Manage 92:119–152

Bliss LC (1981) North American and Scandinavian tundras and polar deserts. In: Bliss LC, Heal OW, Moore JJ (eds) Tundra ecosystems: a comparative analysis. Cambridge Univ Press, Cambridge, pp 8–24

Bliss LC, Svoboda J, Bliss DI (1984) Polar deserts, their plant cover and production in the Canadian High Arctic. Holarct Ecol 7:305–324

Bölter M (2001) Preliminary results about contributions of soil algae and cyanobacteria in microbial communities of arctic soils – a microscopic approach. Nova Hedwigia Beih 123:307–315

Bölter M (2004a) Soil: an extreme habitat for microorganisms? Pedosphere 14:137–144

Bölter M (2004b) Ecophysiology of psychrotrophic and psychrophilic microorganisms. Cell Mol Biol 50:563–573

Bölter M, Blume H-P (2002) Soils as habitats for microorganisms. In: Beyer L, Bölter M (eds) Geoecology of Antarctic ice-free coastal landscapes. Ecological Studies 154. Springer, Berlin Heidelberg New York, pp 285–302

Bölter M, Höller C (1996) Hygienic problems in using permafrost soils for organic waste disposal. Polarforsch 66:61–65

Bölter M, Kanda H (1997) Preliminary results of biological and microbiological investigations on Severnaya Zemlya (1995). Proc NIPR Symp 10:169–178

Bölter M, Pfeiffer E-M (1997) Bacterial biomass and properties of Arctic desert soils (Archipelago Severnaya Zemlya, Northern Siberia). In: Iskandar IK, Wright EA, Radke JK, Sharratt BS, Groenevelt PH, Hinzman LD (eds) Proceedings of the international sympsoium on physics, chemistry, and ecology of seasonally frozen soils, Fairbanks, Alaska, 10–12 June 1997, CRREL Spec Rep 97-10:481–487

Bölter M, Bloem J, Meiners K, Möller R (2002) Enumeration and biovolume determination of microbial cells – a methodological review and recommendations for applications in ecological research. Biol Fertil Soils 36:249–259

Bölter M, Müller-Lupp W, Takata K, Yabuki H, Möller R (2003) Potential CO_2-production in aerobic conditions from a Siberian tundra environment. Polar Biosci 16:70–85

Bremer DJ, Ham JM, Owensby CE, Knapp AK (1998) Response of soil respiration to clipping and grazing in a tall grass prairie. J Environ Qual 27:1359–1348

Broady PA (1996) Diversity, distribution and dispersal of Antarctic terrestrial algae. Biodiv Conserv 5:1307–1355

Broersma K, Krzic M, Newman R, Bomke A (2000) Effects of grazing on soil compaction and water infiltration in forest plantations in the interior of British Columbia. In: Hollstedt C, Sutherland K, Innes T (eds) From science to management and back: a science forum for southern interior ecosystems of British Columbia. Southern Interior Forest Extension and Research Partnership, Kamloops, BC, pp 89–92

Brown J, Miller PC, Tieszen LL, Bunnell FL (eds) (1980) An arctic ecosystem: the coastal tundra at Barrow Alaska. Dowden, Hutchinson and Ross, Stroudsburg, pp 66–101

Butterfield J (1999) Changes in decomposition rates in Collembola densities during the forestry cycle in conifer plantation. J Appl Ecol 36:92–100

Cao G, Tang Y, Mo W, Wang Y, Li Y, Zhao X (2004) Grazing intensity alters soil respiration in an alpine meadow on the Tibetan plateau. Soil Biol Biochem 36:237–243

Chapin FS III (1980) Nutrient allocation and responses to defoliation in tundra plants. Arct Alp Res 12:553–563

Chapin FS III (1991) Effects of stresses on nutrient availability. In: Mooney HA, Winner WE, Pell EJ (eds) Response of plants to multiple stresses. Academic Press, San Diego, pp 67–76

Chapin FS III, Fetcher N, Kielland K, Everett KR, Linkins AE (1988) Productivity and nutrient cycling of Alaskan tundra: enhancement by flowing soil water. Ecology 69:693–702

Chapin FS III, Bret-Harte MS, Hobbie SE, Zhong H (1996) Plant functional types as predictors of transient responses of arctic vegetation to global change. J Veg Sci 7:347–358

Cheng W, Virginia RA, Oberbauer SF (1998) Soil nitrogen, microbial biomass, respiration along an arctic toposequence. Soil Sci Soc Am J 62:654–662

Chiariello NR, Gulmon SL (1991) Stress effects on plant reproduction. In: Mooney HA, Winner WE, Pell EJ (eds) Response of plants to multiple stresses. Academic Press, San Diego, pp 161–188

Clark LJ, Whalley WR, Barraclough PB (2003) How do roots penetrate strong soil? Plant Soil 255:93–104

Cooper EJ, Wookey PA (2001) Field measurements of the growth rates of forage lichens and the implications of grazing by Svalbard reindeer. Symbiosis 31:173–186

Cooper EJ, Wookey PA (2003) Floral herbivory of *Dryas octopetala* by Svalbard reindeer. Arctic Antarct Alp Res 35:369–376

Cornelissen JHC, Callaghan TV, Alatalo M, Michelsen A, Graglia E, Hartley AE, Hik DS, Hobbie SE, Pres MC, Robinson CH, Henry GHR, Shaver GR, Phoenix GK, Gwynn Jones D, Jonasson S, Chapin FS III, Molau U, Neill C, Lee JA, Melillo JM, Sveinbjörnsson B, Aerts R (2001) Global change and arctic ecosystems: is lichen decline a function of increases in vascular plant biomass? J Ecol 89:984–994

Den Herder M (2003) Impacts of ungulates in boreal and subarctic tundra ecosystems in Finland. Thesis, Faculty of Forestry, University of Joensuu, 38 p + 4 app

Den Herder M, Niemelä P (2003) Effects of reindeer on the re-establishment of *Betula pubescens* subsp. *czerepanovii* and *Salix phylicifolia* in a subarctic meadow. Rangifer 23:3–12

Den Herder M, Kytöviita M-M, Niemelä P (2003) Growth of reindeer lichens and effects of grazing on ground cover vegetation in a Scots pine forest and a subarctic heathland in Finnish Lapland. Ecography 26:3–12

Dixon RK, Brown S, Houghton RA, Solomon AM, Trexler MC, Wisniewski J (1994) Carbon pools and flux of global forest ecosystems. Science 263:185–190

Dugas WA, Heuer ML, Mayeux HS (1999) Carbon dioxide fluxes over Bermuda grass, native prairie, and sorghum. Agric For Meteorol 93:121–139

Edenius L, Danell K, Bergström R (1993) Impact of herbivory and competition on compensatory growth in woody plants: winter browsing by moose on Scots pine. Oikos 66:286–292

Edenius L, Bergman M, Ericsson G, Danell K (2002) The role of moose as a disturbance factor in managed boreal forests. Silva Fennica 36:57–67

Eilertsen SM (2002) Utilization of abandoned coastal meadows in northern Norway by reindeer. Thesis, Dept Arctic Biol Inst Med Biol, Univ Tromsø, Tromsø and Norw Crop res Inst, Holt Res Centre, Tromsø, Norway, 31 pp + 4 app

Elberling B, Brandt KK (2003) Uncoupling of microbial production and release in frozen soil and its implications for field studies of arctic C cycling. Soil Biol Biochem 35:263–272

Evans R (1996) Some impacts of overgrazing by reindeer in Finnmark, Norway. Rangifer 16:3–19

Fierer N, Schimel JP, Holden PA (2002) Influence of drying–rewetting frequency on soil bacterial community structure. Microb Ecol (DOI 10.1007/s00248-002-1007-2)

Frank AB, Liebig MA, Hanson JD (2002) Soil carbon fluxes in northern semiarid grasslands. Soil Biol Biochem 34:1235–1241

Frank DA, Groffman PM (1998) Ungulate vs. landscape control of soil C and N processes in grasslands of Yellowstone National Park. Ecology 79:2229–2241

Gebauer RLE, Tenhunen JD, Reynolds JF (1996) Soil aeration in relation to soil physiological properties, nitrogen availability, and root characteristics within an arctic watershed. Plant Soil 178:37–48

Gilmanov TG, Oechel WC (1995) New estimates of organic matter reserves and net primary productivity of the North American tundra ecosystems. J Biogeogr 22:723–741

Grellmann D (2001) Top-down and bottom-up effects in a Fennoscandian tundra plant community. Dissertation, Dept Ecol Environ Sci, Umeå University, 24 pp + 6 app

Grünzweig JM, Sparrow SD, Yakir D, Chapin FS III (2004) Impact of agricultural land-use change on carbon storage in boreal Alaska. Global Change Biol 10:452–472

Hagerman SM, Jones MD, Bradfield GE, Gillespie M, Durall DM (1999) Effects of clear-cut logging on the diversity and persistence of ectomycorrhizae in a subalpine forest. Can J For Res 29:124–134

Hanson PJ, Wullschleger SD, Bohlman SA, Todd DE (1993) Seasonal and topographic patterns of forest floor CO_2 effects from an upland oak forest. Tree Physiol 13:1–15

Havas P, Kubin E (1983) Structure, growth and organic matter content in the vegetation cover of an old spruce forest in Northern Finland. Ann Bot Fenn 20:115–149

Heal OW (1981) Introduction. In: Bliss LC, Heal OW, Moore JJ (eds) Tundra ecosystems: a comparative analysis. Cambridge Univ Press, Cambridge, pp xxvii–xxxi

Heikkinen JEP, Maljanen M, Aurela M, Hargreaves KJ, Martikainen PJ (2002) Carbon dioxide and methane dynamics in a sub-Arctic peatland in northern Finland. Polar Res 21: 49–62

Henry GHR, Svoboda J (1994) Dinitrogen fixation (acetylene reduction) in high-arctic sedge meadows. In: Svoboda J, Freedman B (eds) (1994) Ecology of a polar desert, Alexandra Fiord, Ellesmere Island. Captus University Publ, Toronto, pp 201–206

Horn R (1986) The influence of animal treading on soil physical properties of mountainous soils. Mitteilgn Dtsch Bodenkundl Gesellsch 46:63–71

Huhta V (1976) Effects of clear-cutting on numbers, biomass and community respiration of soil invertebrates. Ann Zool Fenn 13:63–80

Huxman TE, Turnipseed AA, Sparks JP, Harley PC, Monson RK (2003) Temperature as a control over ecosystem CO_2 fluxes in a high-elevation, subalpine forest. Oecologia 134:537–546

Ingerslev M, Mälkönen E, Nilsen P, Nohrstedt H-Ö, Oskarsson H, Raulund-Rasmussen K (2001) Main findings and future challenges in forest nutritional research management in the Nordic countries. Scand J For Res 16:488–501

Insam H (1990) Are the soil microbial biomass and basal respiration governed by the climatic regime? Soil Biol Biochem 22:525–532

Insam H, Rangger A, Henrich M, Hitzl W (1996) The effect of grazing on soil microbial biomass and community structure on alpine pastures. Phyton 36:205–216

IPCC (2001) Climate change 2001: synthesis report. A contribution of working groups I, II, and III to the third assessment report of the intergovernmental panel on climate change (eds: Watson RT and the Core Writing Team). Cambridge Univ Press, Cambridge, 398 pp

Jenkinson DS, Ladd JN (1981) Microbial biomass in soil: measurement and turnover. In: Paul EA, Ladd JN (eds) Soil biochemistry, vol 5. Dekker, New York, pp 415–471

Johnson LC, Shaver GR, Giblin AE, Nadelhofer KJ, Rastetter ER, Laundre JA, Murray JL (1996) Effects of drainage and temperature on the carbon balance of tussock tundra microcosms. Oecologia 108:737–748

Jonasson S (1992) Plant responses to fertilization and species removal in tundra related community structure and clonality. Oikos 63:420–429

Jylhä K, Tuomenvirta H, Ruosteenoja K (2004) Climate change projections for Finland during the 21st century. Boreal Env Res 9:127–152

Kennedy AD (1993) Water as limiting factor in the Antarctic terrestrial environment: a biogeographical synthesis. Arct Alp Res 25:308–315

Kershaw KA, Field GF (1975) Studies on lichen dominated systems. XV. The temperature and humidity profiles in a *Cladina alpestris* mat. Can J Bot 53:2614–2620

Kieft TL (1994) Grazing and plant canopy effects on semi-arid soil microbial biomass and respiration. Biol Fertil Soils 18:155–182

Kucera CL, Kirkham DR (1971) Soil respiration studies in tallgrass prairie in Missouri. Ecology 52:912–915

Kumpula J (2001) Productivity of the semi-domesticated reindeer (*Rangifer t. tarandus* L.) stock and carrying capacity of pastures in Finland during 1960–1990's. Acta Univ Oulu, A 375, 44 pp + 6 app

Kumpula J, Colpaert A (2003) Effects of weather and snow conditions on production and survival of semi-domesticated reindeer (*R. t. tarandus*). Polar Res 22:225–233

Lauer W, Rafiqpoor MD (2002) Die Klimate der Erde. Erdwissenschaftliche Forschung, Bd 40, Franz Steiner Verlag, Stuttgart,271p

Li XZ, Chen ZZ (1998) Effects of different stock rates on C, N, P contents in soil and plant in *Leymus chinensis* steppe. Acta Agresta Sin 6:90–98

Lipson DA, Schadt CW, Schmidt SK (2002) Changes in soil microbial community structure and function in an alpine dry meadow following spring melt snow melt. Microb Ecol 43:307–314

Lloyd J, Taylor JA (1994) On the temperature dependence of soil respiration. Funct Ecol 8:315–323

Lohila A, Aurela M, Regina K, Laurila T (2003) Soil and total ecosystem respiration in agricultural fields: effect of soil and crop type. Plant Soil 251:303–317

Lorenzen CF (1967) Determination of chlorophyll and phaeopigments: spectroscopic equations. Limnol Oceanogr 12:343–346

Maljanen M, Liikanen A, Silvola J, Martikainen PJ (2003) Nitrous oxide emissions from boreal organic soil under different land-use. Soil Biol Biochem 35:689–700

Manseau M, Huot J, Crete M (1996) Effects of summer grazing by caribou on composition and productivity of vegetation: community and landscape level. J Ecol 84:503–513

Manseau M, Huot J, Bergman C (1999) Ecosystem processes characterizing and influencing grazing systems, productivity and sustainability: the caribou herds of the Quebec/Labrador peninsula. In: Proceedings, An International Workshop on The Human Role in Reindeer/Caribou Systems: Coping with Threats to Environmental Security in Northern Landscapes, 10 14 February 1999, Rovaniemi, Finland

McKendrick JD, Batzli GO, Everett KR, Swanson JC (1980) Some effects of mammalian herbivores and fertilization on tundra soils and vegetation. Arct Alp Res 12:565–578

McNaughton SJ (1985) Ecology of grazing ecosystems: the Serengeti. Ecol Monogr 55:259–294

Mikan C, Schimel J, Doyle A (2002) Temperature controls of microbial respiration above and below freezing in Arctic tundra soils. Soil Biol Biochem 34:1785–1795

Miller PC, Webber PJ, Oechel WC, Tieszen LL (1980) Biophysical processes and primary production. In: Brown J, Miller PC, Tieszen LL, Bunnell FL (eds) An arctic ecosystem: the coastal tundra at Barrowm Alaska. Dowden, Hutchinson and Ross, Stroudsburg, pp 66–101

Müller-Wille L (1987) Indigenous peoples, land-use conflicts, and economic development in circumpolar lands. Arct Alp Res 19:351–356

Nadelhoffer KJ, Giblin AE, Shaver GR, Laundre JA (1991) Effects of temperature and substrate quality on element mineralization in six arctic soils. Ecology 72:242–253

Nellemann C, Vistnes I, Jordhøy P, Strand O (2001) Winter distribution of wild reindeer in relation to power lines, roads and resorts. Biol Conserv 101:351–360
Norman JM, Garcia R, Verma SB (1992) Soil surface CO_2 fluxes and the carbon budget of a grassland. J Geophys Res 97:18,845–18,853
Ohtonen R (1994) Accumulation of organic matter along a pollution gradient: application of Odum's theory of ecosystem energetics. Microb Ecol 27:43–55
Oksanen L (1993) Trophic exploitation and arctic phytomass patterns. Am Nat 122:45–52
Oksanen L, Oksanen T (2000) The logic and realism of the hypothesis of exploitation ecosystems. Am Nat 155:703–723
Oksanen L, Virtanen R (1995) Topographic, altitudinal and regional patterns in continental and suboceanic heath vegetation of northern Fennoscandia. Acta Bot Fenn 153:1–80
Olofsson J, Kitti H, Rautiainen P, Stark S, Oksanen L (2001) Effects of summer grazing by reindeer on composition of vegetation, productivity and nitrogen cycling. Ecography 24:13–24
Olsrud M, Christensen TR (2004) Carbon cycling in subarctic tundra; seasonal variation in ecosystem partitioning based on in situ ^{14}C pulse-labelling. Soil Biol Biochem 36:245–253
Owen DF (1980) How plant may benefit from the animals that eat them. Oikos 35:230–235
Parsson AN, Welker JM, Wookey PA, Press MC, Callaghan TV, Lee JA (1994) Growth responses of four sub-Arctic dwarf shrubs to simulated environmental change. J Ecol 82:307–318
Persson I-L (2003) Moose population density and habitat productivity as drivers of ecosystem processes in northern boreal forests. PhD thesis, Swed Univ Agric Sci, Umeå
Persson I-L, Danell K, Bergström R (2000) Disturbance by large herbivores in boreal forests with special reference to the moose. Ann Bot Fenn 37:251–263
Peterson KM, Billings WD (1980) Tundra vegetational patterns and succession in relation to microtopography near Atkasook, Alaska. Arct Alp Res 12:473–483
Phoenix GK, Gwynn-Jones D, Callaghan TV, Sleep D, Lee JA (2001) Effects of global change on a sub-arctic heath: effects of enhanced UV-B radiation and increased summer precipitation. J Ecol 89:256–267
Pietikäinen J, Fritze H (1995) Clear-cutting and prescribed burning in coniferous forest: comparison of effects on soil fungal and total microbial biomass, respiration and nitrification. Soil Biol Biochem 27:101–109
Post ES, Klein DR (1996) Relationships between graminoid growth form and levels of grazing by caribou (*Rangifer tarandus*) in Alaska. Oecologia 107:364–372
Post WM, Emanuel WR, Zinke PJ, Stangenberger AG (1982) Soil carbon pools and world life zones. Nature 298:156–159
Press MC, Potter JA, Burke MJW, Callaghan TV, Lee JA (1998) Responses of a sub-Arctic dwarf shrub heath community to simulated environmental change. J Ecol 86:315–327
Raich JW, Schlesinger WH (1992) The global carbon dioxide flux in soil respiration and its relationship to vegetation to vegetation and climate. Tellus 44B:81–91
Rieger S (1974) Arctic soils. In: Ives JD, Barry RG (eds) Arctic and alpine environments. Methuen, London, pp 749–769
Rodin LE, Bazilevich NI, Rozov NN (1975) Productivity of the world's main ecosystems. Productivity of world ecosystems. Proc Symp 31 Aug – 1 Sept 1972, Seattle, Washington. Nat Acad Sci, Washington, DC, pp 13–26
Ruess RW, McNaughton SJ (1987) Grazing and the dynamics of nutrient and energy regulated microbial processes in the Serengeti grassland. Oikos 49:101–110

Russel WC, Newman C, Williamson DH (1975) A simple cytochemical technique for determination of DNA in cells infected with mycoplasmas and viruses. Nature 253:461–462

Rustad LE, Campbell JL, Marion GM, Norby RJ, Mitchell MJ, Hartley AE, Cornelissen JHC, Gurevitch, GCTE-News, (2001) A meta-analysis of the response of soil respiration and net nitrogen mineralization, abd aboveground plant growth to experimental ecosystem warming. Oecologia 126:532–562

Schleser GH (1982) The response of CO_2 evolution from soils to global temperature changes. Zeitschr Naturforsch 37a:287–291

Schmidt N, Bölter M (2002) Fungal and bacterial biomass in tundra soils along an arctic transect from Taimyr Peninsula, central Siberia. Polar Biol 2:871–877

Schultz J (2000) Handbuch der Ökozonen. Ulmer, Stuttgart

Shaver GR, Chapin FS III (1991) Production:biomass relationships and element cycling in contrasting arctic vegetation types. Ecol Monogr 61:1–31

Siira-Pietikäinen A, Haimi J, Fritze H (2003) Organisms, decomposition, and growth of pine seedlings in boreal forest soil affected by sod cutting and trenching. Biol Fertil Soils 37:163–174

Sjögertsen S, Wookey PA (2002) Climatic and resource quality controls on soil respiration across a forest–tundra ecozone in Swedish Lapland. Soil Biol Biochem 34:1633–1646

Smith VR (2003) Soil respiration and its determinants on a sub-Antarctic island. Soil Biol Biochem 35:77–91

Sommerkorn M, Bölter M, Kappen L (1999) Carbon dioxide fluxes of soils and mosses in wet tundra of Taimyr Peninsula, Siberia: controlling factors and contribution to net system fluxes. Polar Res 18:253–260

Stark S (2002) Reindeer grazing and soil nutrient cycling in boreal and tundra ecosystems. PhD Thesis, Dept Biol, Univ Oulu, Acta Univ Oulu, A382, 31 pp + 5 app

Stark S, Grellmann D (2002) Soil microbial responses to herbivory in an arctic tundra heath at two levels of nutrient availability. Ecology 83:2736–2744

Stark S, Wardle DA, Ohtonen R, Helle T, Yeates GW (2000) The effect of reindeer grazing on decomposition, mineralization and soil biota in a dry oligotrophic Scots pine forest. Oikos 90:301–310

Stark S, Strömmer R, Tuomi J (2002) Reindeer grazing and soil microbial processes in two suboceanic and two subcontinental tundra heaths. Oikos 97: 69–76

Stark S, Tuomi J, Strömmer R, Helle T (2003) Non-parallel changes in soil microbial carbon and nitrogen dynamics due to reindeer grazing in northern boreal forests. Ecography 26:51–59

Sundman V, Huhta V, Niemelä S (1978) Biological changes in northern spruce forest soil after clear-cutting. Soil Biol Biochem 10:393–397

Suominen O, Olofsson J (2000) Impacts of semi-domesticated reindeer on structure of tundra and forest communities in Fennoscandia: a review. Ann Zool Fenn 37:233–249

Suominen O, Danell K, Bergström R (1999) Moose, trees, and ground-living invertebrates: indirect interactions in Swedish pine forests. Oikos 84:215–226

Teucher MW (1986) Abschätzung der benthischen Biomasse über die Analyse von DNA. Diplomarbeit, Institut für Meereskunde, Universität Kiel, 45 pp

Tracy BF, Frank DA (1998) Herbivore influence on soil microbial biomass and nitrogen mineralization in a northern grassland ecosystem: Yellowstone National Park. Oecologia 114:556–562

Vallamil MB, Amiotti NM, Peinemann N (2001) Soil degradation related to overgrazing in the semi-arid southern Calldenal area of Argentine. Soil Sci 166:441–452

Van der Wal R, Brooker R, Cooper EJ, Langvatn R (2001a) Differential effects of reindeer on high-Arctic lichens. J Veget Sci 12:705–710

Van der Wal R, Van Lieshout SMJ, Loonen MJJE (2001b) Herbivore impact on moss depth, soil temperature, and Arctic plant growth. Polar Biol 24:29–32
Van der Wal R, Pearce I, Brooker R, Scott D, Welch D, Woodin S (2003) Interplay between nitrogen deposition and razing causes habitat degradation. Ecol Lett 6:141–146
Vance ED, Chapin FS III (2001) Substrate limitations to microbial activity in taiga forest floors. Soil Biol Biochem 33:173–188
Väre H, Ohtonen R, Mikkola K (1996) The effect and extent of heavy grazing by reindeer in oligotrophic pine heaths in northeastern Fennoscandia. Ecography 19:245–253
Virtanen RJ, Lundberg PA, Moen J, Oksanen L (1997) Topographic and altitudinal patterns in plant communities on European arctic islands. Polar Biol 17:95–113
Vistnes I, Nellemann C, Jordhøy P, Strand O (2001) Wild reindeer: impacts of progressive infrastructure development on distribution and range use. Polar Biol 24:531–537
Webber PJ (1978) Spatial and temporal variation of the vegetation and its production, Barrow, Alaska. In: Tieszen LL (ed) Vegetation and production ecology of an Alaskan Arctic Tundra. Springer, Berlin Heidelberg New York, pp 37–112
Weber GE, Jeltsch F, van Rooyen N, Milton SJ (1998) Simulated long-term vegetation response to grazing heterogeneity in semi-arid rangelands. J Appl Ecol 35:687–699
Weladji RB, Holand Ø (2003) Global climate change and reindeer: effects of winter weather on the autumn weight and growth of calves. Oecologia (DOI 10.1007/s00442-003-1257-9)
White RG, Bunnell FL, Gaare E, Skogland T, Hubert B (1981) Ungulates on arctic ranges. In: Bliss LC, Heal OW, Moore JJ (eds) Tundra ecosystems: a comparative analysis. Cambridge Univ Press, Cambridge, pp 397–483
Widén B (2001) CO2 exchange within a Swedish coniferous forest spatial and temporal variation. Acta Universitas Agriculturae Suecia, Silvestria 184
Williamson DH, Fennel DJ (1975) The use of fluorescent DNA binding agent for detecting and separating yeast mitochondrial DNA. Meth Cell Biol 12:335–351
Wynn-Williams DD (1990) Ecological aspects of Antarctic microbiology. Adv Microb Ecol 11:71–146
Zeller V, Bardgett RD, Tappeiner U (2001) Site and management effects on soil microbial properties of subalpine meadows: a study of land abandonment along a north–south gradient in the European Alps. Soil Biol Biochem 33:639–649
Zhang HQ, Horn R (1996) Einfluß der Beweidung auf die physikalischen Eigenschaften einer Salzmarsch im Deichvorland (I): die gesättigte Wasserleitfähigkeit und das Bodengefüge. Z Kulturtechn Landentw 37:24–31
Zimov SA, Chuprynin VI, Oreshko AP, Chapin FS III, Reynolds JF, Chapin MC (1995) Steppe–tundra transition: A herbivore-driven biome shift at the end of the Pleistocene. Am Nat 146:765–794

14 Hygienic Status of Soils and Surface Waters in Reindeer Herding Areas in Northernmost Europe

N. Kemper, A. Aschfalk, and C. Höller

14.1 Introduction

Enteric organisms - virus, bacteria, and parasites - can possess the potential to cause severe diseases both in humans and animals. Transmission of these pathogens can occur directly from a reservoir to the susceptible animal or indirectly via vectors. The specific aim of this study was to assess the fecal shedding of such pathogens by reindeer and to give information about the hygienic quality of soils and surface waters in the subarctic and boreal ecosystems influenced by reindeer to deduce the potential zoonotic risk to humans of modern reindeer herding.

The main emphasis was on bacteria (*Campylobacter* spp., *Enterococcus* spp., *Escherichia coli*, *Salmonella* spp., *Yersinia* spp.) and parasites (*Cryptosporidium* spp.) that are among the most important pathogenic agents in causing zoonoses, such as enteric and other severe diseases, and that have been isolated from healthy and diseased domestic ruminants (de Rycke et al. 1986; Munoz et al. 1996; Busato et al. 1998, 1999; Tham et al. 1999). In detail, various *Campylobacter* spp. (*C. coli*, *C. jejuni* subspecies jejuni, *C. hyointestinalis*, *C. lari*, and *C. upsaliensis*) were isolated from humans with gastroenteritis (Lawson et al. 1999; Gorkiewicz et al. 2002). *Enterococcus* spp. are known as pyogenic organisms and causative agents for nonspecific hospital infections. They have been isolated from human patients with peritonitis, endocarditis, urinary tract infections, cholecystitis, and wound infections (Hahn et al. 2001).

The virulence and pathogenesis of the bacteria under study depends on different factors. In *E. coli*, for instance, the ability to cause severe disease in humans and animals is associated with the occurrence of several virulence factors, such as shigatoxins. Therefore, the presence of shigatoxin genes can be a warning of the virulence of certain strains, shigatoxin-producing *E. coli* (STEC). The pathologic activity of shigatoxin is triplicate: cytotoxic, entero-

Ecological Studies, Vol. 184
B.C. Forbes, M. Bölter, L. Müller-Wille, J. Hukkinen, F. Müller, N. Gunslay, Y. Konstatinov (Eds.)
Reindeer Management in Northernmost Europe

toxic, and neurotoxic. Aside from botulin, it belongs to the strongest of bacterial toxins (Mainil 1999). Regarding *Salmonella* spp., all 1,443 serotypes of *Salmonella enterica* subspecies enterica can cause gastroenteritis in humans and animals (Rolle et al. 2001). Out of the eleven species of *Yersinia* , *Y. pestis* is the cause of plague, and *Y. pseudotuberculosis* and certain biotypes of *Y. enterocolitica* are of great significance for human health. In Europe, particularly the *Y. enterocolitica* biotypes O:3, O:9 and O:5,27 are associated with human gastroenteritis (Rolle et al. 2001). The parasite *Cryptosporidium*, especially the species *C. parvum*, plays a major role as causative agent for heavy diarrhea in humans and animals worldwide.

All microorganisms examined in this study may be found in the intestinal tract of animals or humans, and many survive in the environment. To focus on these pathogens may serve as a general evaluation of the epidemiological situation in an area. Knowledge of the occurrence of important pathogens shed by reindeer in the environment is necessary for learning of the potential human health risk. As known from other herding or husbandry systems, concentration of animals increases the shedding of infectious microorganisms and the incidence of infectious diseases in animals, and subsequently in humans in direct or indirect contact. This is of special importance as crowding of reindeer for winter feeding is becoming more and more common, particularly in northern Finland; health risks to animals and humans is implicit.

14.2 Reindeer as an Animal Reservoir for Pathogens with Zoonotic Potential

Free-ranging animals may serve as reservoirs or carriers of infectious agents for diseases in domestic livestock and humans. Transmission of infectious agents to humans may occur through direct contacts to free-ranging animals including cervids (Fanning and Edwards 1991), by contamination of the environment through fecal shedding (Aavitsland and Hofshagen 1999) or by con sumption of venison (Keene et al. 1997). In contrast to domestic animals, however, the epidemiological situation in free-ranging animals and in their habitat is difficult to assess. Conclusions that derive from studies focused on domestic animal species, in extensive husbandry systems, can only be applied to free-ranging animals with caution. But it can be assumed that pathogens are transmitted easily through close animal contact and lead to high animal losses as is known from intensive husbandry systems (Woolcock 1991; Rolle et al. 2001).

Regarding the occurrence of possibly zoonotic pathogens in reindeer, data are extremely rare. As far as viruses are concerned, there are several studies about different virus, including the zoonotic *Parapoxvirus* (Tryland et al. 2001), and *Lyssavirus* causing rabies (Prestrud et al. 1992).

Concerning zoonotic bacteria, literature describes *Listeria monocytogenes* (Evans and Watson 1987), *Brucella suis* Biovar 4 (Bollinger and Welch 1994) and *Brucella rangiferi* (Dieterich 1985) associated with mortality in reindeer. Other information is limited to the detection of antibodies in clinically healthy reindeer against pathogens, such as *Leptospira* spp. (Zarnke 1983; Rehbinder and Nikander 1999) and *Francisella tularensis* (Somov et al. 1978).

With regard to the zoonotic enteropathogens examined in this study, *Campylobacter hyointestinalis* subspecies hyointestinalis were isolated in 6 % of 399 healthy Finnish reindeer that were examined (Hänninen et al. 2002). In total, 1,387 fecal samples and 421 meat samples from Finnish reindeer were negative for the virulent strain *E. coli* 0157 (Lahti et al. 2001). *Enterococcus* spp. in reindeer have not been described yet. *Salmonella* spp. have been found associated to mortality in reindeer in Finland (Kuronen et al. 1998). *Yersinia* spp., further analysed as *Yersinia pseudotuberculosis* serotype 1a, were isolated from reindeer earlier (Ueshiba et al. 1998), but in recent samples from 35 Norwegian reindeer no *Yersinia* spp. were found (Aschfalk et al. 2003).

Regarding zoonotic parasites, reindeer can be carriers of *Echinococcus granulosus* cysticerci (Hirvelä-Koski et al. 2003) that cause severe liver and lung infections in humans. Regarding zoonotic protozoa like *Cryptosporidium* (Cr.) spp. in northern European reindeer, no data are available. However, a new genotype of *Cryptosporidium* closely related to *Cr. serpentis*, *Cr. muris* and *Cr. andersoni*, was isolated from 3 out of 49 examined caribou in Canada (Siefker et al. 2002).

14.3 Pathogens in Soils

A high prevalence of bacteria (like *E. coli* and *Enterococcus* spp.) in soils, naturally occurring in the gastrointestinal tract of hematothermal creatures, such as reindeer, indicates an environmental contamination by fecal shedding. These soils represent a health risk to humans, as not only relatively innocuous bacteria but also virulent bacteria from the gastrointestinal tract, such as *Campylobacter* spp., *Salmonella* spp. and *Yersinia* spp., and the parasite *Cryptosporidium parvum* might occur and infect humans.

Due to the extreme climatic conditions in subarctic regions excreted microorganisms, including pathogens, do not survive nearly as well as they do in moderate climates (Bölter 2004a). Temperature and hydration are the main parameters influencing growth and survival of microorganisms (Rolle et al. 2001). Therefore the main limiting factors for maintenance of pathogenic agents in the subpolar zone are the low temperatures during most of the year and the long sunshine duration in summer (Bölter and Höller 1996; Bölter 2004b).

Some of the pathogenic agents examined in this study are more fastidious than the others. *Campylobacter* spp. are extremely sensitive to dryness, heat, and low temperature (Rolle et al. 2001). *Campylobacter* spp. shed in the stools of infected humans can survive for as long as three weeks if maintained at 4 °C (Blaser et al. 1980). However, *C. jejuni* will not replicate below 30 °C, requires micro-aerobic and capnophilic conditions for growth, and is sensitive to drying and heat (Doyle and Jones 1992). On one hand, *Enterococcus* spp. are resistant to high temperatures and high pH values, but on the other hand they die off at temperatures below –4 °C (Hahn et al. 2001). *Escherichia coli* can survive for a long time under humid conditions, keeping an infectious potential for months, and *Salmonella* spp. possess an extremely high ability to survive under dry, cold, or hot conditions. A special characteristic of *Yersinia* spp. is the capacity to multiply at temperatures down to –4 °C (Rolle et al. 2001). *Cryptosporidium parvum* can survive long periods in moist human and cattle feces at 4 °C, but a temperature below –22 °C can inactivate this organism within a few minutes to a few hours (Robertson et al. 1992; Fayer 1994). Even though arctic climate does not favor growth of microorganisms, persistence may be helped by the effect of low temperature to induce delayed rotting of the nutrient-rich feces that keep them alive.

14.4 Pathogens in Surface Waters

Surface waters used for drinking are of high infectious risk to humans. In 2001, 42 % of the drinking water consumed in Finland originated directly from surface water sources (The Finnish Environment Institute 2003). Because of the low human population density (15 inhabitants per km^2), abundant areal water resources (33,000 km^2), and effective sewage treatment of municipal and industrial wastewaters, it might be assumed that the surface waters of Finland are generally of good quality. In contrast to the more contaminated rivers of the coastal area, this assumption should apply especially to the rivers and lakes of northern Finland, as shown by long-term monitoring (including screening for indicator bacteria) since the 1960s (Niemi et al. 1997). However, even minor release of indicator bacteria, through human wastes, "diffuse loading" (indicator bacteria of unknown origin), or fecal shedding of wild animals, e.g., reindeer, may alter this evaluation. In addition, due to the cold average temperature of waters in northern latitudes, persistence of indicator bacteria may be increased, biasing the data.

The importance of *E. coli* and *Enterococcus* spp. in surface waters is attributable to their role as indicator organisms for fecal contamination. They are not able to reproduce in the aquatic environment but have a long duration of survival. Fecal indicator bacteria are enumerated in waters in order to evaluate their hygienic quality. If indicator bacteria are present there is a probability that pathogenic organisms excreted in feces are also present and that the water can transmit waterborne infectious diseases. Waterborne outbreaks of infections in humans caused by pathogenic *E. coli* strains are described in Canada, Japan, Hungary, and the USA (Auckenthaler and Huggenberger 2003).

The survival time of thermophilic *Campylobacter* spp. in water is between a few hours and 14 days (Blaser et al. 1980). Several outbreaks of waterborne campylobacteriosis are documented in northern Norway and Finland (Aho et al. 1989; Melby et al. 1991, 2000; Hänninen et al. 2003). In a Finnish study consisting of 139 surface water samples *Campylobacter* spp. were isolated from 24 (17.3 %) (Hörman et al. 2004).

Cases of gastroenteritis caused by water contaminated with *Salmonella* spp. are known from the USA and Switzerland (Auckenthaler and Huggenberger 2003), but no data exist of waterborne outbreaks of salmonellosis or the occurrence of *Salmonella* spp. in surface waters in subarctic Europe.

Surface waters are a well known reservoir for *Yersinia* spp. (Rolle et al. 2001). A causal connection of *Yersinia enterocolitica* infections in humans and the consumption of water of inadequate hygienic quality was proven by studies in Norway (Saebo et al. 1994; Kapperud et al. 1995).

Cryptosporidium spp. are well recognized as organisms able to cause severe waterborne enteric infections even in small doses. They have not been reported to cause waterborne epidemics in Finland, but were isolated in 14 (10.1 %) of 139 Finnish surface water samples (Hörman et al. 2004).

14.5 Materials and Methods

Table 14.1 gives an overview of all examined fecal, soil, and water samples.

14.5.1 Fecal Samples

To obtain a representative picture of the epidemiological situation in reindeer husbandry areas in northern Finland and Norway, it was necessary to examine a large amount of fecal material from different herds. This study was conducted during different seasons as the excretion of pathogens often follows seasonal patterns. Thus, 2,243 fecal samples from that number of healthy reindeer (adults and calves of both genders) were examined for the occurrence of *Campylobacter* spp., *Enterococcus* spp., *E. coli*, *Salmonella* spp., *Yersinia* spp., and *Cryptosporidium* spp. The samples were taken in the course of eleven months (June 2001 to April 2002) from Finnish and Norwegian free-ranging and corralled reindeer herds, considering parameters such as the intensity of herding, location, and season. Samples were taken off the ground or per rec-

Table 14.1. Origin, date, and number of samples

Type of sample		Origin	Date	Number n	N total
Fecal samples	Free-living reindeer	Näkkälä	June 2001	147	2,243
		Lappi	August 2001	222	
		Lappi	October 2001–January 2002	800	
		Karasjok	September 2001	410	
	Fenced reindeer	Kaamanen	June 2001	40	
		Näkkälä	February 2002	100	
		Sallivaara	April 2002	100	
		Palojärvi	March 2002	325	
		Kiiminki	March 2002	99	
Soil samples	Finnish side	Jauristunturit	August 2002	171	341
	Norwegian side	Jauristunturit	August 2002	172	
Surface water samples	–	Enontekiö	September 2003	50	50

tum from slaughter animals, sent to Kiel, Germany, directly after collection and kept frozen (–4 °C) until further processed within one week.

The examination for *Campylobacter* spp. was carried out by inoculating 1 g fecal material into 9 ml Preston broth (Oxoid, Wesel, Germany). After 24 h incubation in a micro aerophilic atmosphere (5 % oxygen, 10 % carbon dioxide, 3 % hydrogen and 82 % nitrogen) at 37 °C, a loopful of the enriched suspension was plated on Preston agar (Oxoid) and incubated for 48 h under the above-described conditions. *Campylobacter*-like colonies were analyzed by Gram stain, catalase and oxidase tests, and further biochemical reactions (ApiCampy, bioMérieux, Nürtingen, Germany).

To detect *Enterococcus* spp., 1 g fecal material was diluted in 9 ml glucose-azide broth (Merck, Darmstadt, Germany) and incubated for 48 h at 37 °C. A loopful broth was then spread both on kanamycin-aesculin-azide agar (Merck) and Slanetz and Bartley agar (Oxoid). After 48 h at 37 °C suspicious colonies were Gram stained and their biochemical reactions were analyzed further by catalase and oxidase tests.

Escherichia coli was isolated by adding 1 g feces to 9 ml Gram-negative broth (Difco, Becton and Dickinson, Franklin Lakes, USA). After 24 h of incubation at 37 °C a loopful of broth was plated onto Endo-c agar (Merck) and incubated under the above-mentioned conditions for 24 h. Typical shiny metallic colonies were subcultured on blood agar (Oxoid), incubated for 24 h at 37 °C and tested for their biochemical reactions, applying API 20E (bioMérieux). PCR was used to detect the occurrence of shigatoxin 1 and 2 genes (*stx1*, *stx2*), the intimin gene (*eae*) and EHEC-hemolysin gene (*hlyEHEC*). Primers were developed with help of the European Molecular Biological Library database and the oligo 6.0 software (Molecular Biology Insight, Cascade, USA) and produced commercially (Invitrogen, Karlsruhe, Germany).

A loopful of *E. coli* colonies was diluted in 1 ml of treated saline solution (0.85 %) and heated in a thermo block for 15 min at 100 °C. The samples were inserted into an ultrasonic bath for 3 min at 190 W and 50–60 Hz and centrifuged at 9,875 g for 30 s. Five µl of the supernatant was added to one ready-to-go bead (Amersham Pharmacia Biotech, Buckinghamshire, UK) dissolved in 18 µl sterile double-distilled water. EHEC EDL 933 (*stx1,2* positive) was used as a positive control and *E. coli* ATCC 11 229 (*stx1,2* negative) was included as negative control. The conditions for the PCR were 95 °C for 12 min 30 s for initial denaturation, followed by 35 cycles at 95 °C for 20 s (denaturation), 57 °C for 30 s (primer annealing), 72 °C for 40 s (DNA synthesis), and 5 min of final extension at 72 °C performed with a thermal cycler (Perkin-Elmer, Norwalk, USA). The amplified products were analyzed by electrophoresis in a 2 % agarose gel and were visualized following ethidium bromide staining (100 µl/100 ml gel; Sigma-Aldrich, Steinheim, Germany) under UV-light and photographed (AlphaInnotech, Biozym, Hessisch Oldendorf, Germany) using the AlphaImager 1220 software (Biozym).

For the selective enrichment of *Salmonella* spp., 1 g feces was inoculated into 14 ml of tetrathionate broth (Merck) and incubated for 24 h at 37 °C. One ml of this enriched broth was brought into tetrathionate broth the next day and incubated for another 24 h at 37 °C. This enrichment step was repeated one more time. On the fourth day one loopful of the cultured medium was plated both on Salmonella-Shigella agar (Difco) and Leifson agar (Merck). After 24 h of incubation at 37 °C presumptive *Salmonella* spp. colonies were Gram stained and tested by API 20E (bioMérieux).

Cultural examination of *Yersinia* spp. was performed by adding 1 g feces into 9 ml of Gram-negative broth and incubating for 48 h at 21 °C. One loopful of broth was then plated on Yersinia-selective agar (Difco) and incubated for another 48 h at 21 °C. Colonies with the typical bull's-eye appearance were subcultured on blood agar and Gram stained and biochemical tests were subsequently carried out with API 20E (bioMérieux) and Micronaut (Merlin, Bornheim-Hersel, Germany). To detect various *Yersinia* genes, PCR was performed using the primers to detect the genes encoding 16SrRNA, yadA, and v-antigen. The PCR conditions consisted of an initial 94 °C denaturation step of 10 min followed by 30 cycles at 94 °C for 1 min, 57 °C for 1 min, and 72 °C for 1 min. The final cycle was followed by incubation at 72 °C for 10 min. Amplified DNA fragments were visualized as mentioned for the *E. coli* virulence genes.

For the detection of *Cryptosporidium* locusts, immunomagnetic separation was applied using Dynabeads anti-Cryptosporidium (Dynal, Oslo, Norway). Twenty µl of the immunoconcentrate were used for a direct immunofluorescence test (Cryptosporidium-Antigen-IFT, medac, Wedel, Germany; Das Graças C. Pereira et al. 1999).

Cryptosporidium parvum oocysts from a calf (Iowa isolate, USA) served as the positive control. Using a fluorescence microscope at ×400–×1000 magnification, *Cryptosporidium* oocysts appear, 6–10 µm in size, round or oval in shape, and with bright green fluorescence.

For statistical analyses, the data were evaluated with the Statistica 5.0 software (StatSoft GmbH, Hamburg, Germany), following the instructions of Trampisch and Windeler (1997). For all analyses, differences were considered significant at $P \leq 0.05$.

14.5.2 Soil Samples

For the evaluation of the environmental contamination regarding the examined pathogens, 341 soil samples were collected at the Finnish–Norwegian border (Jauristunturit) in August 2002. At this time, no reindeer were present at the slightly-grazed Norwegian side of the border. The Finnish side was more intensively grazed. Due to these conditions a comparison of areas with high and low reindeer density was possible. The number of samples was 171 on the Finnish and 170 on the Norwegian side. A 3 km^2-sized area was sampled every 100 m with the help of a GPS grid. Samples were taken aseptically from the humus layer and examined as described for fecal samples in Sect. 14.5.1.

14.5.3 Surface Water Samples

In total, 50 water samples were collected from waters in the "natural state" located in the reindeer grazing region of Enontekiö (Finland). The samples were collected into sterilized 0.5 L glass bottles. Brooks and small rivers were sampled by holding a bottle by hand. Lakes and ponds were sampled with the help of a device that could be cast 10–20 m from the shore. The samples were immediately placed into insulating boxes provided with cooling devices.

The membrane filtration technique using bacteriological filters (Schleicher & Schuell, Dassel, Germany) and standard agar media as described in Sect. 15.5.1 was employed for quantitative estimation of *Yersinia* spp.. The occurrence of *Enterococcus* spp. and *E. coli* was tested with microtiter plates (Biorad, Munich, Germany) and the use of a fluorogenic metabolite in the MPN technique. Tests for the presence of the pathogens *Campylobacter* spp., *Salmonella* spp., and *Cryptosporidium* spp. in water samples was not done due to their absence or very low prevalence in fecal samples. These pathogens in water, if possibly found, most probably do not originate with reindeer.

14.6 Results

In 2,224 (99.2 %) out of the total number of 2,243 fecal samples one or more of the targeted bacterial species were isolated. Figure 14.1 shows the incidence of all isolated pathogens from feces.

Campylobacter sp., identified as *Campylobacter hyointestinalis*, was detected only in one feces sample (0.04 %). *Enterococcus* spp. were isolated in 2,084 (92.9 %) samples, *Escherichia coli* in 2,123 (94.7 %). However, only a few of the isolated *E. coli*-strains were found to possess genes encoding for virulence, to wit: *stx1* (0.14 %), *stx2* (0 %), *eae* (0.61 %), and *hlyEHEC* (1.08 %). No Salmonella spp. or Cryptosporidium spp. were found in the feces.

In total, 108 (4.8 %) strains of *Yersinia* spp. were isolated, consisting of *Y. enterocolitica* biogroup 1A ($n = 29$), *Y. intermedia* ($n = 2$), *Y. kristensenii* ($n = 72$), *Y. mollaretii* ($n = 3$). and *Y. rhodei* ($n = 2$).

Regarding the degree of intensity of reindeer herding, the season, or the geographic origin, no significant differences were found for *Enterococcus* spp. and *E. coli*, but the incidence of *Yersinia* spp. differed significantly ($P \leq 0.001$) in that the prevalence of *Yersinia* spp. in free-ranging reindeer in summer and autumn was significantly higher than in fenced reindeer during winter.

Out of the total number of 341 soil samples 13 were positive for *E. coli*. No virulence genes were detected by using PCR. Four samples contained *Entero-*

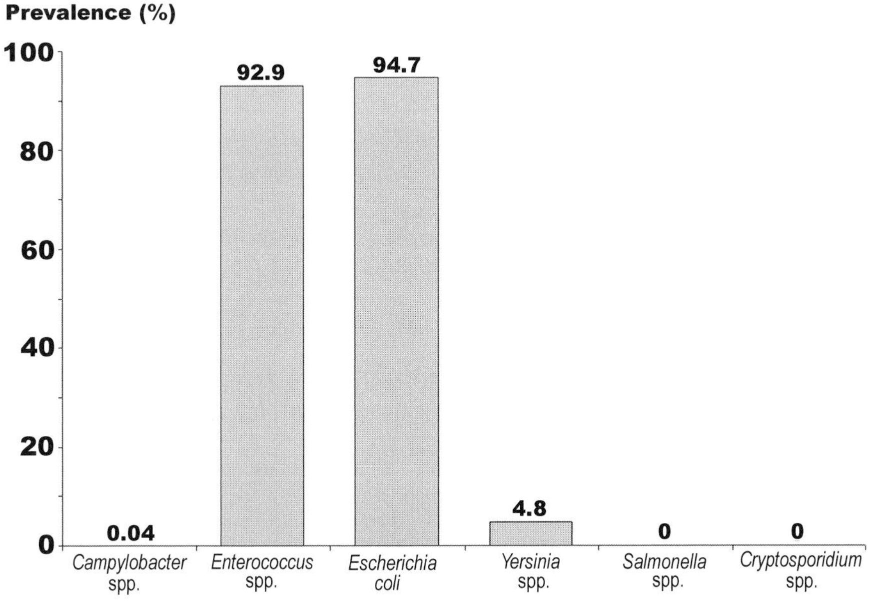

Fig. 14.1. Prevalence of analyzed pathogens in feces of reindeer ($n = 2,243$)

coccus spp. and one sample *Yersinia kristensenii*. Figure 14.2 illustrates the spatial distribution in the research area.

Table 14.2 shows the different incidence of the isolated bacteria on the Norwegian and Finnish side.

Nine (18 %) of the water samples were positive for *E. coli* and two (4 %) for *Enterococcus* spp. Their location is shown in Fig. 14.3.

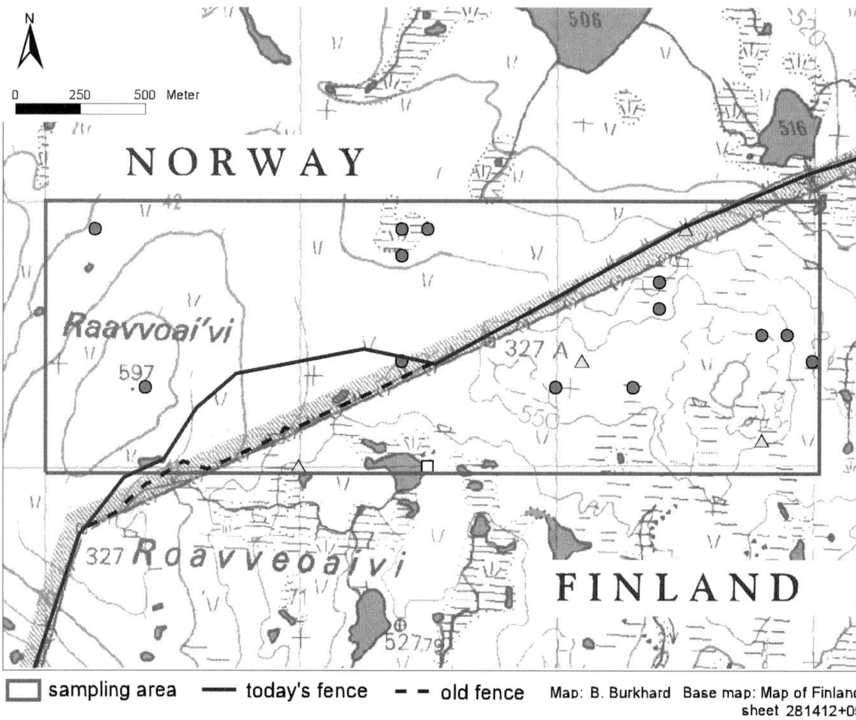

Fig. 14.2. Occurrence of *E. coli* (*circles* ○), *Enterococcus* spp. (*triangles* △) and *Yersinia kristensenii* (*squares* □) in soil samples from the research area Jauristunturit. Sampling was carried out during August–September 2002

Table 14.2. Bacteria isolated from soil samples taken in the Jauristunturit area. Percentages in parentheses refer to total number of soil samples *n*F and *n*N

	Enterococcus spp.	*Escherichia coli*	*Yersinia kristensenii*
Finland (*n*F=171)	4 (2.3 %)	7 (4.1 %)	1 (0.6 %)
Norway (*n*N=170)	0	6 (3.5 %)	0

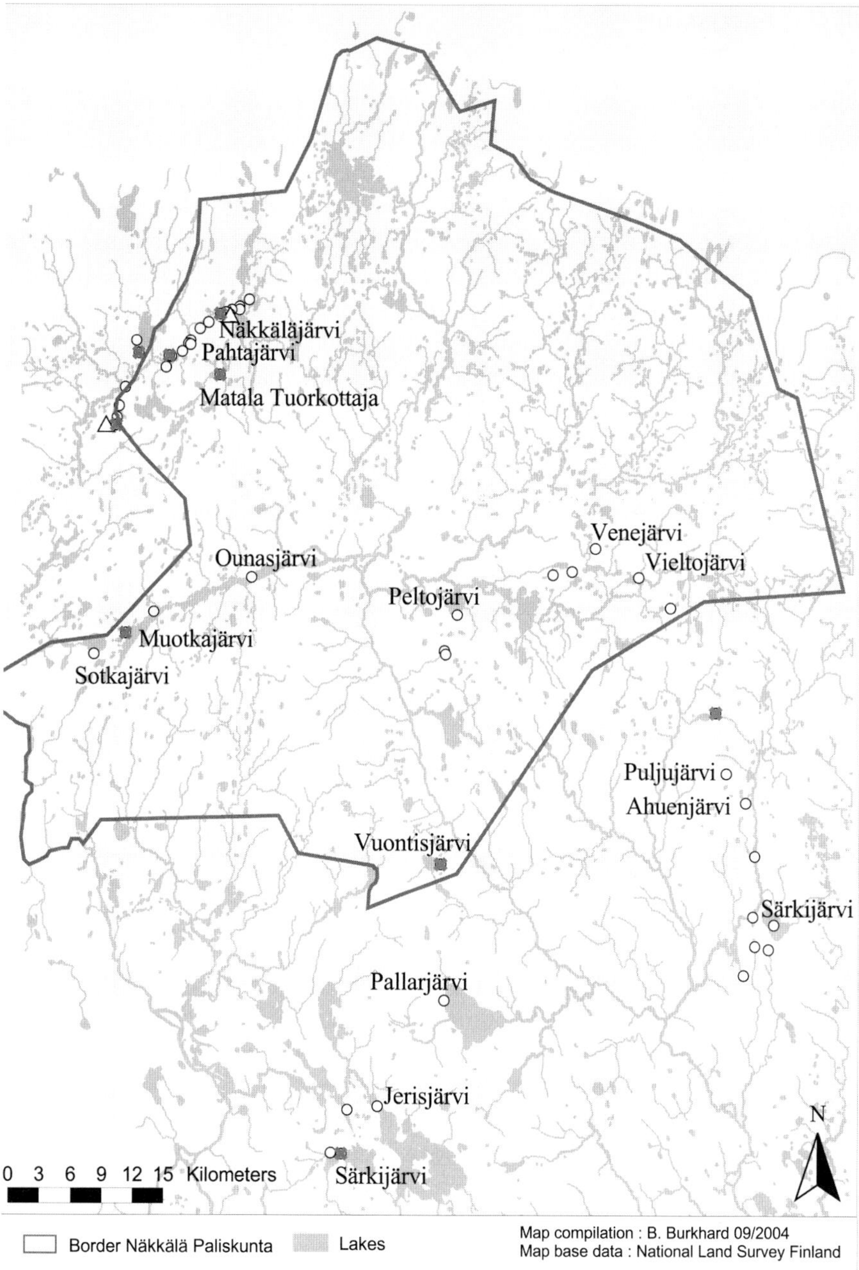

Fig. 14.3. Occurrence of examined bacteria in water samples from Näkkälä paliskunta region. Sampling sites are indicated by *circles* (○), occurrence of *E. coli* is indicated by *crosses* (×), and occurrences of *Enterococcus* spp. by *triangles* (△). Sampling was performed during September 2003 by the Finnish Environment Institute

14.7 Discussion

All bacteria investigated in this study may be found in northern Europe in the environment in aquatic, terrestrial, and animal reservoirs (Kapperud 1981), and they have been isolated from the intestinal tract of healthy or diseased ruminants worldwide (Adesiyun et al. 1998; Busato et al. 1998). Even though most of the isolated bacteria strains do not have the potential to cause severe human or animal health problems, certain strains might be a risk, especially for immunosupressed, old, or very young persons and animals. Therefore, one has to pay heed to the epidemiological impact of transmission of these infectious agents from the environment to reindeer and man and vice versa.

In reindeer, *Enterococcus* spp. and *E. coli* occurred at very high incidence, showing the affiliation of these two species with the normal intestinal flora of healthy reindeer. Concerning *E. coli*, there are only few reports on diseases caused by shigatoxin-producing bacteria in ruminants (Sherwood et al. 1985; Mainil 1999), however, these bacteria are of extreme importance in causing severe diseases in humans (Griffin and Tauxe 1991). As the genes encoding stx1, eae and hlyEHEC were detected only in very low numbers of the isolated *E. coli* strains, the human health risk due to *E. coli* excreted by reindeer can be considered very low at present. These results comply with another study detecting no *E. coli* O157:H7 in 1,387 fecal and 421 meat samples from reindeer (Lahti et al. 2001). It is known, however, that STEC virulence factors are mobile within bacterial populations (Yamamoto et al. 1984; Pupo et al. 1997). Therefore, an increase of the occurrence of toxin genes in *E. coli* from reindeer can not be excluded when influencing parameters such as herding conditions are changed.

Yersinia spp. were isolated in 108 samples. The identified species *Y. intermedia*, *Y. kristensenii*, *Y. mollaretii*, and *Y. rhodei* have been isolated before from various environmental samples (fresh water, soil, etc.), food, healthy animals, and healthy and diseased humans (Baier and Puppel 1981; Sulakvelidze 2000). Even though these species are widely distributed in nature, their actual impact on human health is a matter of controversy. However, as they are isolated from persons with gastrointestinal disorders, the role of these species should not be disregarded (Sulakvelidze 2000). The isolated *Y. enterocolitica* strains belonged to biogroup 1A, which embraces the non-pathogenic European *Y. enterocolitica* strains, often isolated from environmental samples, foods, and animal and human feces (Bottone 1997).

Campylobacter hyointestinalis was isolated from one sample only. As the cultivation of *Campylobacter* spp. is exceedingly difficult, the real prevalence might be higher. Hitherto *Campylobacter hyointestinalis* has been associated only sporadically with human gastrointestinal disorders (Edmonds et al. 1987; Gorkiewicz et al. 2002). Even though the incidence of *Campylobacter* spp. in this study was very low, it shows that reindeer can be carriers. This is confirmed by another study, which detected *Campylobacter hyointestinalis* in a prevalence of 6 % in Finnish reindeer feces (Hänninen et al. 2002). It is surprising that neither *Salmonella* spp. nor *Cryptosporidium* oocysts were detected in reindeer in this study as both pathogens have been isolated from the environment, farm animals, and humans in Fennoscandia (Refsum et al. 2002; Hörman et al. 2004), as were Salmonella spp. from reindeer as well in Finland (Kuronen et al. 1998).

Bacteria isolated from soil samples cannot necessarily be attributed to fecal shedding by reindeer, other possible contamination sources must be considered. However, detection rates of bacteria were higher in the permanent and more intensively grazed Finnish evaluation site than in Norway, which has winter grazing only. These results might be further supported by the fact that better growth conditions for bacteria were found on the Finnish side of the border, due to a number of wetlands. As water is a crucial factor for survival of bacteria, most enterobacteria do not tolerate soil desiccation (Pedersen and Jacobsen 1993).

Regarding the water samples, there was no hint that the detected bacteria originated from reindeer. All positive water samples were taken near human settlements, these being the putative sources of the bacteria found in this environment.

The extent to which an actual health risk exists depends on a number of local factors. Recapitulating, the excretion risk of pathogens by reindeer has to be seen against the background of the extreme climate conditions in the research area. Permafrost soils of the tundra and taiga are a domain of psychrophilic and psychrotolerant organisms (Bölter 2004a), but enteropathogens, usually living in the intestinal tracts of warm blooded animals at temperatures of about 37 °C, are not adapted to these living conditions. On one hand, northern environmental conditions lead to rapid destruction of these sensitive microorganisms. On the other hand, low temperatures may contribute to a prolonged lifespan of introduced microorganisms and thus keeping them infectious for a long time. Those pathogens can show viability at reduced rates of activity at low temperatures for long periods of time, although they have high demands on temperature stability and nutrients (Ravel et al. 1995). The ability to survive in this climate varies, as does the minimum concentration of pathogens necessary to cause an infection. The more pathogens that survive in the environment and the lower the infection rate, the higher is the actual health risk for susceptible creatures.

In reindeer feces, the examined enteropathogens were either not detected at all (*Salmonella* spp. and *Cryptosporidium* spp.), in very small numbers (*Campylobacter* spp.) or if detected, their virulence and pathogenicity was very low (*E. coli* and *Yersinia* spp.). Correspondingly, the incidence of enteropathogens in soils and water was very low.

This study clearly shows that the potential human and animal health risk through reindeer excretion of certain important enteropathogenic bacteria and *Cryptosporidium* spp., is very low at present. However, especially if reindeer are crowded, e.g., for winter feeding, an increased prevalence of enteric pathogens excreted by reindeer and eventually an increased risk to the consumer has to be considered, as is known from other intensive animals husbandry systems worldwide.

References

Aavitsland P, Hofshagen T (1999) Salmonella utbrudd i Herøy kommune. Nytt fra miljø og samfunnsmedisin 3:12

Adesiyun AA, Seepersadsingh N, Inder L, Caesar K (1998) Some bacterial enteropathogens in wildlife and racing pigeons from Trinidad. J Wildl Dis 34:73–80

Aho M, Kurki M, Rautelin H, Kosunen TU (1989) Waterborne outbreak of Campylobacter enteritis after outdoor infantry drill in Utti, Finland. Epidemiol Infect 103:133–141

Aschfalk A, Kemper N, Höller C (2003) Bacteria of pathogenic importance in faeces from cadavers of free-ranging or corralled semi-domesticated reindeer in northern Norway. Vet Res Com 27:93–100

Auckenthaler A, Huggenberger P (2003) Pathogene Mikroorganismen im Grund und Trinkwasser. Birkhäuser, Basel

Baier R, Puppel H (1981) Enteritis durch "atypische" Yersinien. Dtsch Med Wochenschr 106:208–210

Blaser MJ, Hardesty HL, Powers B (1980) Survival of *Campylobacter fetus* subsp. jejuni in biological milieus. J Clin Microbiol 11:309–313

Bölter M (2004a) Ecophysiology of psychrophilic and psychrotolerant microorganisms. Cell Mol Biol 50:563–573

Bölter M (2004b) Soil: an extreme habitat for microorganisms? Pedosphere 14(2):137–144

Bölter M, Höller C (1996) Hygienic problems in using permafrost soils for organic waste disposal. Polarforsch 66:61–65

Bollinger TK, Welch D (1994) A retrospective review of disease diagnoses in caribou from Canada. In: Haugerud RE (ed) The 6th North American Caribou workshop, Prince George. Rangifer Special Issue 9, pp 395

Bottone FJ (1997) *Yersinia enterocolitica*: the charisma continues. Clin Microbiol Rev 10:257–276

Busato A, Lentze T, Hofer D, Burnens A, Hentrich B, Gaillard C (1998) A case control study of potential enteric pathogens for calves raised in cow–calf herds. Zentralbl Veterinarmed (B) 45:519–528

Busato A, Hofer D, Lentze T, Gaillard C, Burnens A (1999) Prevalence and infection risks of zoonotic enteropathogenic bacteria in Swiss cow–calf farms. Vet Microbiol 69:251–263

Das Graças C Pereira M, Atwill ER, Jones T (1999) Comparison of sensitivity of immunofluorescent microscopy to that of a combination of immunofluorescent microscopy and immunomagnetic separation for detection of *Cryptosporidium parvum* oocysts in adult bovine faeces. Appl Environ Microbiol 65:3,236–3,239

De Rycke J, Bernard S, Laporte J, Naciri M, Popoff MR, Rodolakis A (1986) Prevalence of various enteropathogens in the feces of diarrheic and healthy calves. Ann Rech Vet 17:159–168

Dieterich RA (1985) An overview of Alaskan reindeer diseases and current control methods. Roy Soc NZ Bull 22:97–100

Doyle MP, Jones DM (1992) Food-borne transmission and antibiotic resistance of *Campylobacter jejuni*. In: Nachamkin I, Blaser MJ, Tompkins LS (eds) *Campylobacter jejuni*: current status and future trends. American Society for Microbiology, Washington, DC, pp 45–48

Edmonds P, Patton CM, Ward GE, Griffin PM (1987) *Campylobacter hyointestinalis* associated with human gastrointestinal disease in the United States. Clin Microbiol 25:251–263

Evans MG, Watson GL (1987) Septicemic listeriosis in a reindeer calf. J Wildl Dis 23:314–317

Fanning A, Edwards S (1991) *Mycobacterium bovis* infection in human beings in contact with elk (*Cervus elaphus*) in Alberta (Canada). Lancet 228:1253–1255

Fayer R (1994) Effect of high temperature on infectivity of *Cryptosporidium parvum* oocysts in water. Appl Environ Microbiol 60:2,732–2,735

Gorkiewicz G, Feierl G, Zechner R (2002) Transmission of *Campylobacter hyointestinalis* from a pig to human. J Clin Microbiol 40:2,601–2,605

Griffin PM, Tauxe RV (1991) The epidemiology of infections caused by *Escherichia coli* O157:H7, other enterohemorrhagic *E. coli*, and the associated haemolytic uremic syndrome. Epidemiol Rev 13:60–98

Hahn H, Falke D, Kaufmann SHE, Ullmann U (2001) Medizinische Mikrobiologie. Springer, Berlin Heidelberg New York

Hänninen ML, Sarelli L, Sukura A, On SLW, Harrington CS, Matero P, Hirvelä-Koski V (2002) *Campylobacter hyointestinalis* subsp. hyointestinalis, a common Campylobacter species in reindeer. J Appl Microbiol 92:717–723

Hänninen ML, Haajanen H, Pummi T, Wermundsen K, Katila ML, Sarkkinen H, Miettinen I, Rautelin H (2003) Detection and typing of *Campylobacter jejuni* and *Campylobacter coli* and analysis of indicator organisms in three waterborne outbreaks in Finland. Appl Environ Microbiol 69:1,391–1,396

Hirvelä-Koski V, Haukisalmi V, Kilpela SS, Nylund M, Koski P (2003) *Echinococcus granulosus* in Finland. Vet Parasitol 111:175–192

Hörman A, Rimhanen-Finne R, Maunula L, von Bonsdorff C, Torvela N, Heikinheimo A, Hänninen ML (2004) *Campylobacter* spp., *Giardia* spp., *Cryptosporidium* spp., noroviruses, and indicator organisms in surface waters in southwestern Finland. Appl Environ Microbiol 70:87–95

Kapperud G (1981) Survey on the reservoirs of *Yersinia enterocolitica* and Yersinia enterocolitica-like bacteria in Scandinavia. Acta Pathol Microbiol Scand (B) 89:29–35

Kapperud G, Ostroff SM, Nesbakken T, Hutwagner LC, Bean NH, Lassen J, Tauxe RV (1995) Risk factors for sporadic *Yersinia enterocolitica* infections in Norway: a case-control study. Contrib Microbiol Immunol 13:25–28

Keene WE, Sazie E, Kok J, Rice DH, Hancock DD, Balan VK, Zhao T, Doyle MP (1997) An outbreak of *Escherichia coli* O157:H7 infections traced to jerky made from deer meat. J Am Med Ass 277:1,229–1,231

Kuronen H, Hirvelä-Koski V, Nylund M (1998) Salmonella poronvasoissa (Salmonellosis in reincalv). Poromies 4/5:54

Lahti E, Hirvelä-Koski V, Honkanen-Buzalski T (2001) Occurrence of *Escherichia coli* O157 in reindeer (*Rangifer tarandus*). Vet Rec 19:633–634

Lawson AJ, Logan MJ, O'Neill GL, Desa M, Stanley J. (1999) Large-scale survey of Campylobacter species in human gastroenteris by PCR and PCR-enzyme-linked immunosorbent assay. J Clin Microbiol 37:3,860–3,864

Mainil J (1999) Shiga/Verocytotoxins and Shiga/Verotoxigenic *Escherichia coli* in animals. Vet Res 30:235–257

Melby K, Gondrosen B, Gregusson S, Ribe H, Dahl OP (1991) Waterborne campylobacteriosis in northern Norway. Int J Food Microbiol 12:151–156

Melby K, Svenby JG, Eggebo T, Holmen LA, Andersen BM, Lind L, Sjogren E, Kaijser B (2000) Outbreak of Campylobacter infection in a subarctic community. Eur J Clin Microbiol Infect Dis 19:542–544

Munoz M, Alvarez M, Lanza I, Carmenes P (1996) Role of enteric pathogens in the aetiology of neonatal diarrhoea in lambs and goat kids in Spain. Epidemiol Infect 117:203–211

Niemi JS, Niemi RM, Malin V, Poikolainen ML (1997) Bacteriological quality of Finnish rivers and lakes. Environ Toxicol Water Qual 12:15–21

Pedersen JC, Jacobsen CS (1993): Fate of *Enterobacter cloacae* PJ120 and *Alcaligenes eutrohus* AE106 (pRO101) in soil during water stress: effects of culturability and viability. Appl Environ Microbiol 59:1,560–1,564

Prestrud P, Krogsrud J, Gjertz I (1992) The occurrence of rabies in the Svalbard Islands of Norway. J Wildl Dis 28:57–63

Pupo GM, Karaolis DKR, Lan R, Reeves PR (1997) Evolutionary relationships among pathogenic and nonpathogenic *Escherichia coli* strains inferred from multilocus enzyme electrophoresis and mdh sequence studies. Infect Immun 65:2,685–2,692

Ravel J, Knight IT, Monahan CE, Hill RT, Colwell RR (1995): Temperature-induced recovery of *Vibrio cholerae* from the viable but non-culturable state: growth or resuscitation? Microbiology 14:377–383

Rehbinder C, Nikander S (1999) Ren och Rensjukdomar. Studentlitteratur, Lund

Refsum T, Heir E, Kapperud G, Vardund T, Holstad G (2002) Molecular epidemiology of *Salmonella enterica* serovar typhimurium isolates determined by pulsed-field gel electrophoresis: comparison of isolates from avian wildlife, domestic animals, and the environment in Norway. Appl Environ Microbiol 68:5,600–5,606

Robertson L, Campbell A, Smith H (1992) Survival of *Cryptosporidium parvum* oocysts under various environmental pressures. Appl Environ Microbiol 58:3,494–3,500

Rolle M, Mayr A, Büttner M (2001) Medizinische Mikrobiologie, Infektions- und Seuchenlehre. Enke, Stuttgart

Saebo A, Kapperud G, Lassen J, Waage J (1994) Prevalence of antibodies to *Yersinia enterocolitica* O:3 among Norwegian military recruits: association with risk factors and clinical manifestations. Eur J Epidemiol 10:749–755

Sherwood D, Snodgrass DR, O'Brien AD (1985) Shiga-like toxin production from *Escherichia coli* associated with calf diarrhea. Vet Rec 116:217

Siefker C, Rickard LG, Pharr GT, Simmons JS, O'Hara TM (2002) Molecular characterisation of *Cryptosporidium* sp. isolated from Northern Alaskan caribou (*Rangifer tarandus*). J Parasitol 88:213–216

Somov GP, Shubin FN, Gopachenko IM, Leonova GN, Ivanenko TP (1978) Natural focal infections in the north of the Far East. The detection of natural foci of infections in Chukot. Zh Mikrobiol Epidemiol Immunobiol 10:101–104

Sulakvelidze A (2000) Yersiniae other than *Y. enterocolitica*, *Y. pseudotuberculosis*, and *Y. pestis*: the ignored species. Microbes Infect 2000:497–513

Tham W, Bannerman E, Bille J, Danielsson-Tham ML, Eld K, Ericsson H, Gavier-Widen D, Rocourt J, Morner T (1999) *Listeria monocytogenes* subtypes associated with mortality among fallow deer (*Dama dama*). J Zoo Wildl Med 30:545–549

The Finnish Environment Institute (2003) Water services in Finland. http:/www.vyh.fi/eng/environ/state/waterre/waterser/waterser.htm

Trampisch HJ, Windeler J (1997) Medizinische Statistik. Springer, Berlin Heidelberg New York

Tryland M, Josefsen T, Oksanen A, Aschfalk A (2001) Contagious ecthyma in Norwegian semidomesticated reindeer (*Rangifer tarandus*). Vet Rec 149:394–395

Ueshiba H, Kato H, Tohru MA, Tsubokura M, Nagano T, Kanako S, Uchiyama T (1998) Analysis of the superantigen-producing ability of *Yersinia pseudotuberculosis* strains of various serotypes isolated from patients with systemic or gastroenteric infections, wildlife animals and natural environment. Zentralbl Bakteriol 288:277–291

Woolcock JB (1991) Microbiology of animals and animal products. Elsevier, Amsterdam

Yamamoto T, Honda T, Miwatani T, Yokota T (1984) A virulence plasmid in *Escherichia coli* enterotoxigenic for humans: intergenetic transfer and expression. J Infect Dis 150:688–698

Zarnke RL (1983) Serologic survey for selected microbial pathogens in Alaskan wildlife. J Wildl Dis 19:324–329

Part III

Integrative Models for Reindeer Management: The Interface Between Social and Natural Sciences

Photo B. Burkhard

Preface

This last section of the volume contains three synthesizing papers. Thus, it functions as a concluding synopsis of the RENMAN project, integrating many of the single aspects that have been reported in the preceding chapters.

Timo Helle and Ilpo Kojola (Chap. 15) use data sets on reindeer population dynamics from different Northern countries and regions to test different hypotheses that try to explain the observed highly synchronized changes. In their survey they provide statistical analyses, checking the following potential explanations: market disorders, supplementary feeding, changes in calf harvesting strategies, anti-parasite treatments, the Chernobyl nuclear disaster, herding behavior ("tragedy of the commons"), governmental reindeer policies, misperceptions of carrying capacities, and climatic variations. The authors have found a certain gradient of significance in to all these important driving factors of modern reindeer management.

In Chapter 16 Benjamin Burkhard and Felix Müller provide a systems-based analysis of reindeer management in Finland. In their approach most of the parameters described in Chapters 1–15 are integrated. The tools used are landscape analysis, conceptual modeling and geographic-simulation modeling. On the basis of regional land use patterns, indicators are derived to describe the structural and functional states of the ecosystems (ecological integrity), the economic welfare of the herders, and their social welfare. These indicators are used as outcomes of simulation experiments with different land-use intensities. Finally, a potentially sustainable scenario is described.

Chapter 17 provides a synthesis of the focal results from the RENMAN project, summarizing the environmental and sociopolitical conditions modern of reindeer management. Its pages describe the problems arising from competing land use interests and changing mobility potentials and herding practices. Also the changing environmental conditions are discussed, and the consequences for the human dynamics of herding and management are derived. This treatment includes the diverging cultural and philosophical premises, socioeconomic considerations and political and legal aspects. The Chapter ends with some recommendations for sustainable reindeer herding in the future.

15 Population Trends of Semi-Domesticated Reindeer in Fennoscandia – Evaluation of Explanations

T. Helle and I. Kojola

15.1 Introduction

Sustainability is the central issue in modern reindeer management. The relative importance of its different elements, be they ecological, economic or cultural, varies depending on the present status of reindeer management. There exist two different trends. Reindeer herding in Siberia, one of the core areas, shows a declining trend due to the impacts of other land uses and drastic economic changes in the society at large, while in Fennoscandia and Northwest Russia such a decline has not taken place (Jernsletten and Klokov 2002). In areas with regressive trends the goal of the reindeer owners is to revitalize the lost economic and cultural status by re-building the necessary infrastructure and develop collapsed market systems. In Fennoscandia, most attention has been paid to ecological and economic sustainability (e.g., Dahle et al. 1999).

In Finland, Norway, and Sweden, striving towards ecological and economic sustainability has been, to a large extent, a response to wide fluctuations in herd sizes. These are not new phenomena in the history of reindeer herding. The time series from the middle of the 19th century show that wide fluctuations were characteristic even for traditional reindeer herding (Riseth 2000; Moen and Danell 2003). That is not surprising because population fluctuations are a common trait for all northern ungulates (Saether 1997) including wild reindeer and caribou (Reimers 1977; Caughley and Gunn 1993; Post and Forchammer 2002). However, for modern market-orientated reindeer management, fluctuations are a great problem, because they mean unexpected annual variations in income, and during the worst years there is practically no income at all (Heikkinen 2002).

Official statistics for the last four decades show that the reindeer numbers started sharply to increase in each country in the mid-1970s and continued to

Ecological Studies, Vol. 184
B.C. Forbes, M. Bölter, L. Müller-Wille, J. Hukkinen, F. Müller, N. Gunslay, Y. Konstatinov (Eds.)
Reindeer Management in Northernmost Europe

do so into the late 1980s, after which they declined. It is noteworthy that in Finland and Norway the numbers reached higher levels than ever before (Riseth 2000; Ermala 2002). In Sweden they were roughly at the same level as earlier peaks of the last 150 years (Moen and Danell 2003). The subsequent heavy use of winter pastures has been documented in Finland (Kautto et al. 1986; Mattila 1998; Kumpula et al. 2000) and in Norway (Tömmervik and Johansen 1992; Johansen and Karlsen 1998). In Sweden, winter pastures have not been monitored, but the same trend has been obvious there in some areas (Moen and Danell 2003). Density dependency of reproduction, carcass weights, and tooth wear were subjects of reports in northern Finnish Lapland (Kojola et al. 1995, 1998; Kumpula et al. 1998).

Explanations of the impetus for population increase have been extremely variable, even within the same country. If there is no agreement on how to explain the recent history of this livelihood (Chap. 2), it may be asked whether one can speak prudently about sustainable reindeer management as a goal. We approach the question by studying first the synchrony of population fluctuations in different parts of Fennoscandia. In ecology, synchronous population dynamics is referred to as the Moran effect, which may be induced by dispersal mechanisms among neighboring populations (Ranta et al. 1997), predator–prey or parasite–host relationships (Ims and Steen 1990; Ims and Andreassen 2000), climate variation (Post and Forchammer 2002), and synchronous human harvesting. In the case of semi-domesticated reindeer, potential factors are also new management practices, marketing disorders, and the impacts of various institutional systems (Chap. 3). We assume that if the fluctuations are synchronous, area- or country-specific explanations would be not the best ones. Therefore, we assess the relevance of the explanations by within/between country comparisons in order to find the smallest common denominators.

15.2 Material and Methods

We have used data on reindeer numbers in Finland, Norway, and Sweden from a total from seven subpopulations between 1960 and 2000 (Fig. 15.1). These include about 90 % of the total reindeer population, and represent well the existing variations in natural environment, management practices, and ethnic groups (Sámi and Finnish). The southern half of the Finnish reindeer management area is located deep in the coniferous forest belt. Finnish settlers and farmers have practiced reindeer herding there for several hundred years as a part of small-scale agriculture, fishing, and hunting (Kortesalmi 1996); the relationships with agriculture and forestry are still strong. The management units are small in area without seasonal migrations and the average herd size of a household is low. In northern Finnish Lapland, Norway, and Sweden rein-

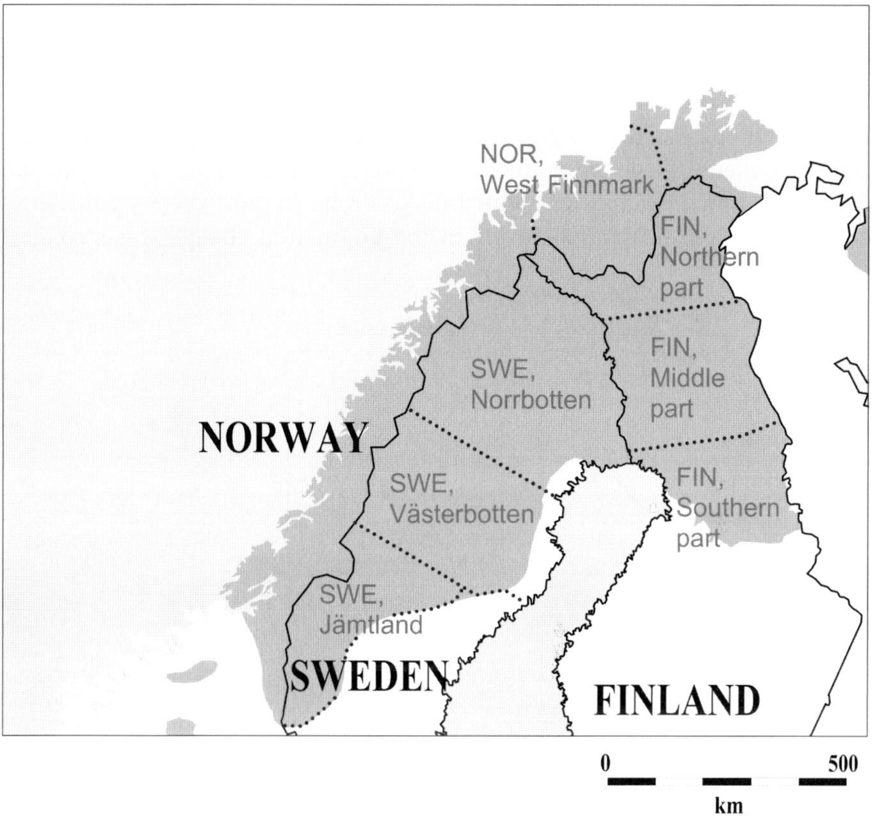

Fig. 15.1. The study area with subpopulation of semi-domesticated reindeer

deer owners are almost exclusively Sámi, to whom reindeer herding is the main or an important additional livelihood. In Norway and Sweden the management is still characterized by seasonal migrations between winter and summer pastures. In Sweden winter pastures are located in coniferous forests and summer pastures on the fells. In West Finnmark seasonal migrations take place between winter pastures inland and summer pastures on the coast of the Arctic Ocean.

We used the Pearson correlation and linear regression in the statistical analysis. The annual change in the population size (P) was calculated as $P=\ln N_{t+1}-\ln N_t$ in which N_t was the number of reindeer in the year t and N_{t+1} was the number of reindeer in the year t+1. All statistical tests were two-tailed and considered significant at $P \leq 0.05$.

15.3 Synchrony in Population Trends

The number of reindeer for each subpopulation in 1960–2000 is given in Fig 15.2. As indicated by highly significant correlations between subpopulations (r value 0.750–0.956) population trends have been highly synchronized. Similarly, annual changes in population size were in most cases significantly correlated (Table 15.1). Synchrony in the population change was strongest

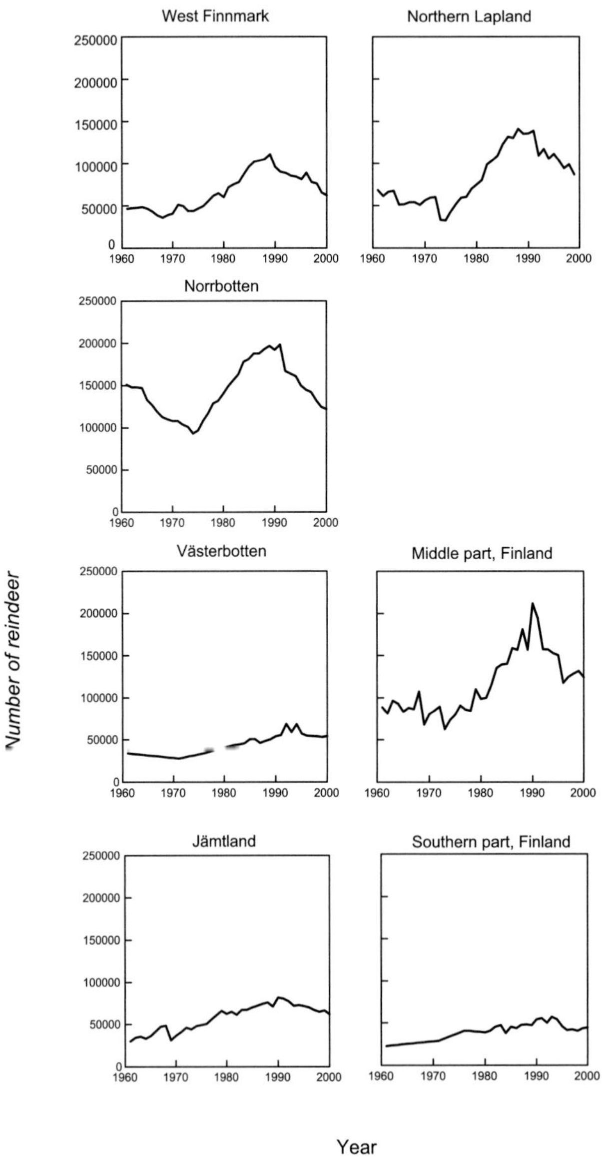

Fig. 15.2. Number of reindeer (summer herd) in different subpopulations 1960–2000

Table 15.1. Intercorrelation of the annual population change in different sub-areas of Norway (N), Finland (F), and Sweden (S) 1960–2000 (calculated from smoothed values). Significances: * P<0.05, ** P<0.01, *** P<0.001

	West Finnmark (N)	Norr-botten (S)	Northern part (F)	Central part (F)	Southern part (F)	Väster-botten (S)
1 West Finnmark (N)						
2 Norrbotten (S)	0.845***					
3 Northern part (F)	0.681***	0.843***				
4 Central part (F)	0.603***	0.845***	0.725***			
5 Southern part (F)	0.298	0.346*	0.053	0.353*		
6 Västerbotten (S)	0.354*	0.574***	0.393*	0.495***	0.192	
7 Jämtland (S)	0.095	-0.161	-0.034	0.390***	0.450***	0.475***

among West Finnmark, northern Finnish Lapland, Norrbotten, and Västerbotten.

The hypothesis that population synchrony is induced by dispersal between neighboring areas is not relevant in the case of semi-domesticated reindeer. The movements across the state borders are prevented by fences, which are also commonly used to separate herding associations and other management units from each other. Similarly, there is no evidence that population fluctuations in Fennoscandia might be the result from predator–prey or parasite–host interactions. In the management of semi-domesticated reindeer, harvest is normally proportional, therefore it cannot explain why the herds have experienced ups and downs. It is also known that fluctuations appear even in wild populations lacking both predators and human harvest (e.g., Reimers 1977; Caughley and Gunn 1993; Aanes et al. 2000).

15.4 Explanations of Population Trends

Figure 15.2 depicts the population trends in the different regions of northern Scandinavia. There are several reason which influence these trends.

Supplemental feeding. Especially in Finland, the population increase since 1975 has been explained by new management practices. Supplemental feeding in winter was adopted in the southern part in the early 1970s and spread from there northward (Helle and Saastamoinen 1979). The need for supplemental feeding was caused by intensive forestry decreasing the area of old forests rich in arboreal lichens and, at the same time, creating great variations in animal numbers and production. The Act on Field Reservation of 1968 made supplemental feeding practical. In order to prevent milk overproduction, the state subsidized farmers to abandon dairy farming, but the act allowed the use of hay for reindeer. Supplemental feeding decreases winter mortality and improves carcass weights and reproduction (Helle and Kojola 1993; Kumpula et al. 1998). The impact was visible in the southern area during the population increase 1975–1990 (105 kg dry hay per reindeer) and to some extent in the central area (44 kg per reindeer). In the northern part supplemental feeding (6 kg per reindeer) was restricted to some few herding associations. It was used to help herding work. In Sweden supplemental feeding has been occasional and restricted to spring migrations and exceptional snow conditions (Moen and Danell 2003).

Calf harvest. This practice was adopted in the southern part of the Finnish reindeer management area in the early 1960s (Alaruikka 1964). Because the calves are most vulnerable to winter mortality, calf harvest reduces overall mortality. It also enhances reproduction in the following year (Kojola and Helle 1993). In winter the female nurtures her calf by sharing the feeding crater, but if the calf is slaughtered before that, the female is capable of investing more in the developing fetus, which will increase the birth weight of the calf and thus enhances its further survival.

Anti-parasite treatments. Since the mid-1970s this application has been used to control the warble fly (*Oedomagena tarandi*) and the nasal bot fly (*Cephenomya trombe*). This treatment improves skin quality of slaughtered reindeer and, in given circumstances, this can increase autumn weight of the calves (Nieminen et al. 1980; Persen et al. 1982; Nordkvist et al. 1983). On the other hand, long-term effects of anti-parasite treatments on the immunity of reindeer are not adequately known.

Market disruptions. It is a common opinion, especially among people working in reindeer administration that, in the late 1980s, higher reindeer numbers were a result of changing market conditions for reindeer meat. In April 1986 the Chernobyl nuclear disaster increased the Caesium 137 content in meat especially in the southern parts of the Swedish and Norwegian reindeer management areas. In Sweden, 90,000 reindeer carcasses were destroyed (Åhman 2002) with an expected reduction in demand for reindeer meat elsewhere; if there were no buyers, it was useless to slaughter reindeer (Anonymous 1986). In 1989 *Poro ja Riista*, the largest buyer of reindeer meat in Finland, filed for bankruptcy, weakening the reputation and marketability of reindeer meat (Huttu-Hiltunen 1989).

The tragedy of the commons. Bad management and unreasonable use of commonly owned property was familiar to Aristotle; its modern version, " the tragedy of the commons", is described by Hardin (1968). Detrimental overuse occurs most likely if the access to resources is free or under conditions where rights and obligations are not determined (Ostrom 1992). With regard to reindeer herding, pastures are not a resource with free access, instead grazing rights in a given area are restricted solely to a given number of herders or households. However, that does not necessarily exclude competition among owners: the larger the herd of an owner, the greater the benefit the owner will derive from the common resource. Competition can result in deterioration of pastures, which is a disadvantage to every owner. The rationality for increasing the herd size is thus also based on the reasoning that if the owner has a large herd, there will be enough reindeer left even after a catastrophe (Ingold 1980).

Common property problems have evoked heavy scientific debate in Norway, but very little interest in Finland and Sweden. In Norway the opinions whether the increase in the number of reindeer since 1975 can be explained by a tragedy of the commons are highly contradictory. Brox (1989) considered that reindeer nomadism is not capable of a zero-sum increase: if one owner or group of owners reduce the number of reindeer in order to improve the condition of pastures, the others are likely to increase their herds. Similarly, Nilsen and Mosli (1994) hold the situation in Finnmark to be a serious indication of a tragedy of commons. According to Björklund (1990), conditions prevail that a tragedy of the commons is likely to occur, although he points out that it has been accelerated by the unexpected side effects of governmental reindeer policy. Riseth (2000) does not take a stance on the tragedy of the commons, but pays attention to differences in natural conditions, ethnic relations, and capacity building between southern (maximizing meat production) and northern (maximizing herd size) reindeer herding.

Governmental reindeer policy. The role of governmental reindeer policy has been discussed especially in Norway. Paine (1992) and Berg (1996) deny the presence of "the tragedy of the commons" in reindeer husbandry in Norway, and blame, instead, the state for a misguided reindeer policy. According to Paine (1992), the population increase was a result of a subsidy system started in 1976. Subsidies were so great that reindeer owners had no need to harvest and sell their reindeer. An additional impetus was the change in the law in 1980, which allowed an increase in the number of reindeer households (Berg 1996). Both authors also criticized the governmental reindeer policy at more general level. Their central concept is "herder responsibility", which the state has taken over, leaving herders external to the central issues in their own livelihood. The discussion is closely associated with the question of property rights, in which both the state and Sámi reindeer herders try to present themselves as being competent and responsible.

Misperceptions of carrying capacity. Another common reason for the overexploitation of resources is associated with difficulties in understanding the dynamic nature of the resource, which is obvious also linked with reindeer–pasture interactions. Moxnes (1998) studied experimentally the capability of test individuals to improve the condition of an overgrazed lichen pasture by reducing the number of reindeer. That appeared to be a difficult task even for professional reindeer herders. He concluded that misinterpretations are the result of the fact that people commonly understand natural processes in a static and linear fashion, although they commonly are, as in the reindeer–pasture relationships, dynamic and non-linear.

Climatic variation. Authors of articles in professional journals focusing on reindeer herding in each of the three Nordic countries commonly agree that winter conditions were favorable to reindeer from the mid-1970s to the late 1980s. After that several snow-rich winters and population declines were reported. Helle et al. (2001) explained a 53 % annual variation in reproduction (calves/100 females) due to snow conditions (snow depth and icing) in the district of the herding association of Käsivarsi, northern Finnish Lapland, between 1960 and 2000.

These findings are congruent with the fact that winter weather is related to large-scale climatic systems expressed by the index of NAO (The North Atlantic Oscillation; Hurrell 1995; Jones et al. 1997) and AO (Arctic Oscillation; Ambaum et al. 2001); despite the different origin they are closely related to each other (r value 0.830 for the period 1960–2000). The NAO index is depicted in Fig. 15.3. Regarding the NAO there are two extreme positions for high and low pressure centers in the North Atlantic region. In the first, the low-pressure center is located near Iceland and the high-pressure center near the Azores, and westerly winds prevail. Winter in northern Europe is mild and snow-rich. In the second, the situation is the opposite; easterly winds dominate and winter in northern Europe tends to be cold and dry. Several recent publications show that both NAO and AO exert correlated effects on local weather, affecting population dynamics of several northern ungulate species (Post and Stenseth 1998; Post et al. 1999; Post and Forchhammer 2002; Weladji et al. 2002). Therefore, NAO and AO are commonly used as proxies.

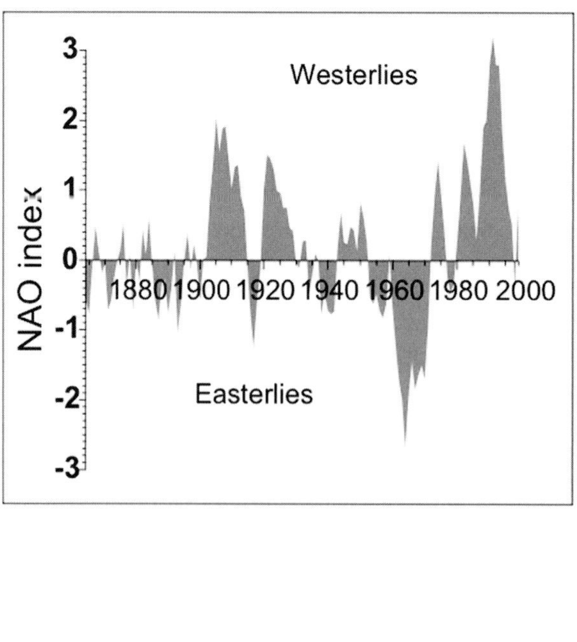

Fig. 15.3. North Atlantic Oscillation (NAO) index 1865–2000 (Jones et al. 1997) with prevailing wind directions in winter

15.5 Evaluation of Explanations

Management practices. The spatial and temporal variation in the distribution of new management practices is summarized in Table 15.2. It shows that these conditions cannot comprehensively explain the synchrony in population trends. The widest distribution of an application is anti-parasite treatment, but its influence on the weights and reproduction of reindeer is rather small and not noticeable every year (Nieminen et al. 1980; Persen et al. 1982; Nordkvist et al. 1983). In the southern and middle parts in the Finnish reindeer management area, supplemental feeding was an obvious reason for population increase, but a corresponding increase took place also in the northern part, West Finnmark (Norway), and Norrbotten (Sweden) without supplemental feeding. Neither could new practices prevent the population decline in the late 1980s or early 1990s, although their intensity increased through the whole study period.

Riseth (2000) suggested that in West Finnmark the herds had begun to increase in the beginning of 1970s due to the use of snowmobiles, which increased the mobility of herding routines and changed the earlier relative stable systems as to the division of pastures among herding districts or individual reindeer owners. However, a similar population increase took place in the southern half of the Finnish reindeer husbandry area without snowmobiles and resultant changes in pasture use.

Market disruptions. Fig. 15.4 presents the number of harvested reindeer in relation to the size of summer herds in West Finnmark and in the northern part of the Finnish reindeer management area. Both 1986 (Chernobyl nuclear disaster) and 1989 (bankruptcy of *Poro ja Riista* in Finland) are depicted in Fig. 15.4. In West Finnmark there was more variation along the regression line, but even there the data do not support the opinion that these disruptions had an essential effect on the harvest.

Table 15.2. Distribution of new management practices in different sub-areas in Norway (N), Finland (F), and Sweden (S) from 1975 to 1990

Location	Calf harvest	Anti-parasite treatment	Supplemental feeding
West Finnmark (N)			
Norrbotten (S)			
Northern part (F)	x	x	
Central part (F)	x	x	x (some)
Southern part (F)	x	x	x
Västerbotton (S)	x	x	
Jämtland (S)	x	x	

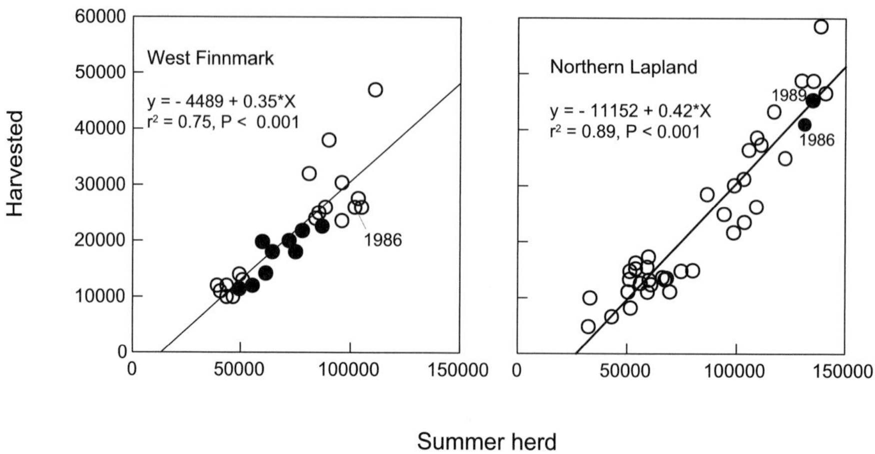

Fig. 15.4. The relationship between the number of harvested reindeer and the size of summer herd in West Finnmark and northern Finnish Lapland 1960–2000. 1986 => Chernobyl nuclear disaster and 1989 => bankruptcy of *Poro ja Riista*. In West Finnmark 1976–1984 depicted by *black circles* (see discussion in text, Paine 1992). Data on harvested reindeer in West Finnmark from Riseth (2000)

The tragedy of the commons. The role of the tragedy of the commons remains unclear, but seemingly it does not explain the population increase in each area. In the southern part of the Finnish reindeer herding area the highest permitted number of reindeer were reached in the late 1970s, and the same happened in many herding associations in the middle part of the area in the early 1980s. By law, if the number is exceeded, each owner is obligated to reduce his herd by a fixed proportion, which makes it impossible to increase the herd size. In addition, the highest permitted reindeer number for each owner was relatively low, commonly 80–100 reindeer. Competition over limiting winter pastures had lost its meaning, because a great majority of reindeer were fed in corrals for three to four months during mid and late winter. In the southern part of the area, furthermore, reindeer herding is characteristically an additional livelihood, therefore the socioeconomic status of a reindeer owner is less dependent on the herd size than in other areas in this study.

Misperceptions of carrying capacity. Developments during the last 40 years show how difficult it is to evaluate the carrying capacity of winter pastures. The population increase in Finland and Sweden since the mid-1970s was a complete surprise to those engaged in pasture research, which started at that time. In 1972, Kärenlampi (1973) performed the first study on lichen biomass. He concluded that lichen vegetation was severely overgrazed and that lichen production was therefore far below the maximum. Some years later Mattila (1981) drew a similar conclusion with a much larger sample size. Mattila

(1979) also found that arboreal lichens, an important food source for reindeer in the coniferous forests in late winter, had declined due to forest renovation. In Sweden, Skuncke (1968) explained that high winter mortality of reindeer was caused by the deterioration of lichen pastures. V. Andrejev, a Soviet classic expert in pasture science, visited Norrbotten in 1969 and found that the height of lichens was only about 10 mm. He suggested that the number of reindeer should be reduced (Andrejev 1977). In West Finnmark, on the contrary, the condition of lichen pastures was essentially far better in the beginning of 1970s; more than 80 % of lichen pastures were relatively lightly grazed (Johansen and Karlsen 1998).

Moxnes (1998) emphasized the difficulty in interpreting reindeer–pasture relationships in a sustainable way on the basis of changes in lichen biomass or lichen height. Therefore, reindeer herders have in recent years tended to evaluate ecological sustainability by other means. Traditionally, the same pasture was not used every year (pasture rotation), or, at minimum, the herd size was kept in balance with winter pastures in the sense that reindeer were provided fresh, non-excavated pasture during the entire winter. There were also reserve areas used only when snow conditions were exceptionally difficult (Itkonen 1948).

Another source for feedback is the condition of reindeer, i.e., mortality, reproduction, and carcass weight (Heikkinen 2002). During peak years feedback citing failure in reproduction or high mortality was almost totally lacking. In the southern half of the Finnish reindeer herding area (including the southern part and the southern portion of the central part) mortality decreased and reproduction and carcass weights increased with heightened density (Helle and Kojola 1993, 1994). In the northern part, mortality remained unaffected and reproduction increased with the population expansion (Fig. 15.7) although the weights of females and calves decreased (Helle and Kojola 1994). The meat production increased as well, and it correlated positively with reindeer density.

Governmental reindeer policy. Paine (1992) considered misguided subsidy policy by the state as the main reason for the population increase and resultant overgrazing. He found that the proportion of subsidies with respect to total incomes increased in Norway almost proportionally between 1976 and 1984. Therefore reindeer owners had no need to obtain income from slaughtering and selling reindeer. However, the population increase in West Finnmark was somewhat slower than in Norrbotten and the northern part of the Finnish reindeer herding area (Fig. 15.2), although in Sweden and Finland subsidies were not paid before 1988 and 1995, respectively, i.e., when the growth period was already over. Clear differences in harvest did not exist from 1960 to 2000 between West Finnmark and the northern part of the Finnish reindeer herding area (Fig. 15.4). These findings do not support Paine's (1992) suggestion about the central role of the subsidies for the population increase in West Finnmark.

Climatic variation. A negative relationship between AO or NAO and population change would be expected, since highly positive AO and NAO are associated with high precipitation and resultant deep snow cover, which in turn affect both winter mortality (increase) and reproduction (decrease; Helle et al. 2001; Aanes et al. 2002). In our data the correlation was significant only in West Finnmark (Fig. 15.5). Instead, the number of reindeer in each of the seven sub-populations correlated significantly and positively with AO and NAO with one exception, namely, Norrbotten (Fig. 15.6). The correlations were significant with time lags of one to four years.

Although the positive correlations between AO or NAO and the number of semi-domestic reindeer were just opposite the expectation, they are difficult to consider as only an artifact. For the period 1960–2000, the behavior of AO and NAO can be roughly divided into three periods (see Fig. 15.3). In the period before the common population increase, AO and NAO indices were negative, suggesting shallow snow cover and good winters for reindeer. However, in the whole Finnish reindeer herding area, at least, the period was characterized by difficult snow winters with resultant high mortality and low reproduction (Kojola et al. 1991; Helle and Kojola 1993).

From the middle 1970's to the late 1980s common population growth was accompanied by low winter precipitation, shallow snow cover, and early onset of summer in Kilpisjärvi in northwestern Finnish Lapland (Helle et al. 2001). High winter mortality or poor reproduction did not occur. Applied to the entire Finnish reindeer herding area, Fig. 15.7 shows that in the northern part of the Finnish reindeer husbandry area calf percentage was higher between 1975 and 1986 than on the average despite increasing animal numbers. During that period AO and NAO indices showed an increasing trend, but they still remained around zero. However, during the third period from 1990 onward, the NAO index was higher than it ever had been throughout its known history since the middle of the 19th century (Jones et al. 1997). Western winds from the Atlantic Ocean were particularly strong and resultant snow-rich winters cut

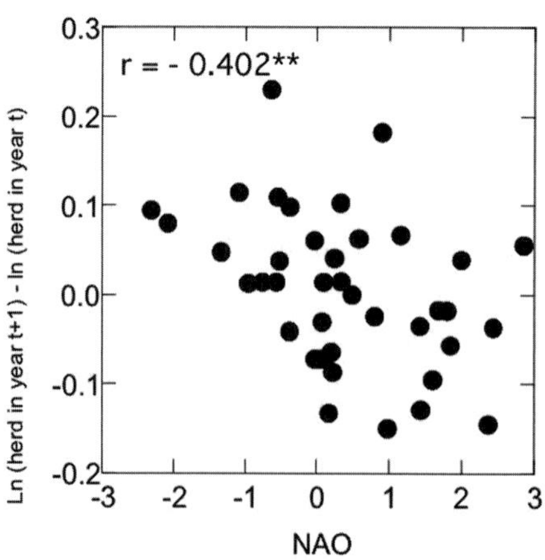

Fig. 15.5. The relationship between annual population change and North Atlantic Oscillation (NAO) index in West Finnmark 1960–2000

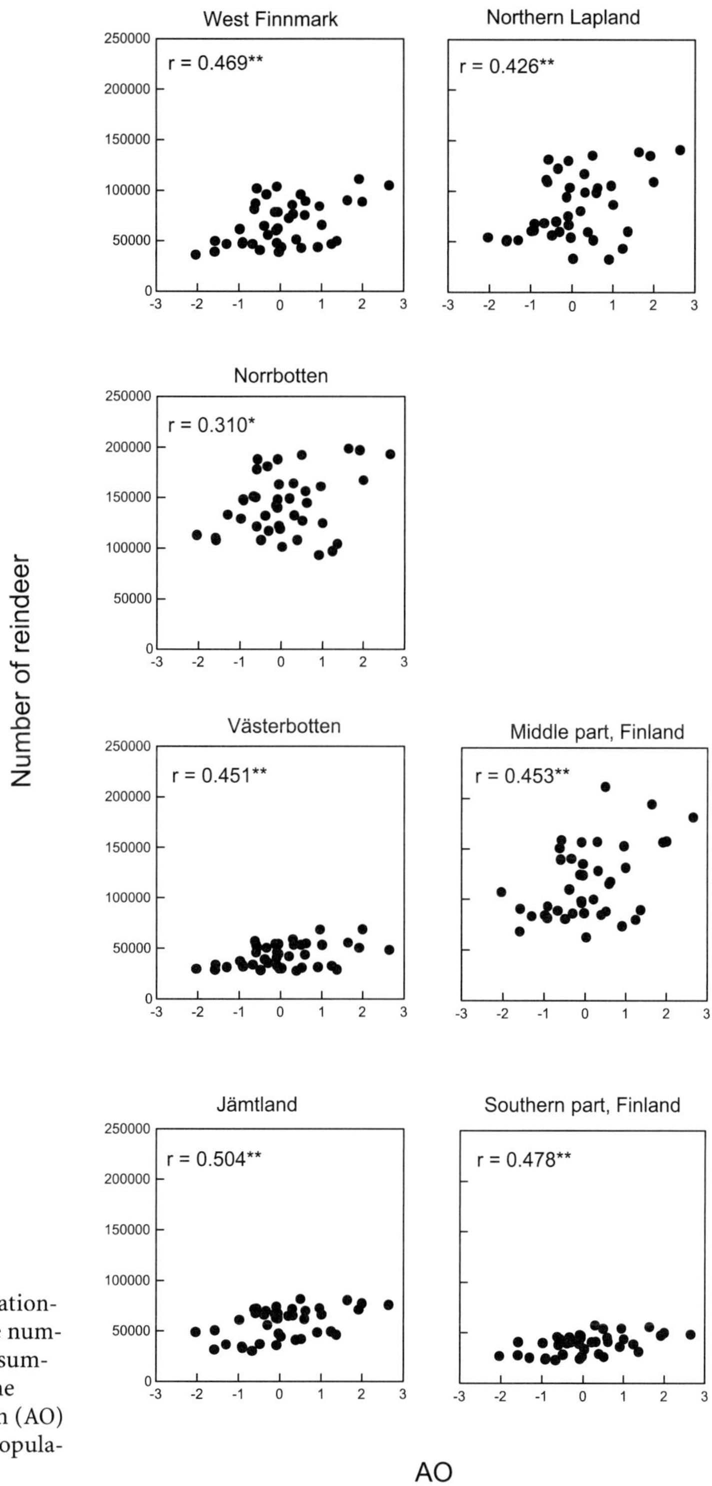

Fig. 15.6. The relationship between the number of reindeer (summer herd) and the Arctic Oscillation (AO) in different subpopulations 1960–2000

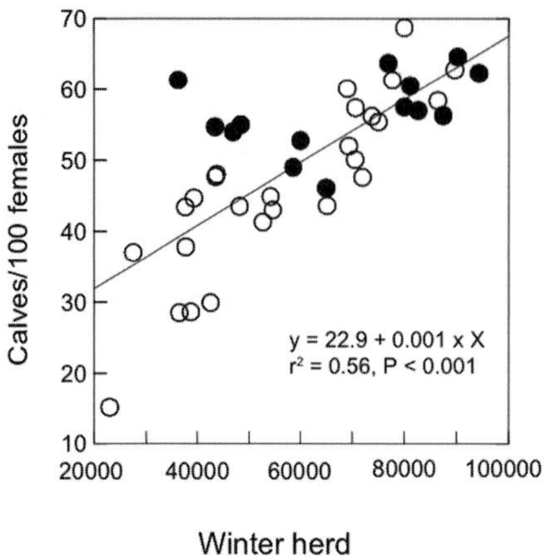

Fig. 15.7. The relationship between the number of reindeer in winter herds and reproduction (calves/100 females) in the northern part of the Finnish reindeer management area 1960–2000. Years of the population increase,1975–1988, are depicted with *black circles*

reindeer numbers everywhere in the Finnish herding area. As shown in Fig. 15.2 the same thing happened simultaneously in other sub-areas, and we suggest that all the decreases were the result of the same phenomenon, i.e., adverse winter conditions.

15.6 Discussion

Numerical data compared among the countries as well as within the borders of each suggest that there has been an overestimation of the importance of the Chernobyl nuclear disaster, the bankruptcy of *Poro ja Riista* in Finland, and the subsidy policy in Norway for the population increase during the period of 1976–1984. This applies also to newly introduced management practices; no doubt they were not insignificant, but they could explain neither the population increase over the whole area nor the decline to follow.

But the importance of the nonquantitated "tragedy of the commons" and the misperception of carrying capacity are more difficult to evaluate. Our data do not give any new information on these topics, but suggest that "the tragedy of the commons" is not the smallest denominator for the population increase that began in the mid-1970s. In fact, it is not known whether the herders at that time had their own institutional systems to control herd size or not. Ingold (1980) put forward the hypothesis that reindeer nomadism is an inherently unstable system due to competition between communities and/or owners resulting in pasture deterioration and great fluctuations in animal numbers: Increasing the herd size was reasonable from the viewpoint of an individual owner, because thereby one would have a much better chance of owning a sufficient herd even after a population crash.

However, in northern Norway strong pasture competition and resultant overuse of pastures seem to be a rather recent phenomenon triggered by the "snowmobile revolution" (Riseth 2000). The increased mobility has changed earlier practices in pasture division and herding techniques. Snowmobiles are expensive with high maintenance costs, increasing management costs and forcing herders to expand their herds. Furthermore, many other aspects have changed in the herders' society, including consumption behavior and expectations, working opportunities external to reindeer herding, as well as transfers of incomes. However, very little attention has been paid to the impact of these changes. Moreover, issues over a common approach to continuing growth under the ecologically sustainable utilization of reindeer pastures are discussed intensively.

As the onset of the population increase in the mid-1970s was unexpected, another surprise was experienced during the population peak. Food intake per reindeer was approximately one fourth of that in early 1970s, but density effects were found only with respect to (lower) carcass weights (Helle and Kojola 1994; Kojola et al. 1995). "Suddenly, all places were full of reindeer", Heikkinen (2002) was told by reindeer herders in Käsivarsi. They were also aware of the deterioration of winter pastures. Despite exponentially increasing harvest rates the decline in animal numbers was caused primarily by several poor winters before and after 1990. One may assume that the changes functioned in combination with density-dependent food limitations; due to the low growth rate of reindeer lichens, a time lag of several years may occur before one sees the influence of the pasture condition on reindeer.

In Finland supplemental fodder has lessened the focus on carrying capacity. The purpose of supplemental feeding was initially to compensate for the reduction of arboreal lichens that resulted from forest renovation (Helle and Saastamoinen 1979), but later it spread to areas where forestry has not been practiced. A clear intensification of supplemental feeding in the northern part of the area took place after severe snow winters before and after 1990.

Some institutional mechanisms also favor supplemental feeding (Helle et al. 1985). If the highest permitted number of reindeer is reached in a herding association, each owner must reduce his herd by a fixed proportion, calculated using official statistics based on the counting of reindeer during round-ups the previous autumn/winter. Not taken into account is the distinct possibility that in natural ranges winter mortality will be higher and reproduction lower than in herds with supplemental feeding. Consequently, maintaining one's herd requires using supplemental fodder if the other owners are doing so. In the southern half of the area "overfeeding" is commonly solved by establishing a household-specific quota. In northern Finnish Lapland some herding associations have decided that supplemental feeding requires a common agreement, because otherwise herders who do it would be capable of increasing their relative herd size. The decision is based on economic rationality, to wit, the high cost of supplemental feeding in relation to the producer's price of reindeer meat, which in 2004 was only 30–50 % of that in mid-1980s. Therefore, there is a rising trend to return reindeer "back to nature" – while still maintaining the readiness to apply supplemental feeding in order to prevent annual fluctuations in animal numbers due to severe snow conditions.

Our hypothesis is that the population trends were primarily a result of the variability of winter weather. This statement is in agreement with Gaughley and Gunn (1993), who showed that even short-term weather events could have marked effects on population dynamics of herbivores. Because weather data over the whole area were not available, we used AO and NAO indices as proxies.

In semi-domesticated reindeer, an increasing NAO index had a negative effect on body weight of calves in southern Norway (Weladji and Holand 2003). In this study, the expected negative correlation between NAO and the annual population change was significant only in West Finnmark. Several explanations are possible. Weladji and Holand (2003) suggested that human interference might mask the impacts of NAO, whereas the impacts are strengthened by density dependency. The correlation between NAO or AO and local weather may vary. Moreover, the impacts of NAO and AO on actual food availability of reindeer are very likely dependent on the habitat structure (open fell vs. coniferous forests) providing different kind of food for reindeer (ground vegetation vs. arboreal lichens). Paradoxically, the smallest common denominator we found, the correlations between NAO and AO and the number of reindeer, is difficult to interpret. The most obvious discrepancy deals with the period before the population increase. Animal numbers were fluctuating in each sub-area at a relatively low level, although proxies used here predicted favorable winter conditions. Therefore Helle et al. (2001) put forwards a hypothesis that when AO and NAO indices are negative, eastern winds, instead of bringing dry air from the continent, might blow over the Barents Sea bringing moisture and snowfall to northern Fennoscandia.

The essential role of the winter climate for the population dynamics of semi domesticated reindeer can explain, at least partly, the discrepancy in interpretations among biologically orientated researchers and authorities and reindeer herders (see other chapters in this book). The discrepancy is commonly related to how and to what extent the impact of other land uses should be taken into account when explaining the declining condition of winter pastures. However, the discrepancy is apparent also in areas where the impact is marginal. In Finland, the conflict emerged in the late 1980s when the reindeer numbers peaked, but it was after this that the condition of winter pastures showed signs of deterioration. Nevertheless, furthermore, the production was maintained at a high level (Kojola et al. 1991), and reindeer owners had difficulty understanding why they were asked to cut highly productive herds. As we have shown in this chapter, the exceptionally favorable period between the mid-1970s and late 1980s changed the earlier considerations of carrying capacity among reindeer herders. For herders, carrying capacity varies, while for authorities it is a fixed point on the curve describing the relationship between a standing crop and production.

In the management of reindeer herds, the concept of carrying capacity (Caughley 1976) or maximum sustained yield is based on the idea that maximal lichen production will automatically result in maximal meat production (Skogland 1985). According to a rule of thumb that happens if the biomass is about half of the maximum (Caughley 1976), which in northern Finland means a lichen biomass of about 1,000–1,500 kg ha^{-1} (Kärenlampi 1973; Helle and Kojola 1993) or even 2,600–2,800 kg ha^{-1} in dry sites (Kumpula et al. 2000). However, the functional response (describing the relationship between the average lichen biomass and the proportion of lichens in the diet of the reindeer) indicates that reindeer can fulfill their requirement in normal snow conditions with a lichen biomass of about 350–500 kg ha^{-1} (Helle and Kojola, unpubl.). The present solution to sufficient food supply during winters with difficult snow conditions is emergency or supplemental feeding.

One of the great challenges when striving toward sustainability is to combine ecological and social aspects (Berkes and Folke 1998). For research this demand is even more important. This study demonstrates, however, how rare such an approach has been in reindeer research. For social scientists reindeer management is a social construction, while for ecologists it is one form of prey–predator relationships, an attempt to maximize meat production in a sustainable way in stochastically varying environments. The roots of this discrepancy are deeply embedded in the history and traditions of social and biological sciences. Reindeer management is an example of the "slippery slope" (Takacs 1996) between social construction and essentialism or realism; it includes both of them at the same time. Our main finding, which links population fluctuations with large-scale climatic variations, represents post-wisdom, hindsight in the sense that most explanations and analyses in this chapter had been presented before the late 1990s when the first papers on the impacts of NAO and AO were published.

NAO explains a considerable proportion of the variation in winter weather, including temperature, precipitation, snow depth, and the speed and direction of prevailing winds, over an extensive area of temperate and cold climate zones in the northern hemisphere (e.g., see Hurrell 1995; Hurrell and van Loon 1997), and its contribution to observed warming is obvious (Wallace et al. 1995; Kerr 1997; Corti et al. 1999; Hurrell et al. 2001). That comprehension also seems to apply to AO in arctic and subarctic areas (Aanes et al. 2002; Ambaum et al. 2001). An increasing body of research is showing that NAO and AO influence, through impact on local weather, numerous ecological processes in terrestrial and aquatic ecosystems (Post and Stenseth 1999; Stenseth et al. 2002; Walther et al. 2002). In addition, it is obvious that summer weather is mediated by large-scale air mass movements affecting local weather and thus quality and quantity of vegetation, which is in turn reflects on the condition of herbivorous animals (Mysterud et al. 2001). However, much research is needed, because great variation exists between local and regional weather and large-scale climate indices.

Acknowledgements. We are grateful to Päivi Soppela, Nicolas Gunslay, and Ludger Müller-Wille for their comments on the earlier draft of the manuscript.

References

Aanes R, Saether B-E, Öritsland N (2000) Fluctuations of an introduced population of Svalbard reindeer: the effects of density dependence and climatic variation. Ecography 23:437–443

Aanes R, Saether B-E, Smith F, Cooper E, Wookey P, Öritsland N (2002) The Arctic Oscillation predicts effects of climate change in two trophic levels in a high-arctic ecosystem. Ecol Lett 5:445–453

Åhman B (2002) Radiocesium in reindeer July 2000 to June 2002. Annual report to the Swedish National Food Administration and the Swedish Board of Agriculture, Swedish University of Agricultural Sciences, Uppsala, 9 p (in Swedish)

Alaruikka Y (1964) Reindeer husbandry in Finland (in Finnish). Rovaniemi

Ambaum M, Hoskins B, Stephenson D (2001) Arctic Oscillation or North Atlantic Oscillation. J Climate 14:3,495–3,507

Andrejev V (1977) Reindeer management in the Soviet Union is based on pasture research (in Finnish). Suomen Luonto 2:142–144

Anonymous (1986) Radioactive contamination of reindeer meat (in Finnish). Poromies 4:1

Berg B (1996) Government intervention into Sami Reindeer-management in Norway. Has it prevented or provoked "Tragedies of the Commons". Acta Borealia 2):69–89

Berkes F, Folke C (eds) (1998) Linking social and ecological systems. Management practices and social mechanisms for building resilience. Cambridge Univ Press, Cambridge

Björklund I (1990) Sami reindeer pastoralism as an indigenous resource management system in northern Norway. A contribution to the common property debate. Dev Change 21

Brox O (1989) Can rural livelihoods be profitable (in Norwegian). Gyldendal, Oslo

Caughley G (1976) Wildlife management and the dynamics of ungulate populations. In: Coaker TH (ed) Applied biology, vol 1. Academic Press, London

Caughley G, Gunn A (1993) Dynamics of large herbivores in deserts: kangaroos and caribou. Oikos 67:47–55

Corti S, Molteni F, Palmer TN (1999) Signature of recent climate change in frequencies of natural atmospheric circulation regimes. Nature 398:799–802

Dahle H, Danell Ö, Gaare E, Nieminen M (1999) Reindeer husbandry in North-West Europe 1998 – biological opportunities and limitations (in Norwegian). TemaNord 1999:510

Ermala A (2002) Tops and downs in reindeer numbers in Finland (in Finnish). Poromies 69(2):23–25

Hardin G (1968) The tragedy of the commons. Science 162:1,243–1,248

Heikkinen H (2002) Models of adaptation. Adaptation of reindeer husbandry to post-industrial society in the western part of the Finnish reindeer management area 1980–2000 (in Finnish). SKS, Helsinki, 369 pp

Helle T, Kojola I (1993) Reproduction and mortality of Finnish semi-domesticated reindeer in relation and management strategies. Arctic 46:72–77

Helle T, Kojola I (1994) Body mass variation in semi-domesticated reindeer. Can J Zool 72:681–688

Helle T, Saastamoinen O (1979) The winter use of food resources of semi-domesticated reindeer in northern Finland. Communicationes Instituti Forestalis Fenniae 95(6):1–26

Helle T, Kojola I, Timonen M (2001) Impacts of the snow on demography of semi-domesticated reindeer in Käsivarsi – which is the role of the North Atlantic Oscillation (NAO)? (In Finnish with English summary.) Suomen Riista 47:75–85

Helle T, Pöyhönen I, Lotvonen E (1985) Economic evaluation of current trends in Finnish reindeer management. In: Siuruainen E (ed) Ten years of work at Research Institute of Northern Finland, vol 3. Research Institute of Northern Finland, Serie A, pp 50–61

Hurrell JW (1995) Decadal trends in the North Atlantic oscillation. Science 269:676–679

Hurrell JW, van Loon H (1997) Decadal variations in climate associated with the North Atlantic Oscillation. Climate Change 36:310–326

Hurrell JW, Kushnir Y, Visbeck M (2001) The North Atlantic Oscillation. Science 291:603–605

Huttu-Hiltunen V (1989) Annual report on reindeer husbandry in Finland (in Finnish). Poromies 3:9–11

Ims RA, Andreassen HP (2000) Spatial synchronization of vole population dynamics by predatory birds. Nature 408:194–196

Ims RA, Steen H (1990) Geographical synchrony in microtine population cycles: a theoretical evaluation of the role of nomadic avian predators. Oikos 57:381–387

Ingold T (1980) Hunters, pastoralists and ranchers. Cambridge Univ Press, Cambridge

Itkonen T (1948) The Lapps in Finland up to 1945, vol II (in Finnish). Porvoo, Helsinki, 629 pp

Jernsletten JL, Klokov K (2002) Sustainable reindeer management. Arctic Council 2000–2002. Centre for Saami Studies, University of Tromsö, Tromsö, 157 pp

Johansen B, Karlsen SR (1998) Changes in lichen vegetation i Finnmarksvidda in 1987–1996 based on Landsat 5/TM data (in Norwegian). NORUT Informasjonsteknologi

Jones PD, Jönsson T, Wheeler D (1997) Extension to the North Atlantic Oscillation using early measuremental pressure observations from Gibraltar and south-west Iceland. Int Climatol 17:1,433–1,450

Kärenlampi L (1973) Condition, biomasses and production of lichen pastures in Finnish reindeer husbandry area in 1972 (in Finnish). Poromies 40(2):15–19

Kautto A, Kärenlampi L, Nieminen M (1986) The change in condition of lichen pastures in Finnish reindeer husbandry area 1972–1983 (in Finnish). Poromies 53(3):28–34

Kerr RA (1997) A new driver for the Atlantic's mood and Europe's weather. Science 275:754–755

Kojola I, Helle T (1993) Calf harvest and reproductive rate of reindeer in Finland. J Wildl Managem 57(3):451–453

Kojola I, Helle T, Aikio P (1991) Productivity of semi-domesticated reindeer in Finland. Rangifer 11(2):53–63

Kojola I, Helle T, Niskanen M, Aikio P (1995) Effects of lichen biomass on winter diet, body mass and reproduction of semi-domesticated reindeer *Rangifer t. tarandus* in Finland. Wildl Biol 1:33–38

Kojola I, Helle T, Ranta E, Niskanen M (1998) Foraging conditions, tooth wear and herbivore body reserves: a study of female reindeer. Oecologia 117:26–30

Kortesalmi J (1996) Peasant reindeer breeding in northernmost Russian Karelia: its origin and development up to 1922 (in Finnish with English summary). Vammalan Kirjapaino Oy, Vammala

Kumpula J, Colpaert A, Nieminen M (1998) Reproduction and productivity of semi-domesticated reindeer in northern Finland. Can J Zool 76:269–277

Kumpula J, Colpaert A, Nieminen M (2000) Condition, potential recovery, and productivity of lichen (*Cladina* spp.) ranges in the Finnish reindeer management area. Arctic 53(2):152–160

Mattila E (1979) Characteristics of mineral soil forests with arboreal lichens in the Finnish reindeer management area, 1976–1978. Folia Forestalia 417: 1–39 (in Finnish with English summary)

Mattila E (1981) Survey of reindeer winter ranges as a part of the Finnish National Forest Inventory in 1976–1978. Communicationes Instituti Forestalis Fenniae 99(6):1–74

Mattila E (1998) Quantitative data on reindeer winter pastures from 50 years (in Finnish). Metsäntutkimuslaitoksen tiedonantoja 678:67–83

Moen J, Danell Ö (2003) Reindeer in the Swedish mountains: an assessment of grazing impacts. Ambio 32:397–402

Moxnes E (1998) Not only tragedy of commons: misperceptions of bioeconomics. Manage Sci 44:1,234–1,248

Mysterud A, Stenseth NC, Yoccoz N, Langvatn R, Steinheim G (2001) Nonlinear effects of large-scale climatic variability on wild and domestic herbivores. Nature 410:1,096–1,099

Nieminen M, Timisjärvi J, Laitinen M (1980) The effects of antiparasitic treatment on the condition of the semi-domestic reindeer (*Rangifer tarandus*). Rep Kevo Subarct Res Sta 16:23–26

Nilsen R, Mosli J (1994) Socioeconomic adaptation of reindeer husbandry in Guovdageaidnu 1960–1993 (in Norwegian). BAJOS Utviklingsselskap AS/NORUT Samfunnsforskning AS, Trömsö

Nordkvist M, Rehbinder C, Cristensson D (1983) A comparative study of the efficiency of four antihelminthics on some important reindeer parasites. Rangifer 3:19–38

Ostrom E (1992) Governing the commons. Cambridge Univ Press, Cambridge

Paine R (1992) Social construction of the "tragedy of commons" and Saami reindeer pastoralism. Acta Borealia 2):3–20

Persen E, Jacobsen E, Lenvik D, Skjenneberg S (1982) Weight development and survival rate of reindeer calves treated against *Oedomagena tarandi* L. and *Cephenomyia trombe* L (in Norwegian with English summary). Rangifer 2(1):39–49

Post E, Forchammer M (2002) Synchronization of population dynamics by large-scale climate. Nature 420:168–171

Post E, Stenseth NC (1998) Large scale climatic fluctuations and population dynamics of moose and white-tailed deer. J Anim Ecol 67:537–543

Post E, Stenseth NC (1999) Climatic variability, plant phenology, and northern ungulates. Ecology 80: 1,322–1,339

Post E, Forchammer MC, Stenseth AN (1999) Population ecology and the North Atlantic Oscillation (NAO). Ecol Bull 47:117–125

Ranta E, Kaitela V, Lindström J, Helle E (1997) The Moran effect and synchrony in population dynamics. Oikos 78:136–142

Reimers E (1977) Population dynamics in two subpopulations of reindeer in Svalbard. Arct Alp Res 14:295–300

Riseth JÅ (2000) Sami reindeer management under technological change 1960–1990: implications for common-pool resource use under various natural and institutional conditions. Thesis, Dept Economics and Social Sciences, Agricultural University of Norway, Ås, Norway

Saether BE (1997) Environmental stochasticity and population dynamics of large herbivores: a search for mechanisms. Trends Ecol Evol 12:143–149
Skogland T (1985) The effects of density-dependent resource limitations on the demography of wild reindeer. J Anim Ecol 54:359–372
Skuncke F (1968) A reindeer catastrophe caused by an overutilization of a good winter pasture. Zool Revy 82:105–112
Stenseth NC, Mysterud A, Ottersen G, Hurrell JW, Chan K, Lima M (2002) Ecological effects of climate fluctuations. Science 297:1,292–1,296
Takacs D (1996) The idea of biodiversity: philosophies of paradise. Johns Hopkins University Press, Baltimore
Tömmervik H, Johansen B (1992) Environmental studies in Elgå reindeer district. Vegetation map and condition of reindeer pastures (in Norwegian). FORUT Information Technology, 97 pp
Wallace J, Chang Y, Renwick J (1995) Dynamic contribution to hemispheric mean temperature trends. Science 270:780–783
Walther G-R, Post E, Convey P, Menzel A, Parmesan C, Beebee TJC, Fromentin J-M, Hoegh-Guldberg O, Bairlein F (2002) Ecological responses to recent climate change. Nature 416:389–395
Weladji R, Holand Ö (2003) Global climate change and reindeer: effects of winter weather on the autumn weight and growth of calves. Oecologia (published online 18.04.2003), (DOI 10.1007/s00442-003-1257-9)
Weladji R, Klein D, Holand Ö, Mysterud A (2002) Comparative response of reindeer/caribou and other northern ungulates to climatic variability. Rangifer 22:33–50

16 Systems Analysis of Finnish Reindeer Management

B. BURKHARD and F. MÜLLER

16.1 Introduction

One of the targets of the RENMAN project was development of new tools and models of participatory research and planning in reindeer management that will facilitate integrated and sustainable use of semi-domesticated reindeer and related living resources in subarctic/boreal regions of Europe (Forbes et al. 2004). To achieve this interdisciplinary goal, systems analytical procedures were carried out within the presented work package. For this, integrative, systems-oriented analysis of reindeer management in northernmost Europe, different demands for landscape utilization were investigated from social, economic, and ecological aspects. Quantitative and qualitative data from the other RENMAN working groups were combined and applied in conceptual and simulation models of socioeconomic and ecological interactions. Finally, the results from exemplary studies were used for the simulation of different future scenarios for the reindeer management system. Based on these results, several recommendations for sustainable future reindeer herding strategies have been elaborated.

Semi-domesticated reindeer husbandry, as can be found in all countries in the sub-polar zone in northern Fennoscandia and Asia, has been an appropriate way to exploit the sparse natural resources of the fjeld and forest limit areas for a long time (Aikio 1987). Besides fundamental economical interests, reindeer have always played a major role in the livelihood and culture of local communities. Thus, there has always been a strong connection between ecological, social, and economic aspects in reindeer management.

For the systems analysis and modeling tasks, the main challenge can be seen in the integration of these socioeconomic and ecological factors to depict a figure of reindeer management as a holistic entity. This holistic view may help to promote a future-oriented and sustainable concept of reindeer management (Fig. 16.1). The application of systems-analysis methods like

Ecological Studies, Vol. 184
B.C. Forbes, M. Bölter, L. Müller-Wille, J. Hukkinen, F. Müller, N. Gunslay, Y. Konstatinov (Eds.)
Reindeer Management in Northernmost Europe

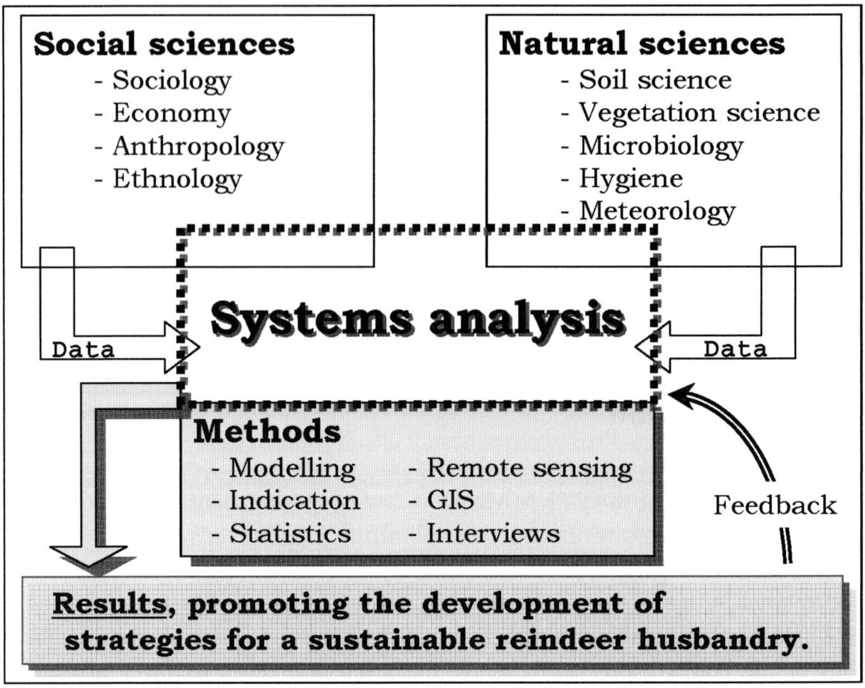

Fig. 16.1. Perception of systems analysis in the RENMAN project

indicator derivation, scenario development, and expert consultations in combination with state-of-the-art instruments such as simulation models, remote sensing techniques, geographical information systems (GIS), and data base tools enables an integrative analysis of complex systems and the management and analysis of comprehensive data sets (Jensen and Bourgeron 2001).

The key problems identified in today's reindeer management are the competition for natural resources with other forms of land use (such as forestry, hydropower, tourism, mining, nature preservation, and predator protection), difficulties concerning pasture management, environmental problems, and the challenges caused by the demanding adaptation to the contemporary market and living conditions (Forbes et al. 2004; RENMAN 2004; Frei 2002; Kumpula 2001).

Referring to these problems, the following key questions will be discussed:

- Is it possible to analyse the key problems of the Finnish reindeer management with the systems-analysis techniques available today?
- Can the results be applied to support a sustainable reindeer management and to improve the living conditions of the local reindeer herders?
- What kind of future is conceivable for Finnish reindeer management?

16.2 Methods and Data Basis

Large amounts of the data acquired by the different RENMAN disciplines were provided to the systems-analysis working group for the integrative analyses. For the application of the particular results, an evaluation scheme based on 32 indicators was developed regarding land use structure, social and economic welfare, and ecological integrity in different reindeer herding systems (Burkhard et al. 2003; Burkhard and Müller 2004). The comparison of different investigation sites with varying management strategies has provided the base for the estimation of the effects that reindeer herding can have on the related subsystems and of the effects that changes in their compositions can have on reindeer herding. Based on these assessments, various scenarios for future development within the reindeer management region were simulated.

For the respective data management and analysis, data base and statistical software packages [MS Excel (Microsoft, Redmond, WA, USA), Visual dBASE (dataBased Intelligence, East Vestal, NY, USA), SPSS (SPSS Inc., Chicago, IL, USA, Statistica (StatSoft, Tulsa, OK, USA)] were used. Spatial data found entry into Geographical Information Systems [ArcGIS and ArcView (ESRI, Redlands, CA, USA)] that enabled proper management of locally obtained data in combination with the Global Positioning System (GPS), topographical map information, and data originating from remote sensing sources (Burkhard et al. 2003; Kumpula 2003).

For the ecological part of this systems analysis, various data from the collaborating RENMAN working groups (e.g., M. Bölter, T. Kumpula, S. Peth, C. Uhlig, A. Zink; see Forbes et al. 2004) were used directly or found entry into information systems and simulation models. For instance, a high number of input data was needed to run the water-and-substance simulation model WASMOD (DigSyLand, Husby, Germany; Reiche 1996). Figure 16.2 gives an overview of the WASMOD data organisation and the model structure. Extensive climate records from the investigation areas applied in WASMOD were purchased from the Finnish Meteorological Institute (FMI). In situ weather measurements that took place during the summer seasons 2002 and 2003 were used to complete and calibrate the climate data. Certain WASMOD input data that were not investigated within the RENMAN project, for instance, plant-specific evaporation rates and leaf-area indices, were taken from comparable investigations (e.g., Bliss et al. 1981; Riess 1993; Dierssen 1996; Scurlock et al. 2001) and, if necessary, adjusted to the local conditions at the RENMAN research areas.

Concerning the socioeconomic part of this systems analysis, various statistical data sources, interviews, and results from RENMAN social science working groups (e.g., Hukkinen et al. 2002; Laakso 2002) were explored. However, due to difficulties in adjusting the varying data sources to the indicator-based evaluation scheme and to the proposed scenarios, additional expert inter-

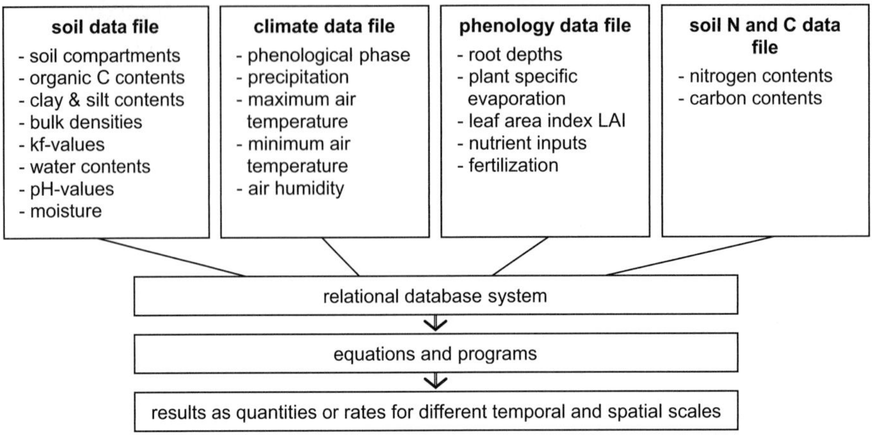

Fig. 16.2. WASMOD model structure

views with particularly designed questionnaires were carried out for the improvement of the scenario derivations. During the interviews, selected experts were asked to estimate how the reindeer management system, including social, economic, and land use-pattern factors, would develop over a 25-year time period in accordance with three predetermined future scenarios. The estimations were specified on a scale ranging from a strong decrease or worsening of the respective indicators (-5), no change compared to today (0), to a strong increase or improvement (+5) of the element represented by the particular indicator.

The spatial scales considered during the RENMAN investigations ranged from fine- and meso-scaled local field studies to coarse-scaled studies at the level of reindeer herding cooperatives. The results of the project will be applied for recommendations on the regional scale of the whole northern Fennoscandian reindeer management region.

16.3 The Investigation Sites

For several reasons, including the availability of expertise and resources within the RENMAN project, systems analysis was applied primarily to the Finnish reindeer management system. Most of the data applied for the ecological simulations were gathered in the two intensive RENMAN research areas within Finland, in Jauristunturit and Lappi.

Jauristunturit (68°49' N 23°48' E) is situated along the border between Norway and Finland and is part of the Norwegian West Finnmark reindeer area and the Finnish reindeer herding district Näkkälä (see map in Chap. 1). Located in the mountainous northern part of Finland at 550 meters above sea level, Jauristunturit belongs to the subarctic tundra zone containing huge fjeld areas. The vegetation are dominated by lichen (*Cladina stellaris, Cetraria nivalis*), dwarf birch (*Betula nana*), dwarf shrubs (e.g., *Empetrum nigrum*) and numerous peatlands (Oksanen and Virtanen 1995; Olofsson et al. 2001).

A reindeer fence that was installed in the 1950s in this region to prevent animal migrations between Norway and Finland has been used as a basic experimental arrangement. The differently managed reindeer pastures on either side of the fence provided possibilities for comparing impacts of reindeer grazing on ecosystem properties. In the Norwegian part of the area, the pastures are used for winter grazing or left as emergency reserve pasture. The area on the Finnish side of the fence has been used primarily as summer pasture for decades. Thus, this area has been subject to both grazing and trampling by reindeer, whereas the Norwegian area has been affected less due to curtailed grazing periods and the protective cover of the snow in winter (see Chaps. 8, 9). A third, relatively small area with differing grazing intensity was detected in a place were the reindeer fence was relocated from the Finnish side to the Norwegian side about 15 years ago (referred to hereafter as the transition zone). As a consequence, this location has been part of the Finnish management system during the last 15 years. This fact was considered for scenario simulations in which the data from the variously managed pastures corresponded to different modifications of reindeer herding (Sect. 16.6).

The investigations in the Lappi district were concentrated on the effects of changing land use patterns on the reindeer management system. Competing land use activities are special issues in this boreal forest zone. Whereas in the Jauristunturit research area, reindeer herding is the dominating form of land use, in Lappi the natural resources are also exploited by forestry, the hydroelectric energy industry, and tourism. The effects that reindeer grazing and trampling have on the environment were studied at the Tuorekolla forest site (68°17' N 26°60' E; Chaps. 11–13).

16.4 Modeling and Indicator Derivation

The systems analysis started with the development of a conceptual model which supports the understanding of a complex system like reindeer management. The reindeer herd can be seen as the central element that follows the seasonal rhythm of reindeer herding with winter and summer pastures, calf marking, separations, and slaughtering (Fig. 16.3).

The different process components are associated with certain inputs ranging from labor, reindeer fences, and herding equipment to energy and supplementary fodder. The main outputs are meat products, antlers, and furs that

Fig. 16.3. Conceptual model of the reindeer management system

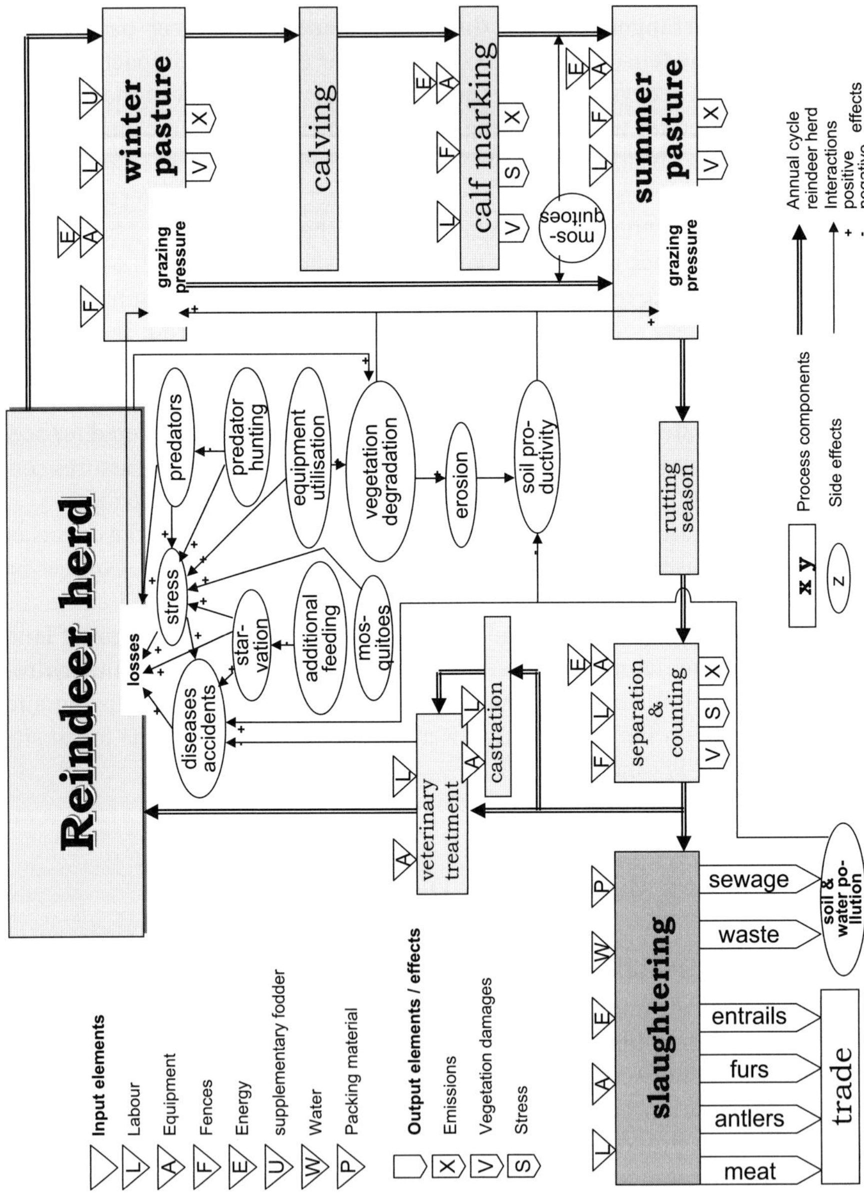

are marketed, but also waste and further by-products. Examples of effects, caused by the action of reindeer herding, include vegetation changes and gaseous emissions (e.g., methane from the animals and CO_2 from the machines). Furthermore, potential interactions and side effects like disease or predation that can cause reductions in the reindeer herd are shown in the diagram. Of course there are variations among different herding areas in the dominant management practices, but the basic pattern is quite similar. This rather simple model gives an illustrative basic overview of the reindeer management system.

The points of main interest within the system shown in Fig. 16.3 are changes and developments. Sample questions include: What happens if certain elements are changing? if for instance the size and quality of the pastures are shrinking? or if the size of the reindeer stock is changing as was the case during the last few decades (Filppa 2003)? What is the reaction of the different stakeholders? Would inputs into the system be increased in the form of supplementary feeding or an increased amount of work? What economical, social, and ecological consequences are to be expected? The model can be used to conceive potential reactions of the system to meet any one or all of the various possible changes.

For the evaluation of the state and development of a system, it is convenient to develop functional indicators (Wiggering and Müller 2003) that also can provide useful information concerning the achievement of certain political objectives. According to the basic concept for sustainable development, which demands modern society to treat economic, social, and ecological issues in an equivalent way (World Commission on Environment and Development 1987; UNCED 1992), we have fashioned indicators for economic welfare, social welfare, and ecological integrity (Fig. 16.4). Based on the assumption that most of the changes in human-environmental systems are caused by human actions, which are conducted due to certain driving forces (according to the Drivers-Pressure-State-Impact-Response approach, EEA 1999), the established land use pattern has to be perceived as a forcing function and is therefore described in an individual indicator group.

The derivation of the socioeconomic indicators was basically influenced by investigations on the quality of life (Fürntrapp-Klopp 1995; McMahan 2002; Alcamo and Bennett 2003) and the geographical concept of "basic functions of human life" (Ruppert and Schaffer 1969). The concept of ecological integrity has been described in Barkmann et al. (2001), and the respective indicators were developed during the research project INΔECO[2] (Baumann 2001; Müller 2005).

The four indicator subgroups can be aggregated into a set which encompasses a broad scale of different factors, and thus affords the opportunity for an evaluation and illustration of different reindeer management systems. Based on these 32 indicators, the analysis of past, current, and future conditions of reindeer management was carried out. For the modeling of past and

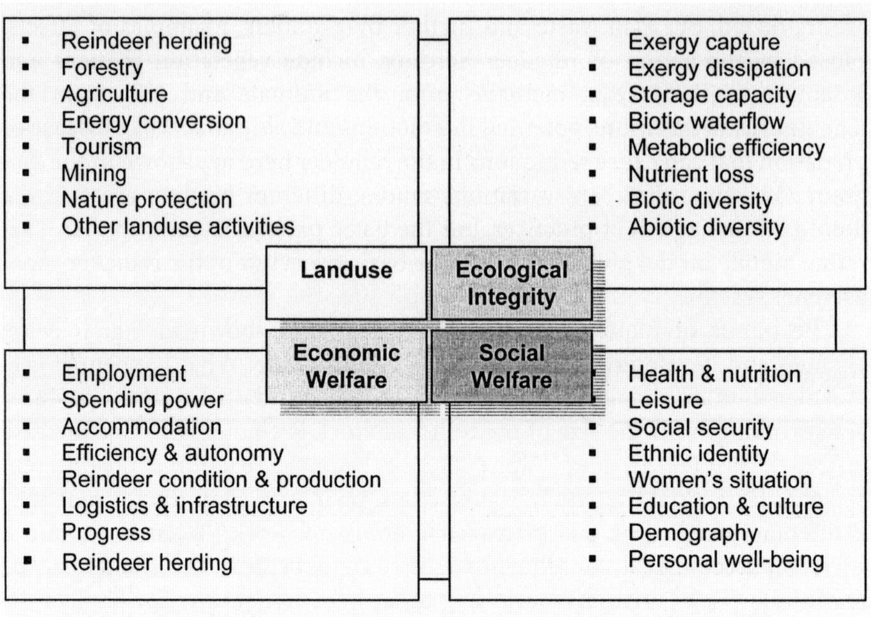

Fig. 16.4. Indicator set for the assessment of the reindeer management system

current conditions, a huge amount of statistics and literature-derived data were used. The data were complemented by new research results produced during the RENMAN project. Concerning the socioeconomic components of the analysis, we were able to integrate the results of expert interviews carried out among reindeer herders, researchers, and decision makers. The indicator groups for land use and ecological integrity were processed with the results from spatial data analyses carried out with Geographical Information Systems (GIS), remote sensing methods, and ecological modeling. Various ecological parameters could be explored by using the water and substance simulation tool WASMOD (Reiche 1996) and data from field investigations and measurements within RENMAN.

16.5 Analysis

Following the derivation of the indicator set, the construction of the conceptual model, data collection, and the installation of WASMOD, the individual sections were applied and integrated. Due to the high number of different components, the presentation within this text is concentrated on a few representative results only. Further information can be found in Burkhard (2004).

16.5.1 Land Use Pattern

In the Lappi district research area, the competition of different forms of land use for natural resources has been a key concern in recent decades (e.g., Hukkinen et al. 2002). Therefore, special focus of the analysis was placed on the investigation of the current land use distribution, its historical development, and possible future developments within this region. The results illustrate that in the last decades forestry, tourism, and energy conversion have increased significantly (two large water reservoirs, Lokka and Porttipahta, constructed in 1967 and 1970, are situated within the Lappi district; Cyffka et al. 1999; Tilastokeskus 2002).

For the study of the current land use pattern, the GIS of the RENMAN research areas was applied to generate a rough calculation of the individual land use types' spatial extension. The integration of remote sensing data (LANDSAT Thematic Mapper 5 image with seven channels and 30 m spatial resolution) enabled the identification of old-growth pine- and spruce-dominated forests with high amounts of arboreal tree lichen as suitable areas for reindeer herding. Furthermore, areas with intensive forestry in form of clear cuts were classified (see Chap. 8). These Landsat data originated in the year 1990. Hence, recent developments like forest cuts in the year 2003 (pers. obs.) were not taken into account. The topographical base data were provided as digital GIS layers by the Finnish National Land Survey and included information about the extension of the artificial lakes for energy conversion, infrastructure installations, agricultural areas and nature protection areas. The calculation of buffer zones around the tourism areas, summer hiking, and winter skiing trails, and around certain other land use structures (like roads or power lines) were calculated and applied as land use types. Reindeer herding in nature protection areas or tourism on artificial lakes are examples of spatially overlapping land use forms that partly impacted the accuracy of this analysis.

landuse	area in km²	data
Reindeer herding	1846	old forests, according to LANDSAT image classification (T. Kumpula)
Nature protection	1818	GIS-data (National Land Survey of Finland)
Energy conversion	462	GIS-data (National Land Survey of Finland)
Other activities	263	buffer zones around roads and power line
Forestry	237	clear cut areas, according to LANDSAT image classification (T. Kumpula)
Tourism	170	buffer zones around touristical facilities
Mining	2	rough field estimation
Agriculture	0,27	GIS-data (National Land Survey of Finland)
Whole district	4453	

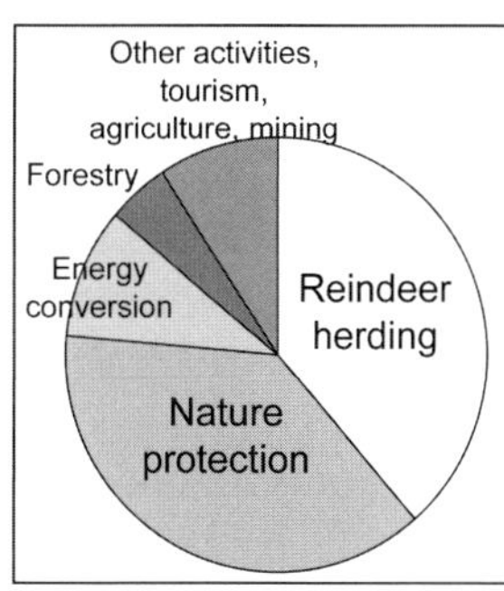

Fig. 16.5. Land-use distribution in the Lappi herding district

Figure 16.5 shows the current land use pattern in the Lappi district, with reindeer herding and nature protection as the land use forms with the largest spatial extensions. The areas of intensive forestry (clear-cut areas) correspond to at least 5.3 % of the whole district. However, the impact of forest clear-cuts or installation of large water reservoirs on local ecosystems must be considered to be many times higher than the impact of reindeer herding or tourism.

16.5.2 Ecological Integrity

Human societies receive benefits from the services provided by nature in different ways (de Groot 1992; Costanza et al. 1997; Alcamo and Bennett 2003). At the same time, the activities carried out by humans cause pressures ecological systems (pressure-state component in the DPSIR-approach; EEA 1999). To ensure a sustainable provision of these ecosystem services, the maintenance of ecophysiological functioning of ecosystems, the ecological integrity (Barkmann et al. 2001) is essential.

Reindeer herding as a traditional form of land use has exerted various pressures on northern boreal and subarctic ecosystems for centuries. To assess the degree of this impact, the data and verbal information from the relevant scientific working groups in RENMAN were integrated into the analysis. The comparison of selected sites that were used with different seasonal access by reindeer herding enabled an estimation of the effects of reindeer grazing and trampling on important ecosystem properties in the two main research areas.

In the Tuorekolla area of the Lappi district, an almost undisturbed forest site was compared with a forest site that bore visible signs of reindeer grazing and a narrow reindeer-trampled path in the same forest. Neither the soil's physical properties (Peth and Horn 2004), microbial activities (Bölter and Möller 2004) nor nutrient and water cycling (Burkhard 2004) showed significant differences among these three sites. Hence, the investigations were concentrated in the Jauristunturit area at the three differently managed sites close to the reindeer fence. Table 16.1 gives an overview of the parameters used for the indication of ecological integrity at these three. Apart from plant primary production values taken from Olofsson et al. (2001), all values were obtained from RENMAN research activities and WASMOD model outputs.

In Fig. 16.6, a sample output from the WASMOD model, showing the annual course and the sums of interception (amount of precipitation stored on the vegetation's surface and eventually evaporated back into the atmosphere) 1983–1999 at the three Jauristunturit sites is given. Slight differences in the amounts of interception are recognizable among the three. The less intensively managed Norwegian site with a less grazed and trampled vegetation (mainly lichen and dwarf birch) shows a higher interception. However, these differences as well as the variations concerning the other ecological parame-

Table 16.1. Indicators used in the model, their units and sources

Indicator	Applied parameter	Unit	Source
Exergy capture	Above ground plant biomass production	g m^{-2} $year^{-1}$	Olofson et al. (2001)
Exergy dissipation	Soil CO_2 production	g CO_2 h^{-1} g^{-1} soil	Chap. 13
Storage capacity	Soil organic carbon content	$^1/_2$ loss on ignition (LOI)	Chap. 12
Biotic water flow	Intercept by evaporation	mm (sums 1977–1999)	Statistics Finland (WASMOD)
Metabolic efficiency	Soil respiration per bacterial biomass	g CO_2 h^{-1} g^{-1} soil g^{-1} bacterial dwt	Chap. 13+
Nutrient loss	Nitrogen discharge by leaching	Kg ha^{-1} (sums 1983-1999)	WASMOD
Biotic diversity	Number of plant species	n	Chap. 8
Abiotic diversity	Soil humus heterogeneity index	cm (Range of soil humus layer thickness)	Chap. 12, A. Zink (pers. comm.)

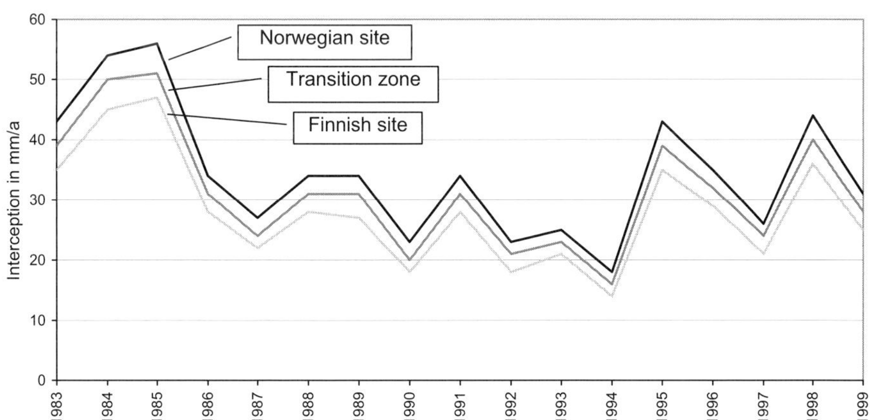

Fig. 16.6. Interception in mm 1983–1999 at three sites in the Jauristunturit research area modeled with WASMOD

ters are rather small (Zink 2003; Bölter and Möller 2004; Burkhard 2004; Peth and Horn 2004).

The most apparent alterations at these three sites are changes in the lichen cover, which attains a height of several centimetres on the Norwegian side, and is visibly diminished within the transition zone and on the Finnish side of the reindeer fence. These differences are clearly visible in satellite images (see Chap. 8). Changes in the vegetation cover at tundra sites frequented by herbivores have the potential to affect the general vegetation distribution, primary production, and nutrient cycling. These animal–plant–soil interactions have been the subject of several investigations (e.g., Henry and Svoboda 1994; Oksanen and Virtanen 1995; Dierssen 1996; Wielgolaski 1997; Löffler 1999; Kumpula 2001; Olofsson et al. 2001; Stark 2002; Chap. 9). Most of these works have demonstrated that moderate reindeer grazing can lead to a transition of lichen and moss-rich tundra heaths into graminoid dominated steppe-like vegetation with higher net productivity rates and higher interspecific species diversity, but reduced total plant biomass. At the RENMAN sites, only minor alterations in the functional states of the ecosystems were detected. Hence, it is difficult to correlate them with differing reindeer grazing and trampling intensities and to draw reliable conclusions about the effects of reindeer herding on ecological integrity on a broader spatial scale.

16.5.3 Economic Welfare

Reindeer management as a livelihood is based on the utilization of natural resources whose state is indicated by the land use pattern and the ecological integrity. How do these factors impact the socioeconomic circumstances in the reindeer herding society (impact component in the DPSIR approach; EEA 1999)? Based on the economic welfare indicator set (Fig 16.4), the development of economic conditions and welfare during the last few decades and the current situation of the reindeer management system were explored using statistical and literature data, interview results, and the input of expert consultants.

Of course, the personal economical situation of individual reindeer herders is highly dependent on the balance between costs and income from reindeer herding (Fig 16.7). The data from 1987 to 2001 reveal relatively similarly developing costs and incomes of reindeer owners and numbers of reindeer slaughtered. It is apparent that incomes have always exceeded costs, except during the late 1980s. Interestingly, these were years characterized by exceptionally high numbers of reindeer in Finland (Kumpula 2001; Paliskuntain Yhdistys 2002).

The composition of the various sources of income in reindeer management is manifold and not everything is directly derived from earnings from reindeer herding. In recent years, subsidies, compensation, and additional employment account for a relatively a higher portion of incomes within reindeer management (Nieminen and Kemppainen 1999; Jernsletten and Klokov 2002). The transition from a self-sustaining nomadic livelihood into a modern subsidized market economy can be seen to have been the greatest challenge for reindeer management for a long time. Finland's contemporary sub-

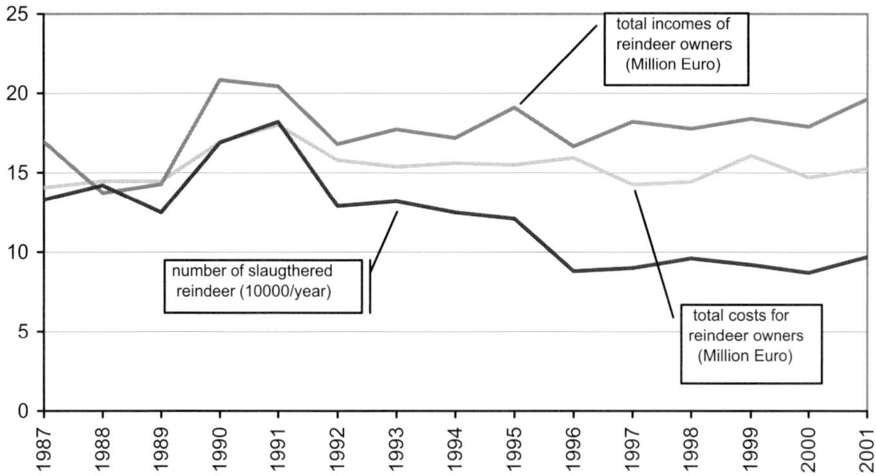

Fig. 16.7. Costs and incomes of reindeer owners and number of reindeer slaughtered in Finnish reindeer management in the years 1987–2001

sidy system, which requires ownership of a minimum number of reindeer in order to obtain support, has led to the consequence that herders with small herds do not receive any financial support. How far these support structures are suitable for establishing a sustainable reindeer management system is one critical question about the future structure of this type of land use.

16.5.4 Social Welfare

The social welfare of the local reindeer herding communities is connected to the economic welfare and thus also to the land use pattern and the state of the ecosystem. This close connection of all the elements within the reindeer management system with the social welfare of the reindeer herders became obvious during the analysis, and was discussed more extensively in the first chapters of this volume.

One main issue hampering local development in "remote" northern regions is the demographical situation (Achenbach 1993). As illustrated in Fig. 16.8, both research areas show typical developments with high emigration rates in combination with low birth rates. These tendencies result in negative population growth, hindering local development and causing generational challenges for reindeer management. This negative trend is evident in the numbers of reindeer owners as well. More and more herders give up their livelihoods and change the business or move away from Lapland. Nevertheless, reindeer management has major importance in keeping rural regions

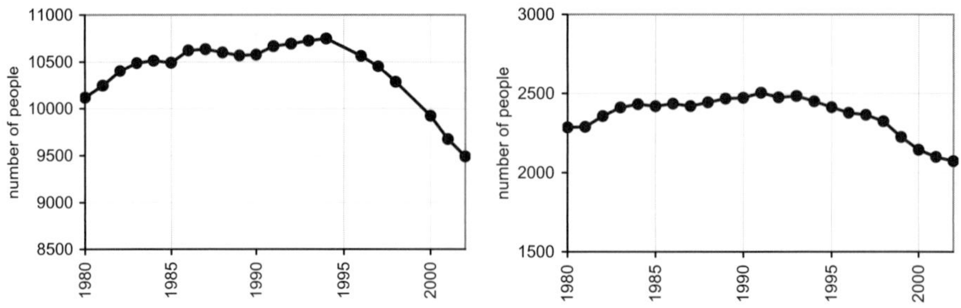

Fig. 16.8. Population developments in the municipalities of Sodankylä and Enontekiö 1980–2002

populated, and it represents an adequate way for exploiting the sparse natural resources of the northern regions and safeguarding the local cultures and traditions (Chaps. 1, 2).

16.6 Future Scenario Derivations

The most applicable (and exciting) simulations are those that consider future developments. One of the main tools for the assessment of assumed future developments in complex systems that often are inherently unpredictable, insufficiently understood, and have high scientific uncertainty, is the designing of scenarios (McCarthy et al. 2001). It is clear that scenarios have a high capacity for giving an idea about possible future developments, but they are not equivalent to prognoses.

The systems analysis in RENMAN was concentrated on three different predetermined scenarios. At first, two extreme developments were chosen as worst-case contrast scenarios marking the positive or negative final states of respectively increasing or decreasing the intensity of reindeer management. These scenarios define the constraints or the bandwidth for all imaginable directions of system developments ("scenario funnel" after Anders et al. 2002). The third conceived scenario, the "business-as-usual scenario", describes the development of the reindeer management system while maintaining today's general conditions in the future, and can be seen as an explorative or trend scenario.

In creating these scenarios no attempt was made to discern the reasons for the respective alterations in the reindeer management system. Rather, the various consequences were analyzed, and this analysis was then used to develop a potentially sustainable scenario as a prime recommendation for actual future Finnish reindeer management. For the simulation of the future land use distributions, the economic welfare, and the social welfare, specific experts were consulted (Burkhard 2004). Applying the indicator set

(Fig. 16.4), different future developments according to the three scenarios were estimated by the individual experts. The ecological-integrity parameters were obtained from the RENMAN field data as shown in Table 16.1. The data from the three differently managed reindeer pastures at the reindeer fence in the Jauristunturit area were applied to correspond to the scenarios. Consequently, the more intensively (during summer) used Finnish area represents an intensification of reindeer management, the less intensively used Norwegian side represents a reduction of reindeer management, and the transition zone can be assigned to "business-as-usual".

The results of the simulations are expressed in "amoeba" diagrams (after Ten Brink et al. 1991), which facilitate the illustration of complex interactions. The land use and social and economic welfare simulations are scaled to a ref-

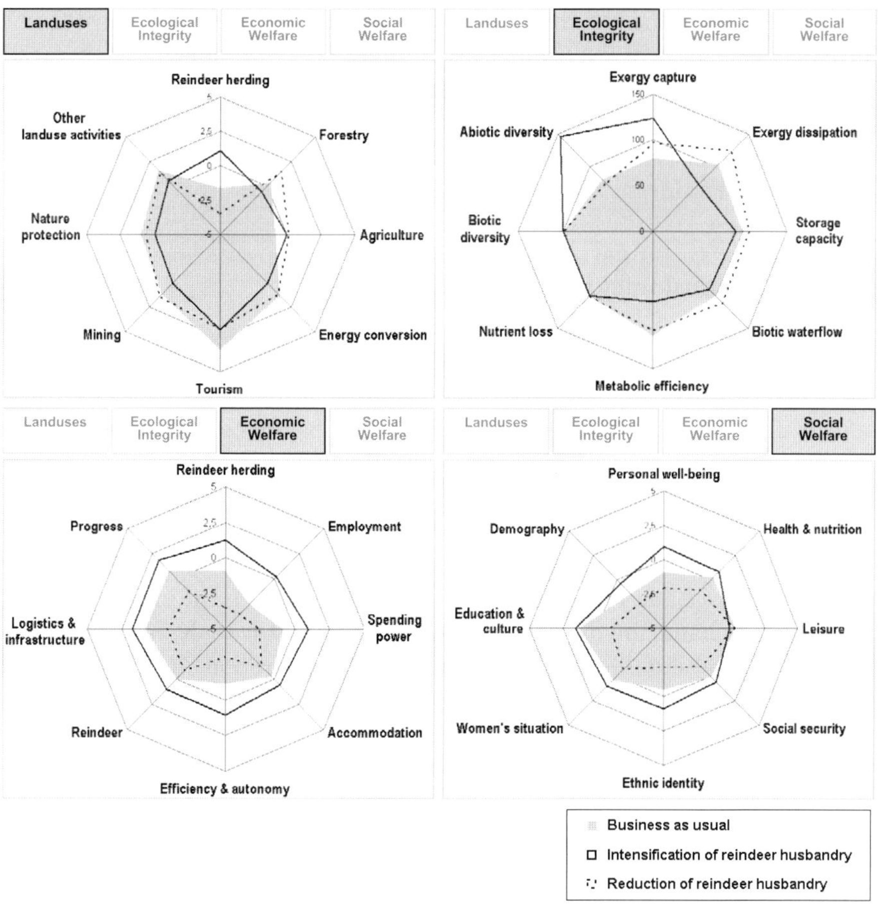

Fig. 16.9. Results of the future scenario simulations for the Finnish reindeer management system

erence value of 0 that denotes no change compared to today, while –5 denotes the greatest conceivable decrease or worsening of the respective indicator and +5 the greatest conceivable increase or improvement. The ecological-integrity simulation is represented in % deviations from the 100 % reference values calculated from the mean values of all three simulations. Apart from potential subjective perceptions in the experts' answers, the scenario derivations presented in Fig. 16.9 gives an illustrative outlook on the simulated future developments.

16.6.1 Land Use Pattern

The land use patterns show clear interconnections between the intensities of reindeer herding and the other land use types. In the scenario "reduction of reindeer management", all forms of land use (besides reindeer herding) increase, while in the scenario with more reindeer management, most forms of land use remain at the same level as today. Forestry, showing high variations according to the degree of reindeer herding, seems to be the major competitor. Tourism, which has been greatly increasing in Lapland in recent decades (Tilastokeskus 2002), is still increasing in all three scenarios, at different rates depending on the intensity of reindeer herding.

16.6.2 Ecological Integrity

Regarding the simulations of ecological integrity, an intensification of reindeer herding with higher numbers of animals leads to an increase in the effects of reindeer grazing and trampling on ecosystems. The most obvious simulated effects concerned abiotic diversity, the exergy capture, and the exergy dissipation (Fig. 16.9).

By eating parts of plants, reindeer diminish the standing stock (plant biomass) but under certain circumstances can increase plant productivity (Henry and Svoboda 1994; Olofsson et al. 2001) and thus, the exergy capture. On the other hand, the diminished plant biomass leads to decreased deposition and availability of dead organic matter and alterations in the water and temperature budgets of the soils (Peth and Horn 2004). These factors are important for the soil microbial activity (Bölter 2004; Bölter and Möller 2004) and thus soil respiration, which indicates exergy dissipation. Moreover, the altered availability of organic matter becomes apparent in the distribution of the soil humus layer, which was generally reduced and more heterogeneous on the more intensively grazed area (Zink 2003). By trampling, reindeer also promote the heterogeneity of the humus layer. This heterogeneity was applied to indicate the abiotic diversity. The storage capacity, the biotic water flow, and the metabolic efficiency decrease in the scenario with more reindeer management. Not any or only very slight differences were proven concerning the biotic diversity (data from T. Kumpula) and the nutrient loss (WASMOD simulation; Burkhard 2004).

Summarizing, it is apparent that the differences among the three sites in Jauristunturit used for the simulation of the ecological effects of the three reindeer herding intensity scenarios have to be regarded as rather modest. Taking into account that the three sites were selected as representing extreme conditions rather than the surrounding environment (Burkhard 2004), the influences within the whole area are probably even lower. The location near the reindeer fence might additionally confound the representativeness because reindeer herds often migrate along fences, and thus their density is usually higher close to them (Olofsson et al. 2001).

In the context of ecosystem dynamics, the results can be interpreted in such a way that by grazing and trampling, the reindeer keep the system permanently at an earlier stage of succession in the more intensively used summer pasture area. Less intensively used winter pasture systems, like the Norwegian pasture area adjoining Jauristunturit, are therefore closer to a climax stage. Hence, the three sites show the typical behavior of herbivory-mediated ecosystems. Additionally, it has to be mentioned that, based on the data and information available, only the status quo of the ecological conditions was captured. However, long-term effects, caused by alterations within the different systems on the soil-developing processes, are conceivable. Due to the slow development and the ostensibly high fragility of these subarctic ecosystems, further attention should be paid to these issues.

16.6.3 Economic Welfare

The close connection between the economic welfare of the reindeer management system and its intensity (as simulated by the three selected scenarios) is reflected by the supposed future developments of the individual economical features (Fig. 16.9).

A reduction of reindeer herding would have major effects, resulting in reduced employment, decreased spending power of the population, and weakened efficiency and autonomy of reindeer management. The same happens in the "business-as-usual" scenario, which indicates a strong need for a changed future management system. Although both the reindeer herders' welfare and the condition and production of reindeer seem to be in a good state at the moment, neither would enjoy such benefits in the future without change. Slightly positive developments were modeled for progress, logistics, and infrastructure in the "business-as-usual" scenario as well as in the scenario with more reindeer herding. Both components may rather be connected to the general regional development than directly to reindeer management. The overall picture implicates that a reduction of reindeer herding results in a general worsening of the economic welfare whereas intensification leads to an improvement in almost all components. The economic welfare feature that is growing least is employment, but more reindeer herding has the potential to stop that negative development.

16.6.4 Social Welfare

Concerning social welfare, an image comparable to the economic welfare picture emerges. A reduction of reindeer herding leads to a worsening of the situation whereas more reindeer herding causes an improvement in almost all indicator values. The demographic development, which is closely linked to the employment situation, is negative in all scenarios, but with different amplitudes. People would leave the region even if almost all social-welfare related components were but little affected – and although education and culture, the situation for women, social security, and health and nutrition show positive developments in two of the three scenarios.

Leisure might be reduced due to the higher labor input needed if reindeer management is intensified. In general, health and nutrition, leisure, and social security are at a relatively high level throughout Finland today (Militz 2002; STAKES Tieto 2002). The values related to the personal well-being and the ethnic identity of the reindeer herders decrease in the "business-as-usual" scenario as well as in that of reduction of reindeer herding. Hence, new strategies are needed for sustaining the reindeer herding culture.

16.6.5 Potentially Sustainable Scenario

From taking into consideration all results from the systems analysis, the RENMAN research activities, and the experiences gathered during the project, a potentially sustainable scenario was developed as a recommendation for future planning in the Finnish reindeer management area. Again, the formulated indicator set was applied, and the results are presented in amoeba diagrams (Fig. 16.10). The scale in the diagrams ranges from –5 for the greatest decrease or worsening of the respective indicator parameter, to +5 for the largest increase or improvement. If maintenance of the current state seems to be appropriate to support sustainable reindeer management, the indicator value is 0. The diagrams are designed to contain certain bandwidths illustrating the scope of actions recommended. The object was to make realistic estimations of the future possibilities and to define accessible targets instead of creating illusions. However, even if subjective perceptions have influenced the definition of the individual components of sustainability, this scenario clearly shows the key problems of reindeer management and has thus the potential for supporting future decisions.

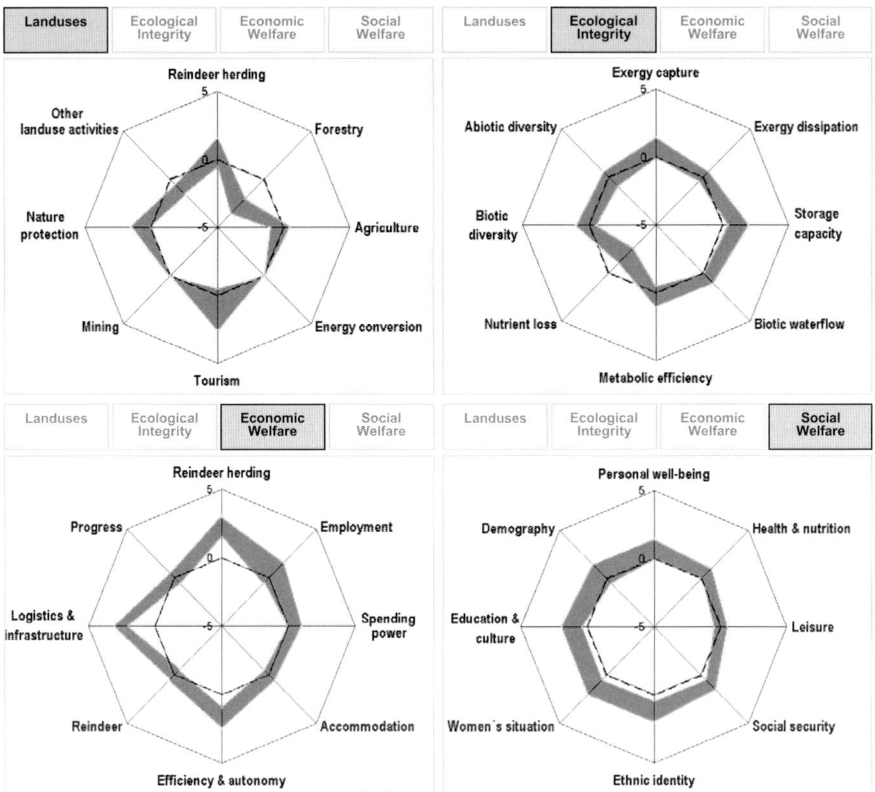

Fig. 16.10. Potentially sustainable scenario for the Finnish reindeer management system

The competition among the different forms of land use is a key issue threatening sustainable reindeer management. Intensive forestry is especially important in that it causes reductions in the quality and quantity of important reindeer pasture areas. Therefore, in the potentially sustainable scenario forestry is explicitly diminished (Fig. 16.10). Energy conversion, mining, and agriculture do not seem to be appreciably reducible, but an increase or the establishment of new areas should be avoided. Synergetic effects are conceivable between reindeer herding, tourism, and nature protection, at least as long as they cooperate (Kemppainen et al. 2003). Therefore, a moderate increase of these land use forms is acceptable. To sustain reindeer management, the herding area has to be kept on at least the same level as today, and other land use activities have to be reduced if they are counteractive.

Crucial factors relating to ecological integrity are the prevention of nutrient losses due to erosion or nutrient discharge and the maintenance of the storage capacity. This demands herding strategies that include prevention of irreversible destruction of the natural systems on which reindeer are dependent. Special attention should be paid to areas where damage to the vegetation cover and upper organic layers of the soils are visible (Chap. 12). From the point of view of reindeer management, an increase of exergy capture via increased biomass production as a result of the grazer-induced conversion of tundra lichen heath areas into grasslands, must be seen as improving the functioning of the system. Nevertheless, the importance of intact lichen pastures as an essential reindeer winter forage resource has to be kept in mind. The development of appropriate pasture management strategies that take into account the importance of both suitable winter and summer food resources is required.

The economic welfare of reindeer herding communities should be promoted by an improvement of efficiency and autonomy, logistics and infrastructure, and reindeer herding economy in general. Improved and direct marketing of reindeer products; more efficient division of labor, equipment, and resources; a decentralization of decision making; and modifications in the current subsidy and support systems are all needed urgently. An improvement of the employment situation within reindeer management, in related businesses, and in the whole region would help sustain livelihoods and the reindeer-management culture. Today's satisfactory conditions of the reindeer herders' accommodations, spending power, and the reindeer themselves have to be safeguarded in the future as well.

The social welfare of reindeer herders as Finnish citizens is on a relatively high level today because Finland's social welfare and healthcare system is one of the best in Europe (Militz 2002). The maintenance of the current state or a general improvement regarding all involved features should be a goal. In any event, the demographical development, the personal well-being, the ethnic identity, and the situation for women have to be considered as sensitive subjects.

16.7 Conclusion

The main focus of our systems analysis was the integration of socioeconomic and ecological factors and data to obtain a holistic assessment of the complex system of reindeer management. The results have confirmed that contemporary reindeer management is in urgent need of future-oriented management and development plans. Due to the complexity of the reindeer management system and its special social, economic, and ecological values, the concept of integrated, sustainable development appears capable of acting as a guide for future decision making. An appropriate definition of sustainability in reindeer management is given in Box 16.1.

According to this definition of sustainability in reindeer management, the functioning of the social, economic and ecologic systems is prerequisite for

Box 16.1. Definition of sustainable reindeer management as used for this presentation

Sustainability in reindeer management means preservation of those structures and processes that are essential prerequisites for the ecological, social, and economical function of the whole system of reindeer management in order to satisfy the needs of present and future generations. Hence, future policy and decision making has to take into account scientific and societal insights and include participation and consensus reaching among the stakeholders.

safeguarding the continuation of reindeer management as a form of livelihood. Hence, no simple solution will be available if all the different factors and interests are considered. For example, an expansion of reindeer herding with more animals and increased herding areas might be a conceivable option for consolidating the reindeer management economy, but conflicts with the other forms of land use as well as ecological and social impacts would have to be expected. The results available from RENMAN revealed that most of the problems in today's reindeer management are connected to the challenging transition from a nomadic subsistence economy to a contemporary subsidized market economy. The loss of pasture areas causes serious additional problems. Although significant ecological problems were not evident in our study, at certain places the effects of reindeer herding were visible and potential long-term effects must be taken into account.

The applied integrative-systems analytical approach proved to be a suitable method to assess the system of reindeer management. The derivation and application of appropriate indicators helped to evaluate the current state, past and future developments, and was applied to define future targets. The chosen models and methods were effectively integrated into the systems analysis. Methodological improvements are needed regarding: (1) the transferability of results into different spatial scales; (2) the development of more dynamic models that enable the integration of altering input parameters; and (3) more standardized methods for acquisition of data concerning socioeconomic as well as nature-and-science questions.

Based on results from the RENMAN analysis, some key recommendations have been formulated and can be applied to future-oriented development:

1. Reindeer management has to be seen as a traditional livelihood in rural subarctic areas. Thus, it should be treated as such instead of being intensified and mechanized further.
2. Strong potential lies in the production and direct marketing of high quality natural products, as well as in linkage to other economies such as tourism, handicrafts, or small-scale trade.
3. Due to the special character of reindeer management, local and national decision making has to take into account social, economic, and ecological

issues equally. Participation and information of local stakeholders is a good basis for reaching consensus.

4. Systems analysis, such as we have undertaken, illustrates the enormous complexity, adaptability, and potential of reindeer management, and hopefully such methods can be used to sustain it for the generations to come.

References

Achenbach H (ed) (1993) Suomi/Finnland. Naturpotentiale und Lebensräume im hohen Norden Europas. Kieler Arbeitspapiere zur Landeskunde und Raumplanung 27, Kiel

Aikio P (1987) Reindeer herding in Norden. In: Uuno V, Tietze W (eds) Norden – Man and environment. Gebrüder Borntraeger, Berlin, pp 332–337

Alcamo J, Bennett EM (2003) Ecosystems and human well-being. Millennium Ecosystem Assessment (MA). Island Press, Washington, DC

Anders K, Prochnow A, Schlauderer R, Wieglab G (2002) Die Szenariomethode: Grundlagen und Anwendung auf das Management von Offenlandstandorten ehemaliger Truppenübungsplätze. In: Wallschläger D, Mrzljak J, Wieglab G (eds) Offenland und Sukzession/Openland and Successsion. Aktuelle Reihe der BTU Cottbus 8, pp 95–102

Baumann R (2001) Indikation der Selbstorganisationsfähigkeit terrestrischer Ökosysteme. Dissertation. University of Kiel, Kiel

Barkmann J, Baumann R, Meyer U, Müller F, Windhorst W (2001) Ökologische Integrität: Risikovorsorge im nachhaltigen Landschaftsmanagement. GAIA 10(2):97–108

Bliss LC, Heal OW, Moore JJ (eds.) (1981) Tundra ecosystems: a comparative analysis. International biological programme 25. Cambridge Univ Press, Cambridge

Bölter M (2004) Soil: An extreme habitat for microorganisms? Pedosphere 14:137–144

Bölter M, Möller R (2004) Changing microbial ecology with changing management. Mitt Kieler Polarforsch 20:17–24

Burkhard B (2004) Ecological assessment of the reindeer husbandry system in northern Finland. EcoSys [Suppl] 43, 117 pp

Burkhard B, Müller F (2004) Integrative analysis and modelling in the RENMAN project. Mitt Kieler Polarforsch 20:46–48

Burkhard B, Kumpula T, Müller F (2003) Landscape assessment for a sustainable reindeer husbandry in Northern Scandinavia. EcoSys 10:116–124

Costanza R, D´Arge R, de Groot RS, Farber S, Grasso M, Hannon B, Limburg K, Naeem S, O´Neill RV, Paruelo J, Raskin RG, Sutton P, van den Belt M (1997) The value of world's ecosystem services and natural capital. Nature 387:253–260

Cyffka B, Zierdt M, Will N, Rauchhaus U (1999) Anthropogen bedingte Sukzessionsstadien des borealen Nadelwalds in Finnisch- und Russisch-Lappland und die Möglichkeiten ihrer nachhaltigen Nutzung. Schriftenreihe des Arbeitskreises vol 13, Norden, Bremen, pp 317–332

De Groot RS (1992) Functions of nature: evaluation of nature in environmental planning, management and decision making. Groningen

Dierßen K (1996) Vegetation Nordeuropas. Ulmer, Stuttgart

EEA (European Environment Agency) (1999) Information for improving Europe's environment. EEA, Copenhagen

Filppa J (2003) Reindeer husbandry in Finland. Rangifer Rep 7, pp 69

Forbes B (ed) (2004) RENMAN – final report. The challenges of modernity for reindeer management: integration and sustainable development in Europe's subarctic and boreal regions. Arctic Center Reports 41. Rovaniemi

Frei B (2002) Challenging the reindeer tragedy: modernisation and rationalisation in Finnish reindeer husbandry. Diploma thesis at the Geographical Institute of the University of Zurich

Fürntrapp-Klopp EF (1995) Quality of life – from a common people's point of view. PappyRossa-Verlag, Köln

Henry GHR, Svoboda J (1994) Comparison of grazed and non-grazed high-arctic sedge meadows. In: Svoboda J (ed) Ecology of a polar oasis. Captus Press, Toronto

Hukkinen J, Jääskö O, Laakso A, Müller-Wille L, Raitio K (2002) Poromiet puhuvat. Helsinki University of Technology

Jensen ME, Bourgeron PS (eds) (2001) A guidebook for integrated ecological assessments. Springer, New York

Jernsletten JL, Klokov K (2002) Sustainable reindeer husbandry. Centre for Saami Studies, University of Tromsø

Kemppainen J, Kettunen J, Nieminen M (2003) Economic research of the Finnish reindeer husbandry 2003–2007. Rangifer Rep 7, Tromsø

Kumpula J (2001) Productivity of the semidomesticated reindeer (*Rangifer t. tarandus* L.) stock and carrying capacity of pastures in Finland during 1960–1990s. Oulu Univ Press; Oulu

Kumpula T (2003) Does high resolution satellites bring new aspects to the pasture research? Rangifer Rep 7:26

Laakso AM (2002) The anomalies of contemporary reindeer herding management and supporting reindeer research: the need for a new paradigm. Pro gradu – tutkielma Sosiologia. University of Lapland, Rovaniemi

Löffler J (1999) Neuere Ergebnisse geoökologischer Kartierungen in Nord-Norwegen. Norden 13:243–265

McCarthy JJ, Canziani OF, Leary NA, Dokken DJ, White KS (2001) Climate Change 2001: impacts, adaptation and vulnerability – contribution of working group II to the Third Assessment Report of IPCC. Arendal

McMahon SK (2002) The development of quality of life indicators – a case study from the City of Bristol, UK. Ecol Indicat 2:177–185

Militz E (2002) Finnland. Perthes Länderprofile. Klett-Perthes Verlag, Stuttgart_Gotha

Müller F (2005) Indicating ecosystem and landscape organisation. Ecol Indicat (in press)

Nieminen M, Kemppainen J (1999) Economical importance of Finnish reindeer husbandry. Rangifer Rep 3:49–50

Oksanen L, Virtanen R (eds) (1995) Topographical, altitudinal and regional patterns in continental and suboceanic heath vegetation of northern Fennoscandia. Acta Botan Fenn 153:1–80

Olofsson J, Kitti H, Rautiainen P, Stark S, Oksanen L (2001) Effects of summer grazing by reindeer on composition of vegetation, productivity and nitrogen cycling. Ecography 24:13–24

Paliskuntain Yhdistys (2002) Information about reindeer husbandry 2002. Rovaniemi

Peth S, Horn R (2004) Impacts of Grazing on Soil Physical Properties as a Result of Reindeer Herding in Northern Scandinavia. Mitt Kieler Polarforsch 20:25–28

Reiche EW (1996) WASMOD. Ein Modellsystem zur gebietsbezogenen Simulation von Wasser- und Stoffflüssen. Darstellung des aktuellen Entwicklungsstandes. EcoSys 4:143–163

RENMAN (2004) RENMAN project homepage : http://www.ulapland.fi/home/renman/

Riess F (1993) Untersuchungen zur Nitratauswaschung nach mineralischer und organischer Düngung von Acker- und Grünland mittels der Saugkerzenmethode. Dissertation TU München-Weihenstephan

Ruppert K, Schaffer F (1969) Zur Konzeption der Sozialgeographie. Geogr Rundsch 21:52–55

Scurlock JMO, Asner GP, Gower ST (2001) Worldwide Historical Estimates of Leaf Area Index, 1932–2000. Oak Ridge National Laboratory, Oak Ridge, Tennessee

STAKES Tieto (2002) Sosiaali- ja terveydenhuollon tilastollinen vousikirja 2002, Helsinki

Stark S (2002) Reindeer grazing and soil nutrient cycling in Boreal and Tundra ecosystems. Oulu

Ten Brink B, Hosper S, Colijn F(1991) A quantitative method for description and assessment of ecosystems: the AMOEBA-approach. Mar Poll Bull 23:265–270

Tilastokeskus (2002) Suomen tilastollinen vuosikirja. Statistics Finland, Helsinki

UNCED (United Nations Conference for Environment and Development) (1992) Agenda 21. United Nations, New York

Wielgolaski FE (1997) Polar and alpine tundra. Ecosystems of the world, vol 3, Kluwer, Amsterdam

Wiggering H, Müller F (eds) (2003) Umweltziele und Indikatoren. Springer, Berlin Heidelberg New York

World Commission on Environment and Development (1987) Our common future (The Brundtland Report). World Commission on Environment and Development, Oxford

Zink A (2003) Verbreitung und physikalische Bodenfunktionen von Tundraböden in Nordskandinavien unter dem Einfluss der Rentierwirtschaft. Diploma thesis in Geography at the Christian-Albrechts-University Kiel

17 Synthesis: Environmental and Sociopolitical Conditions for Modern Reindeer Management in Europe's North

L. Müller-Wille, J. Hukkinen, F. Müller, M. Bölter,
and B.C. Forbes

17.1 Introduction: Linking Practical and Scientific Knowledge

The intricate relationship between reindeer (*Rangifer t. tarandus* L.) and humans in the boreal, subarctic and arctic regions has attracted the keen attention of modern sciences since the early 20th century. Since the 1950s, various scientific traditions have shaped the emergence of separate foci dealing with the natural conditions of the animal and its environment on the one hand and the cultural and socioeconomic dimensions of human–reindeer relations on the other. This divergence of biological and anthropological interpretations of the circumstances of reindeer herding in specific ecosystems and cultural settings has also had an impact on management policies and practices. The measures had tended to rely mainly on results of research conducted by natural scientists. Such studies had focused mainly on natural status of pastures and reindeer, losses to predators, carrying capacity, "overgrazing", and meat production. This type of research had to a large degree neglected the human element in the equation among people, livelihood, animals, and the socioeconomic and cultural significance. Until recently, little emphasis had been given to the intricate knowledge that the practitioners closest to the reindeer have of the animal, the environment, and the ensuing utilization by humans (Aikio 1978).

At the beginning of the 21st century, in the circumpolar North, *Rangifer* populations and their users, herders, and hunters alike are under extensive sociopolitical and economic pressures that have altered and will continue to alter their relationship with reindeer and their own position within northern nation states and their economies (Chap. 2; Forbes and Kofinas 2000; Anderson and Nuttall 2004). These encompassing transformations require extensive

Ecological Studies, Vol. 184
B.C. Forbes, M. Bölter, L. Müller-Wille, J. Hukkinen,
F. Müller, N. Gunslay, Y. Konstatinov (Eds.)
Reindeer Management in Northernmost Europe

and detailed attention by major research efforts joining practitioners and social and natural scientists (Chap. 3; Turi 2000).

The RENMAN project ("The Challenges of Modernity for Reindeer Management – Integration and Sustainable Development in Europe's Subarctic and Boreal Regions") was designed to overcome the separation of methodologies and approaches in research on reindeer herding and create an interdisciplinary context linking natural and social scientists – and the scientists with the practitioners, i.e., reindeer herders – under the auspices of this project funded by the European Union (Forbes 2004). The project initiators felt strongly that these integrating steps had to be taken to understand the emergent processes and the impact of rapid cultural and socioeconomic changes, as well as climatic change, on peoples and the environment in the circumpolar north. Such steps were clearly needed. Peoples relying even if only partially on herding reindeer for their livelihood are currently under continuous internal and external pressures to sustain the development and integrity of their cultural and socioeconomic practices (Jernsletten and Klokov 2002; Anderson and Nuttall 2004).

The objectives of the project were to introduce models and tools of participatory research that had already been successfully implemented in anthropological, educational, and medical research (Hall 1979; Hiebert and Swan 1999). This approach was feasible and manageable and seen as an important baseline for the participation of various actors in these studies and the ensuing integration of different sets of knowledge in and around reindeer herding and management (Chap. 3). The research thrust and problems pursued by the individual workpackages (Chap. 1) were aimed at fostering the integrated sustainable use of reindeer and related resources in northernmost Europe (Chaps. 4–14). Furthermore, recommendations and scenarios were to be developed and formulated for future reindeer herding practices and management in coexistence and conjunction with other expanding and competing land uses in the same area (Chaps. 3, 15, 16).

17.2 Issues of Sustainable Reindeer Herding and Management

The studies conducted under RENMAN (Chaps. 2–16; Forbes 2004) identified a number of issues and concerns that both practitioners and scientists found important without establishing priorities. These topics reoccurred during the research process, analysis, and discussions of results in which all project participants took part (February 2001 to January 2004). One cannot say there was always full agreement on all points. However, the topics discussed here represent the range of problems that reindeer herders and scientists detected in the contact and conflict among diverse land uses currently practiced in the reindeer herding areas throughout northernmost Europe, particularly in northern Finland.

17.2.1 Competing Interests: Space, Mobility, and Land Use Practices

Since at least the 15th or 16th century intensive reindeer herding by Sámi and reindeer keeping by Finns as elements of agriculture (Chap. 2) have been exposed to the increasing expansion of other land uses in their territories. Over the years, these introduced land uses have been supported by public, commercial, and industrial interests, attracted by and founded on diverse resources in the herding areas. In Finland, the "paliskunta" system of reindeer herding cooperatives, established by law in 1898, established a rigorous legal, administrative, and territorial system. Since its inception, this system has regulated herding practices in space, mobility, herd size, and pasture resources (Palkeinen 1913; Aikio 1978; Magga 2003).

Throughout the 20th century economic and industrial forestry, mainly practiced by the modern state in these regions, with variously applied felling techniques such as clear and strip cutting, intensive selective thinning, deep ploughing, spraying of "undesired" species, and construction of logging roads and loading areas (infrastructure), is the major competitor of reindeer herding for space and resources. The expansion of forestry has resulted in the removal of more and more area from reindeer movements and grazing. At the same time, legislation aimed at protection from predators (Beach 1997) has limited herders' options in the realm of control at a time when the number of predators appears to be on the rise (Turi 2000). In addition, the loss of vegetation, e.g., arboreal and ground lichens and thus the decrease of productive grazing lands, has triggered supplemental feeding of reindeer as a consequence. These changes have occurred throughout the economic forestry area as well as on open fell, where soil covers vanish due to increased trampling with consequences for soil structure and soil biology (Chaps. 11–13). Lichen cover and status are regarded as indicators for pasture conditions (Chaps. 7, 9). For the detection of such conditions, modern high-resolution satellite imagery is a useful tool to monitor changes and variations. However, such analyses and interpretations need to be complemented and tested by constant and guided investigations, "ground-truthing" in the field, and comparisons with existing low-resolution imagery, all conducted in close cooperation between practitioners and scientists (Chap. 8).

Similarly, hydroelectric power developments have resulted in the complete loss of certain grazing areas in northernmost Europe. For example, the Lappi Reindeer Herding Cooperative in northern Finland lost 11 % of its grazing lands to flooding in support of the Lokka-Porttipahta hydroelectric complex and its reservoir constructed in the 1960s. The consequence was the reduction of the cooperative's reindeer quota by 2,500 head and the necessity to completely realign its herding strategies within the reduced area (Magga 2003).

Open-pit mining for specific minerals (gold, nickel, or other ore) has caused some problems in specific, but rather restricted and limited areas. However, exploration, stakes, claims and exploitation (particularly of gold) are on the rise and spatial conflicts are anticipated in some reindeer cooperatives in northern Finland.

Tourism and recreation have emerged as the most rapidly expanding land use practices in the reindeer herding areas of Europe's North. Today, activities continue practically throughout all seasons with peaks in late summer and early fall (e.g., sight-seeing, cabin life, hiking, sports hunting) and mid- and late winter (skiing and snowmobiling). These leisure pursuits require an extensive infrastructure and complex facilities for lodging, recreation, entertainment, and mobility (roads, trails, and slope development). Both the fixed infrastructure and the movement of large numbers of visitors along roads and in so-called "wilderness" areas and parks interfere and compete with herding practices and management. Particularly problematic times are the calving period and the round-up season, which are highly sensitive and crucial for the success of reindeer herding (Helle and Särkelä 1993; Vistnes and Nellemann 2001; Jääskö 2003; Magga 2003).

In northernmost Europe, during the last few decades the general infrastructure of settled areas and transport corridors including logging roads have expanded due to a higher concentration of population and services in the few centers with an urban character. Another factor is the rapidly increasing transport and trade links between Central Europe and the Kola Peninsula in Northwest Russia. Traffic corridors have become wider spatially and, in turn, attract reindeer in higher frequency, which has resulted in a higher number of collisions inflicting losses to reindeer owners.

Since the 1960s, the rigorously territorial *paliskunta* system in Finland and its revised Reindeer Law of 1990 have led to the construction of more and more "reindeer fences" along practically all cooperatives' boundaries and the international borders with neighboring countries. These fences have limited movement and required adaptations in reindeer management and rotation strategies within the cooperatives' territory (Magga 2003). Furthermore, fences continue to have an influence on the intensity and density of grazing within the cooperatives (Chaps. 7–9, 11–13).

17.2.2 Potential Impact of Climatic Change and Radiation

Climatic change has had noticeable effects on the environmental conditions of the vegetation cover, concomitant with other physical parameters such as snow depth and animals' constitution and behavior. Previous studies on climatic fluctuations (Vibe 1967) and the research conducted under RENMAN

(Chaps. 7, 11–13, 15) have established that, among other elements, recent winter seasons had less precipitation, shorter freezing periods, less snow cover and duration, longer periods of soil activity, and lower accumulation of litter and humus.

Since climatic change has also caused longer vegetation periods with higher summer temperatures, there is the possibility of higher plant productivity in grazed species and even perhaps among non-preferred forage species (Chap. 9; Turi 2000). On the other hand, in the autumn, due to higher soil temperatures and short duration of sunshine, a more rapid process of humus mineralization might be supported. Furthermore, dry periods favor increases in barren regions by erosion of the thin organic soil horizon (Chaps. 8, 11–13). Moreover, population variations in individual reindeer herds might be explained by climatic fluctuations related to the North Atlantic Oscillation and the Arctic Oscillation (Chap. 15).

Ultraviolet radiation seems to foster the production of phenolic compounds in plants without changes in nitrogen, lignin, and cellulose. Vegetation thus could have a lower level of digestibility as potential fodder plants. However, further studies are needed to investigate the importance of the observed UV effects on plants for reindeer nutrition (Chap. 10).

17.2.3 Environmental Conditions and Herding Strategies

A number of environmental conditions influence strategies developed in reindeer herding and management. The RENMAN studies have identified real and potential circumstances, which are a cause for concern to reindeer herders as they react to changes in the physical environment that might have an impact on the vegetation cover and thus pasture conditions (Chaps. 7, 9–14).

Besides arboreal lichens, *Salix* vegetation is also influenced by grazing, but not by trampling by reindeer. The effects of grazing and trampling in wetter habitats seem to be reversible within rather short periods of time, thus this impact does not seem to be an environmental risk in summer pastures (Chap. 9). Allowing access to winter pastures during the summer, even when the animals move mainly on trails, leads to the reduction of ground lichens, soil degradation, and the destruction of the humus layer via trampling (Chaps. 8–13). This process leads to the reduction of nutrients as well as soil erosion by wind and water, which in turn causes loss of natural capital for the reindeer herds and herders (Chaps. 11, 13, 15). Trampling also induces soil compaction and changes bulk density and aeration as well as hydrological parameters (Chap. 11). These emerging conditions have consequences for nutrient cycling, soil stability, heat flow, and water budgets (Chaps. 12–14). Therefore, these processes have an impact on the productivity of the pasture vegetation. The more serious effects are locally confined, although lichen cover has been reduced over a significant portion of northernmost Finland and also in Finnmark, Norway.

Grazing and trampling by reindeer during snow-free periods can reduce the thickness of the lichen cover locally (Chaps. 8, 9). This results in the deterioration of the lichen cover, the O horizon, and a compaction of the humus layers (Chap. 12). Anthropogenic trampling and traffic by all-terrain vehicles also compact the vegetation and soil cover locally (Chaps. 8, 9). Such impact effects chemical changes in nutrient levels available in plants relative to areas with less intensive grazing and trampling. The levels of degradation in the humus cover were identified by a system of stages based on specific criteria (Chap. 12).

The microbial turnover was found to be quite different as a consequence of changing levels of grazing intensity. This effect was visible in the three subareas of the research sites in the northern tiers of the Näkkälä Reindeer Herding Cooperative on the Finnish side of the Finnish–Norwegian border (Chaps. 11–13). Adaptations in herd management such as rotation (Chaps. 5, 7) were seen to have resulted in vegetation changes that were noticeable in the soil biota because there is a strong relation between microbial activities, organic soil matters, and humus content. This condition was also noticed in the microbiological parameters, e.g., soil respiration. These results, however, partly contradict those from other studies, probably because of local properties as discussed in detail in Chap. 13. Analyses of soils, surface water, and feces concluded that there is no risk of pathogens under normal herding conditions. Risks could occur under crowded conditions of reindeer, for example, at sites where supplemental feeding is applied during winter (Chap. 14).

Since the 1960s, but more widespread since the 1980s, supplemental feeding of hay, birch leaves, vegetable-based pellets (in Finnish "poronherkku", reindeer sweets) and also lichen, if available, has been an important measure in managing the fluctuating size of reindeer populations within Finland's herding districts. Calving success is also an important parameter in the dynamics of reindeer populations. The application of veterinarian treatment is now a regular feature in herd management and has resulted in healthier herds (Chap. 15).

It seems, taking the Näkkälä sites as an example, that environmental and structural changes do not influence the ecosystem's functions in an alarming manner. Nevertheless, if the loss of lichen cover continues, more consequences can be expected for herding practices. The herders have realized such a possibility and have excluded reindeer from some of the disturbed pastures to reverse the trend of lichen depletion by modifying their herding strategies (Chap. 16). The resulting increase in lichen cover was recently detected during a reanalysis of low-resolution satellite imagery from this region (Kumpula et al. 2004).

17.3 Herders and Reindeer: The Human Dynamics of Herding and Management

The modernity of reindeer management is expressed through the intricate, dynamic, and adapting relationships by which herders and herded reindeer are bound, functioning in the boreal and subarctic ecosystems in northernmost Europe. These relationships are shaped not only by the prevailing and changing environmental conditions discussed above, but also in particular by cultural and philosophical premises, approaches to knowledge and science as well as by specific socioeconomic, political, and legal dimensions and circumstances. Several studies conducted under RENMAN paid specific attention to the various aspects of the human dynamics of herding and management (Chaps. 2–7). In these studies the detailed assessment of the dynamic interrelationships was attained through equal partnerships with herders in order to integrate their points of view in a holistic approach to reindeer management.

17.3.1 Divergent Cultural and Philosophical Premises

Reindeer herding is practiced by different peoples throughout the circumpolar North (Jernsletten and Klokov 2002). In Europe's northernmost regions reindeer are herded and utilized by Sámi and Finns in northern Finland (Chaps. 2–5, 7), by Sámi in Norway (Chap. 4), by Sámi and small groups of Finnish speakers in Sweden (Chap. 5), and by Sámi and Komi on the Kola Peninsula in Northwest Russia (Chap. 6). Considerable variations in practices exist among these peoples and throughout the area from east to west and south to north. An important aspect of these variations is the maintenance of cultural identity whose attributes, particularly for the Sámi, include reindeer herding as a core cultural value.

The social organization of each of these cultures faces the pressures of the expanding and encompassing dominant societies of the surrounding nation states (Finland, Norway, Russia, and Sweden) and the forces of transnational institutions such as the European Union through the processes of modernization and globalization. It is noticeable that such pressures and forces can have an integrating influence on the local and regional level. However, disagreements and conflicts continue to occur over how reindeer herding should and would emerge into a modern management system. Divergent views exist with respect to the basic premise of how the "environment" (nature) is perceived and who owns the "land" and has rights to its "use".

For the Sámi there is no strict separation and difference between "nature" and "culture", the way the distinction is made in the Judaeo-Christian and western European tradition. Thus areas that are designated and separated as "wilderness" and protected under modern conservation policies and laws are, in fact, an integral part of Sámi perception, livelihood and land use pattern – the Sámi cultural landscape in its totality (Länsman 2004). These spaces are not excluded from practices such as reindeer herding, rather they are reserved for herding activities and thus do not fall under the premises of everybody's right to free access to public lands as they are treated under current state laws. These divergent positions have created conflicts among the actors in modern reindeer management.

17.3.2 Knowledge and Science: Bridging Diverse Approaches

Reindeer herders exude a holistic approach to the assessment of the physical environment and the human condition. They do not base their decisions on just one fact or another. Scientists and, in particular, environmental managers tend to concentrate mainly on ecological aspects and often underestimate the economic conditions and social problems of herders with respect to the utilization of available local resources. Furthermore, herders have shown that scientists need to learn first from them as the practitioners closest to the animal and the environment. Herders have expressed the strong view that their evaluation of the herding environment is more detailed and encompassing for management decisions; scientists often neglect or fail to recognize elements that are crucial for the understanding of herding practices (Chap. 4; Jääskö 2003; Magga 2003). Examples are the herders' analysis of the impact of insects and wind directions on herding strategies and the decisive elements that are part of the complex decision-making process to rotate and move reindeer for grazing (Chap. 7).

In today's information society, herders feel that there is an accelerating loss of the practitioners' knowledge and that the position of their knowledge based on experience is eroded by the pervasive application of science-based knowledge. This loss of herders' knowledge is also caused by the growing dominance of introduced technologies and techniques as well as the changes that have occurred in the livelihood and living standards of northern residents and herders generally (Chap. 4). For the herders, constant and close contact with the animals has become less important than in earlier times.

Among herders there clearly is the sense that their knowledge needs to be respected and integrated appropriately into a more inclusive knowledge network that combines both the knowledge and science of practitioners and scientists (Chap. 3; Aikio and Müller-Wille 2005). This integrative approach would allow more realistic and rational solutions for the future of reindeer management (Chap. 3).

17.3.3 Socioeconomic Considerations

In the 1960s, the rapid introduction of technological innovations such as the versatile snowmobile (Pelto 1973) made herders change their techniques and practices from year-round intensive herding to extensive, sporadic herding. Herders spent less time with the animals and moved through the district's area intermittently. Still, regular and systematic scanning of the districts by vehicles, small fixed-winged aircraft, and helicopters with instant telecommunications emerged as strategies with which herders would know the movements, locations, and conditions of their reindeer at all times throughout the seasons. The mechanization of reindeer herding increased the demand for money flow considerably to cover the related expenses for hardware and upkeep. In turn, this resulted in higher slaughter rates at the beginning of the "snowmobile revolution" (Pelto 1973), gradually depleting the resources of some herders. Herders with smaller numbers of reindeer could not compete in this economic spiral and many reduced their herds or even sold all their animals and left herding altogether (Pelto and Müller-Wille 1973).

By the beginning of the 21st century reindeer herding had experienced constant changes that saw the formulation of revised legislation, the institutionalization of veterinarian and hygienic measures, quota systems for districts and individual owners, supplemental feeding, compensation payments for loss to road accidents and predators, marketing procedures, accounting requirements, retirement plans, and still other activities. As a result, a web of administrative and bureaucratic frameworks and procedures was created that brought reindeer herding into the complexities of the modern age (Chaps. 3–6). Herders had to integrate these developments into their regular herding work. In addition, due to a denser infrastructure and higher volume of traffic, more reindeer were lost to road accidents than to natural predators. Furthermore, expenditures for equipment, vehicles, maintenance and fuel continue to increase and are not recoverable anymore by increasing the slaughter rates that have their natural limits. Taken together, these changes in fact carry a monetary loss that needs to be covered by income from other sources such as entrepreneurial activities in fishing, berry gathering, multipurpose tourist activities, and additional wage labor income earned by the herders or their family members.

Today, reindeer herding is under strong pressure to adapt to the free market economy that surrounds it. The producers – reindeer herders and owners – do not establish the price for reindeer meat. Instead, larger independent external companies and even monopolies fix the meat price annually. The price has not been stable and has decreased substantially over the last few years (Chaps. 5-6, 15–16). Thus herders' incomes have generally declined (Chap. 2). Reindeer meat originating in Russia, bought and distributed by Swedish companies, has entered markets in Finland and Sweden and competes heavily with local meat (Chap. 6). Reindeer herding, which since 1995 is also embedded within the European Union, cannot easily be transformed from a "traditional" livelihood to meet the demands and conditions of the modern free market economy. The numbers of reindeer simply cannot be increased at will to achieve a higher margin of profits. In fact, it is understood that individual monetary optimization would have a negative impact on the whole system (Chap. 15). To maintain its status as a separate and distinct livelihood in coexistence with the free market economy, appropriate public subsidies, such as in other economic fields, are appropriate and necessary. Still, statements have been made that a system of subsidies would further the dependence of the herders on the state and public support.

The noted and observed loss of competence and skills in reindeer management is also a consequence of the rapid demographic change in the reindeer herding areas. It seems that the herding profession does not attract or keep young people at the required levels. This situation is due partly to the limited economic prospects in herding, but mainly due to the decline in economic opportunities in northern areas since the late 1990s in general. For example, the population of the Province of Lapland in Finland has decreased since the late 1990s because of changing economic opportunities and political constellations in Europe's North, with attracting centers in southern regions as a strong pull factor. Unless there are changes in this trend, it is foreseeable that northern populations will age considerably in the near future. The sector of reindeer herding is not excluded from this development (Chap. 3).

17.3.4 Political and Legal Aspects

At the beginning of the 21st century, the political and legal dimensions of reindeer herding and management in Europe's North hover around several unresolved issues that shape the decision-making and prospects of herders. The issues are different from country to country because the individual nation state's legal framework defines the context within which the relationship between herders and reindeer functions.

For the Sámi, the continuation of reindeer herding is directly related to aboriginal rights to land and water, sovereignty issues, and the secured use of resources in their homeland under international agreements such as the Indigenous and Tribal Peoples Convention No. 169 issued by the International Labour Organization in 1989. Norway has already signed and ratified the convention. Finland, Russia, and Sweden, however, have not done so. In Finland, the Sámi Parliament, representing all registered Sámi, argues that Sámi land ownership also includes the Parliament's competence and authority to manage the intensive use of public lands through, for example, reindeer herding. Land use would thus be administered under the constitutionally defined political autonomy of the Sámi Parliament (Chaps. 2–3). The Sámi Parliament in Finland stresses that such political restructuring and realignment of responsibilities are needed because reindeer herding, by nature, has a low degree of competitiveness versus other land use types, and needs to be guided and secured through structures of direct local decision-making (Chap. 2).

There is also a prevailing attitude described as "eco-colonialism" that propagates legal conservationist measures in areas whose populations, e.g., reindeer herders, need not be consulted over the exclusion of lands from use by reindeer (Chap. 5). This issue concerns the designation of national parks, nature preserves, and wilderness areas whose territories are legally defined in space. In some areas this development has led to the decrease of available space to reindeer herding and inevitably to the reduction of reindeer numbers in certain districts, often without the direct involvement or participation of the practitioners (Magga 2003).

17.4 Anticipated Issues in Herding and Management Practices

In comparison with past situations that had been described as sustainable, the space available to reindeer herding and grazing has gradually been reduced over time as a consequence of various competing land uses. In northern Finland, this spatial contraction has effectively increased the relative number of reindeer per available square kilometre and has caused crowding of reindeer, although the districts' territorial size and borders have not been altered (Magga 2003). As a reaction to higher reindeer density and intensifying grazing and trampling pressure, the quotas set by the Ministry of Agriculture and Forestry for each reindeer herding cooperative have been decreased. This environmental condition has been termed "overgrazing" with the focus on the depletion of the arboreal and ground lichen cover in the pastures caused by browsing reindeer (Chap. 15).

The ecological as well as sociopolitical concept of "overgrazing" is contentious in the discussion among scientists as well as herders (Chaps. 4, 7, 12, 15). In fact, for herders this concept does not exist and is therefore not understood by them (Jääskö 2003; Magga 2003). Some scientists consider it the correct term describing the assumed overuse of vegetation by reindeer (Chap. 12). The use of "overgrazing" still provokes heated scientific and political debates. In Sweden herders have been blamed for transforming the "mountains into a rocky desert" by their herding practices (Beach 2000). The issue and its solution cannot solely be placed with the herders and their management practices. Rather the responsibility lies also with the other expanding land users to find sustainable solutions for the future, because it is they who have introduced the disturbance to the practiced "grazing peace" (Chaps. 4, 7) that exists in reindeer herding through the rotation of reindeer from area to area and leaving specific areas "fallow", to use an agricultural analogy. The depletion of the lichen cover is not the only problem that exists in reindeer management. The reduction of reindeer herds ostensibly decreases pressures on pastures. However, a slight reduction in animals per unit area will not promote lichen growth nearly as effectively as leaving certain winter pastures "in peace", thus eliminating foraging temporarily and preventing related destructive trampling in summer pastures. At the same time, enforced herd reduction also decreases the sector's economic viability. In this respect, the integration of reindeer management into other economic sectors such as appropriate and amenable types of tourism and recreation, as well as into management of ecosystems and protected areas, seem to be feasible prospects that offer socioeconomic opportunities for the future (Chap. 16).

Reindeer herding is faced with the constant loss of live capital ("capital on the hoof") to predation and road accidents. Herders have taken major efforts to conduct surveys on and pursue preventive measures against predators such as bear, wolf, glutton (wolverine), and eagle, to name the major ones. They have also convinced state authorities to pay compensation for the documented predatory loss of reindeer. Still predation and road accidents are a major risk factor in reindeer management for which a secure protection has yet to be found.

Reindeer herding throughout Europe's North is characterized by high diversities of management concepts and practices under different geographical conditions on the local, regional, and international level (Chaps. 2, 3). These practices function according to specific norms and conventions and offer different opportunities such as concession herding (Chap. 5), state-controlled management (Chap. 6), and participatory decision-making regimes, now emerging, that involve primarily herders and other stakeholders (Chap. 3).

It is apparent from the studies conducted that herders will have to show extremely diversified entrepreneurial skills and flexibility in the future. In order to maintain the opportunity to continue reindeer herding as one of several economic sectors, herders have to be versatile in their management strategies to be able to attain an appropriate standard of living and level of well being in their home regions in northernmost Europe.

17.5 Conclusions: Recommendations for Reindeer Management

In the early 1960s, reindeer herding could still be understood as an internal system based on local resources and livelihoods that had a localized cultural and socioeconomic focus, i.e., the *siida* system of the Sámi communities

(Chap. 2). Since then rapid changes have been accompanied by the introduction of technological innovations, monetary systems, management practices, administrative structures, and laws. Reindeer herding has evolved more and more into a commercially defined enterprise that emphasizes the increase of meat production for export beyond the local market network (Lewis 1998). Furthermore, reindeer herding has gradually become accustomed to new constraints introduced by multiple land users functioning in the same space. These rapid changes have received responses that warrant a question about the future of reindeer management in northernmost Fennoscandia and adjacent Northwest Russia.

It is apparent that a holistic approach is needed to manage northern ecosystems, animals, and herding in conjunction with the expectations of modern societies, taking into account cultural, socioeconomic, and political considerations to cope with future developments. As part of its studies the RENMAN project applied the methodological tool of workshops, or roundtables, to discuss issues around "land use practices", bringing together a broad range of stakeholders as a realistic participatory approach to deal with the diversity of opinions, positions, and conflicts (Chap. 3; Hukkinen 2001). Other studies also involved herders directly in the assessments of pasture conditions, affording direct comparisons with standard scientific protocols (Prologue, Chaps. 7, 8, and Epilogue).

It was felt that these applied approaches were successful in alleviating tensions and increasing the level of social awareness, information, and knowledge in the reindeer herding communities. The workshops' discussions centred on the sustainability of reindeer management with respect to the "... maintenance of sociocultural identity supported by economic activities and land use rights in relation to external interests" (Chap. 2) and led, with contributions from other workpackages, to the detailed formulation of models, scenarios, and recommendations for future reindeer management practices (Chaps. 3, 16). In summary the following key points are included in the recommendations:

- The rights of aboriginal peoples to land, resources, and utilization must be recognized legally and implemented politically and administratively, complying to international conventions.
- Decision-making and control of the management of resources and territories by local residents must be enhanced meaningfully.
- Reindeer herding has a broad societal significance and represents a symbol whose cultural, monetary, and economic importance needs to be recognized and maintained.
- Marketing of reindeer products needs to take into account the seasonality of reindeer meat as a delicacy item; this includes the production of high quality meat based on organic feeding, which is attractive to consumers.
- Public subsidies and investment are of paramount importance and need to be made to secure the societal significance of reindeer herding.

- Strong professional institutions are necessary to represent reindeer herders and their interests fully, directly, and effectively.
- The synergetic potential of connecting reindeer herding, nature protection, cultural landscape management, and tourism is a viable opportunity to secure reindeer herding and its territory in the face of the expanding interests of other land users in the same space.
- The role of herders as partners in research projects dealing with reindeer has to be ensured, as well as the acceptance of herders' knowledge in the scientific analysis and deliberations, in order to safeguard the holistic approach to reindeer herding and management.

These selected recommendations highlight the underpinning function of reindeer herding in the successful management of socioecological systems in the boreal and subarctic regions of northernmost Europe. The implementation of the recommendations requires strong integrative application of practical and scientific knowledge. The RENMAN research scheme aimed to initiate a broad range of studies that would lead to extended discussions and debates among the holders of practical and scientific knowledge over the future of the relationship between herders and reindeer. The chapters in this book have made contributions to this end to further these discussions, which would result in the implementation of practical steps that will safeguard the future existence of reindeer management as a viable economic sector in communities in the circumpolar North.

Acknowledgments. The authors are grateful to the many contributions that the members of the RENMAN research consortium have made through their own studies and through their comments that led to the formulation of this synthesis chapter.

References

Aikio P (1978) The breakdown of a Lappish ecosystem in northern Finland. In: Müller-Wille L, Pelto PJ, Müller-Wille Li, Darnell R (eds) Consequences of economic change in circumpolar regions. Univ Alberta Boreal Inst Northern Studies Occas Publ 14:91–104

Aikio MS, Müller-Wille L (2005) Sámi approaches to mountain birch utilization in Northern Sápmi (Finland and Norway). In: Wielgolaski F-E (ed) Plant ecology, herbivory and human impact in northern mountain birch forest. Ecological studies 180. Springer, Berlin Heidelberg New York, pp 255–268

Anderson DG, Nuttall M (eds) (2004) Cultivating arctic landscapes. Knowing and managing animals in the circumpolar North. Berghahn Books, New York

Beach H (1997) Negotiating nature in Swedish Lapland: ecology and economics of Saami reindeer management. In: Smith EA, McCarter J (eds) Contested Arctic. University of Washington Press, Seattle, pp 122–149

Beach H (2000) Reindeer–pastoralism politics in Sweden: protecting the environment and designing the herders. In: Hornburg A, Pálsson G (eds) Negotiating nature: culture, power and environmental argument. Lund Univ Press, Lund, pp 179–211
Forbes BC et al. (eds) (2004) The challenges of modernity for reindeer management: integration and sustainable development in Europe's Subarctic and Boreal Regions. RENMAN – final report. Arctic Centre Rep 41:1–100
Forbes BC, Kofinas G (eds) (2000) Proceedings of the human role in reindeer/caribou systems workshop. Polar Res 19(1):1–142
Hall BL (1979) Participatory research: an approach for change. Convergence 8(2):24–33
Helle T, Särkelä M (1993) The effects of outdoor recreation on range use by semi-domesticated reindeer. Scand J For Res 8:123–133
Hiebert W, Swan D (1999) Positively fit: a case study in community development and the role of participatory action research. Commun Dev J 34:356–364
Hukkinen J (2001) Eco-efficiency as abandonment of nature. Ecol Econ 38:311–315
Jääskö O (2003) Ympäristöpaineiden vaikutukset Paadarin-Muotkatunturin paliskunnan poronhoitoon (in Finnish; The impact of environmental pressures on reindeer herding in the Paadar-Muotkatunturi Reindeer Herding Co-operative). Technol Soc Environ 3:105–128
Jernsletten J-L, Klokov K (2002) Sustainable reindeer husbandry. Centre for Saami Studies, Tromsø
Kumpula J, Colpaert A, Anttonen M, Nieminen M (2004) The repeated reindeer pasture inventory in the northernmost part (13 districts) of the reindeer management area during 1999–2003 (in Finnish with English summary). Finnish Game and Fisheries Research Institute, Helsinki, pp 1–39
Lewis D (1998) Indigenous rights claims in welfare capitalist society: recognition and implementation: the case of the Sámi people in Norway, Sweden and Finland. Arctic Centre Reports 24:1–247
Länsman A-S (2004) Väärtisuhteet Lapin matkailussa. Kulttuurianalyysi suomalaisten ja saamelaisten kohtamisesta (in Finnish, English summary; Host–Guest Relations in Tourism in Sápmi, Finland. A Cultural Analysis of the Encounter Between Finns and Sámi). Kustannus Puntsi, Inari
Magga H (2003) Poronhoidon menetykset ja sopeutuminen suuriin ympäristömuutoksiin Lapin paliskunnassa 1950-luvulta lähtien (in Finnish; Loss and adaptations in reindeer herding in the Lappi Reindeer Herding Co-operative due to considerable environmental changes since the 1950s). Technol Soc Environ 3:11–78
Palkeinen J (1913) Über die Renntierzucht in Finnland im Anfang dieses Jahrhunderts (in German). Doctoral dissertation. Ruprecht-Karls-Universität Heidelberg, Heidelberg
Pelto PJ (1973) The snowmobile revolution: technology and social change in the Arctic. Cummings Publ, Menlo Park, CA
Pelto PJ, Müller-Wille L (1973) Reindeer herding and snowmobiles: aspects of a technological revolution (Utsjoki and Sevettijärvi, Finnish Lapland). Folk 14/15:119–144
Turi JM (2000) Native reindeer herders' priorities for research. Polar Res 19(1):131–133
Vibe C (1967) Arctic animals in relation to climatic fluctuations. Medd Grønland 170(5):23–180
Vistnes I, Nellemann C (2001) Avoidance of cabins, roads, and power lines by reindeer during calving. J Wildl Manage 65:915–925

Epilogue

Reindeer Herder Lost in the Jungle of Science

Traditional reindeer research has viewed the reindeer as a kind of wild animal that utilizes pastures. The "carrying capacity" of the pastures has been studied primarily by comparing the number of reindeer with the available pastures. As an example, the maximum number of reindeer allowed in each reindeer herding cooperative has deliberately been set at a low level based on very light scientific arguments. In these research efforts, little or no attention has been paid to the realities of how herding work can influence the overall sustainability of pastures. Reindeer herding is a method created by indigenous peoples and it has been seen as the best way to utilize the 'fragile' northern environment. I do understand that the financier of the research has, of course, the ultimate authority to decide which topics are investigated.

The RENMAN research project has, already in its planning phase, incorporated the reindeer herders and their views. I have been given the opportunity to work within the project from its beginning and I have therefore also been able to bring forward my own ideas into the research design. When discussing the design and focus of the research project we ended up designing a multidisciplinary research project. As a reindeer herder I see that this is the right starting point for a comprehensive reindeer research project.

As one of the intensive research areas for the project, the researchers and representatives of Näkkälä reindeer herding cooperative together agreed on choosing the areas between Roavvoaivi–Maader and Termisvaara–Kaamos. The Roavvoaivi–Maader area is situated within the tundra zone and it encompasses the summer pastures of the Kalkujärvi *siida* (Sámi village community), but also serves occasionally as an important winter pasture area if the availability of lichens is poor in the southern areas of the cooperative. In the Roavvoaivi–Maader area the availability of lichen forage is usually good, meaning that it is easy for the reindeer to dig for the lichens from under the snow cover. The Termisvaara and Kaamos area is used by the Näkkälä *siida* and it contains more trees but is not quite forest. In addition to the two aforementioned areas we, also chose an area north of Nunnanen village for snow measurements. It is the winter pasture area of the Kalkkujärvi *siida*. The deci-

Ecological Studies, Vol. 184
B.C. Forbes, M. Bölter, L. Müller-Wille, J. Hukkinen,
F. Müller, N. Gunslay, Y. Konstatinov (Eds.)
Reindeer Management in Northernmost Europe

sion of which areas to choose was also partly guided by the satellite images the project intended to use. The area had to be approximately 10 km × 10 km and part of the area needed to be on the Norwegian side of the border fence to include the contrasting management regime (no animals present in summer).

I have been able to join the researchers in their fieldwork both during winter, performing snow measurements, and during summer, conducting vegetation studies, mainly in wetlands. In this way I have been able to get familiar with the different research methods and the actual technology used in modern-day research. For me, it has been really interesting and rewarding. I have been thinking afterwards that, during the fieldwork, there would have been good opportunities to ask the reindeer herders their opinion regarding the research being conducted, for example snow conditions and the state of the winter pastures, and to write them down immediately just like all the other measurement results. In this way it would be possible to compare the results of the snow measurements more directly with the information provided by reindeer herders concerning the terminology used regarding the specific conditions of the winter pastures. This comparison of knowledge would be important for me, as a reindeer herder, because we would then be able to discuss and better understand each other regarding lichen rangelands.

I think that the researchers in the project have done a commendable job of becoming acquainted with the different study areas, the reindeer herders themselves, and also achieved a good level of dialogue. In the research project both female and male representatives of the reindeer-herding livelihood have been interviewed and in this way I am sure that a diverse view of the livelihood has been achieved for the purposes of research. In the project I think that one could have given more attention to the *siida* system of organizing the herding of reindeer inside the Näkkälä reindeer-herding cooperative. Many more of the reindeer herders would have wanted to participate in the fieldwork, so that they could better understand what is being studied within their *siida*. As well, communication among the different *siidas* could have been more efficient.

I see that scientific research related to reindeer herding should be conducted in the future according to the multidisciplinary approach exemplified in the RENMAN research project. During the project I have become more and more convinced that reindeer herders should have the opportunity to participate in research regarding their own livelihood, and also research dealing with pastures. A civil servant of the Finnish Ministry of Agriculture and Forestry has already publicly criticized the RENMAN project because "primitive and illiterate reindeer herders" have been admitted access to an area of science to which they do not belong and that "they only promote their own interests, without any regard for the carrying capacity of the land". I sincerely hope that this is not the official view of the Ministry. I do think that transparency should also be practiced in research.

The experience from the RENMAN project should also be utilized in research related to those northern residents other than indigenous peoples, their livelihoods, and also in research related to the Arctic in general. However, I still believe that the most essential knowledge that is needed to ensure that the livelihoods based on arctic nature and resources are sustainable can be found with the indigenous peoples.

Juha Magga, Reindeer Herder
Näkkälä Reindeer Herding Cooperative

Reindeer as draught animals. Drawing by Chukchi, 19th century. After Ingold, T. 1986: Reindeer economies and the origins of pastoralism. Anthropology Today 2: 5-10

Subject Index

E

F

H

I

Ecological Studies

Volumes published since 2001

Volume 146
Inselbergs: Biotic Diversity of Isolated Rock Outcrops in Tropical and Temperate Regions (2000)
S. Porembski and W. Barthlott (Eds.)

Volume 147
Ecosystem Approaches to Landscape Management in Central Europe (2001)
J.D. Tenhunen, R. Lenz, and R. Hantschel (Eds.)

Volume 148
A Systems Analysis of the Baltic Sea (2001)
F.V. Wulff, L.A. Rahm, and P. Larsson (Eds.)

Volume 149
Banded Vegetation Patterning in Arid and Semiarid Environments (2001)
D. Tongway and J. Seghieri (Eds.)

Volume 150
Biological Soil Crusts: Structure, Function, and Management (2001, 2003)
J. Belnap and O.L. Lange (Eds.)

Volume 151
Ecological Comparisons of Sedimentary Shores (2001)
K. Reise (Ed.)

Volume 152
Global Biodiversity in a Changing Environment: Scenarios for the 21st Century (2001)
F.S. Chapin, O. Sala, and E. Huber-Sannwald (Eds.)

Volume 153
UV Radiation and Arctic Ecosystems (2002)
D.O. Hessen (Ed.)

Volume 154
Geoecology of Antarctic Ice-Free Coastal Landscapes (2002)
L. Beyer and M. Bölter (Eds.)

Volume 155
Conserving Biological Diversity in East African Forests: A Study of the Eastern Arc Mountains (2002)
W.D. Newmark

Volume 156
Urban Air Pollution and Forests: Resources at Risk in the Mexico City Air Basin (2002)
M.E. Fenn, L. I. de Bauer, and T. Hernández-Tejeda (Eds.)

Volume 157
Mycorrhizal Ecology (2002, 2003)
M.G.A. van der Heijden and I.R. Sanders (Eds.)

Volume 158
Diversity and Interaction in a Temperate Forest Community: Ogawa Forest Reserve of Japan (2002)
T. Nakashizuka and Y. Matsumoto (Eds.)

Volume 159
Big-Leaf Mahogany: Genetic Resources, Ecology and Management (2003)
A. E. Lugo, J. C. Figueroa Colón, and M. Alayón (Eds.)

Volume 160
Fire and Climatic Change in Temperate Ecosystems of the Western Americas (2003)
T. T. Veblen et al. (Eds.)

Volume 161
Competition and Coexistence (2002)
U. Sommer and B. Worm (Eds.)

Volume 162
How Landscapes Change: Human Disturbance and Ecosystem Fragmentation in the Americas (2003)
G.A. Bradshaw and P.A. Marquet (Eds.)

Volume 163
Fluxes of Carbon, Water and Energy of European Forests (2003)
R. Valentini (Ed.)

Volume 164
Herbivory of Leaf-Cutting Ants: A Case Study on *Atta colombica* in the Tropical Rainforest of Panama (2003)
R. Wirth, H. Herz, R.J. Ryel, W. Beyschlag, B. Hölldobler

Volume 165
Population Viability in Plants: Conservation, Management, and Modeling of Rare Plants (2003)
C.A Brigham, M.W. Schwartz (Eds.)

Volume 166
North American Temperate Deciduous Forest Responses to Changing Precipitation Regimes (2003)
P. Hanson and S.D. Wullschleger (Eds.)

Volume 167
Alpine Biodiversity in Europe (2003)
L. Nagy, G. Grabherr, Ch. Körner, D. Thompson (Eds.)

Volume 168
Root Ecology (2003)
H. de Kroon and E.J.W. Visser (Eds.)

Volume 169
Fire in Tropical Savannas: The Kapalga Experiment (2003)
A.N. Andersen, G.D. Cook, and R.J. Williams (Eds.)

Volume 170
Molecular Ecotoxicology of Plants (2004)
H. Sandermann (Ed.)

Volume 171
Coastal Dunes: Ecology and Conservation (2004)
M.L. Martínez and N. Psuty (Eds.)

Volume 172
Biogeochemistry of Forested Catchments in a Changing Environment: A German Case Study (2004)
E. Matzner (Ed.)

Volume 173
Insects and Ecosystem Function (2004)
W. W. Weisser and E. Siemann (Eds.)

Volume 174
Pollination Ecology and the Rain Forest: Sarawak Studies (2005)
D. Roubik, S. Sakai, and A.A. Hamid (Eds.)

Volume 175
Antarctic Ecosystems: Environmental Contamination, Climate Change, and Human Impact (2005)
R. Bargagli

Volume 176
Forest Diversity and Function: Temperate and Boreal Systems (2005)
M. Scherer-Lorenzen, Ch. Körner, and E.-D. Schulze (Eds.)

Volume 177
A History of Atmospheric CO_2 and its Effects on Plants, Animals, and Ecosystems (2005)
J.R. Ehleringer, T.E. Cerling, and M.D. Dearing (Eds.)

Volume 178
Photosynthetic Adaptation: Chloroplast to Landscape (2005)
W.K. Smith, T.C. Vogelmann, and C. Chritchley (Eds.)

Volume 179
Lamto: Structure, Functioning, and Dynamics of a Savanna Ecosystem (2005)
L. Abbadie et al. (Eds.)

Volume 180
Plant Ecology, Herbivory, and Human Impact in Nordic Mountain Birch Forests (2005)
F.E. Wielgolaski (Ed.) and P.S. Karlsson, S. Neuvonen, D. Thannheiser (Ed. Board)

Volume 181
Nutrient Acquisition by Plants: An Ecological Perspective (2005)
H. BassiriRad (Ed.)

Volume 182
Human Ecology: Biocultural Adaptations in Human Cummunities (2006)
H. Schutkowski

Volume 183
Growth Dynamics of Conifer Tree Rings: Images of Past and Future Environments (2006)
E.A. Vaganov, M.K. Hughes, and A.V. Shashkin

Volume 184
Reindeer Management in Northernmost Europe: Linking Practical and Scientific Knowledge in Social-Ecological Systems (2006)
B.C. Forbes, M. Bölter, L. Müller-Wille, J. Hukkinen, F. Müller, N. Gunslay, Y. Konstantinov (Eds.)

Printing and Binding: Stürtz GmbH, Würzburg